MATHEMATICS
INTEGRAL CALCULUS

&

DIFFERENTIAL CALCULUS

(12th , Engineering)

Content: —

- ➢ **Indefinite Integral**
- ➢ **Definite Integral**
- ➢ **Differential Equation**

Integration

Indefinite Integral: – Some Basic Integrals in Standard Form: –

(i) $\int x^n \, dx = \dfrac{x^{n+1}}{n+1}$, $\dfrac{d}{dx}\left(\dfrac{x^{n+1}}{n+1}\right) = x^n$ (ii) $\int \sin x \, dx = -\cos x$, $\dfrac{d}{dx}(\cos x) = -\sin x$

(iii) $\int \cos x \, dx = \sin x$, $\dfrac{d}{dx}(\sin x) = \cos x$ (iv) $\int \sec^2 x \, dx = \tan x$, $\dfrac{d}{dx}(\tan x) = \sec^2 x$

(v) $\int \operatorname{cosec}^2 x \, dx = -\cot x$, $\dfrac{d}{dx}(\cot x) = -\operatorname{cosec}^2 x$

(vi) $\int \sec x \tan x \, dx = \sec x$, $\dfrac{d}{dx}(\sec x) = \sec x \tan x$

(vii) $\int \operatorname{cosec} x \cot x \, dx = -\operatorname{cosec} x$, $\dfrac{d}{dx}(\operatorname{cosec} x) = \operatorname{cosec} x \cot x$

(viii) $\int \dfrac{1}{x} \, dx = \log x$, $\dfrac{d}{dx}(\log x) = \dfrac{1}{x}$ (ix) $\int e^x \, dx = e^x$, $\dfrac{d}{dx}(e^x) = e^x$

(x) $\int \dfrac{dx}{\sqrt{1-x^2}} = \sin^{-1} x$, $\dfrac{d}{dx}(\sin^{-1} x) = \dfrac{1}{\sqrt{1-x^2}}$

(xi) $\int \dfrac{dx}{-\sqrt{1-x^2}} = \cos^{-1} x$, $\dfrac{d}{dx}(\cos^{-1} x) = \dfrac{1}{-\sqrt{1-x^2}}$

(xii) $\int \dfrac{dx}{1+x^2} = \tan^{-1} x$, $\dfrac{d}{dx}(\tan^{-1} x) = \dfrac{1}{1+x^2}$

(xiii) $\int \dfrac{dx}{-(1+x^2)} = \cot^{-1} x$, $\dfrac{d}{dx}(\cot^{-1} x) = -\dfrac{1}{1+x^2}$

(xiv) $\int \dfrac{dx}{x\sqrt{x^2-1}} = \sec^{-1} x$, $\dfrac{d}{dx}(\sec^{-1} x) = \dfrac{1}{x\sqrt{x^2-1}}$

(xv) $\int \dfrac{dx}{-x\sqrt{x^2-1}} = \operatorname{cosec}^{-1} x$, $\dfrac{d}{dx}(\operatorname{cosec}^{-1} x) = -\dfrac{1}{x\sqrt{x^2-1}}$

Some Standard Formula: –

(i) $\int \dfrac{dx}{\sqrt{x^2+a^2}} = \log\left|x + \sqrt{x^2+a^2}\right|$ or $\sinh^{-1}\left(\dfrac{x}{a}\right)$

(ii) $\int \dfrac{dx}{\sqrt{x^2-a^2}} = \log\left|x + \sqrt{x^2-a^2}\right|$ or $\cosh^{-1}\left(\dfrac{x}{a}\right)$ (iii) $\int \dfrac{dx}{\sqrt{a^2-x^2}} = \sin^{-1}\dfrac{x}{a}$

(iv) $\int \dfrac{dx}{x^2-a^2} = \dfrac{1}{2a} \log\left|\dfrac{x-a}{x+a}\right|$, when $x > a$ (v) $\int \dfrac{dx}{a^2-x^2} = \dfrac{1}{2a} \log\left|\dfrac{a+x}{a-x}\right|$, when $x < a$

(vi) $\int \dfrac{dx}{x^2+a^2} = \dfrac{1}{a} \tan^{-1}\dfrac{x}{a}$

(vii) $\int \sqrt{x^2+a^2} \, dx = \dfrac{x}{2}\sqrt{x^2+a^2} + \dfrac{a^2}{2}\log\left|x + \sqrt{x^2+a^2}\right|$ or $\dfrac{x}{2}\sqrt{x^2+a^2} + \dfrac{a^2}{2}\sinh^{-1}\left(\dfrac{x}{a}\right)$

(viii) $\int \sqrt{x^2-a^2} \, dx = \dfrac{x}{2}\sqrt{x^2-a^2} - \dfrac{a^2}{2}\log\left|x + \sqrt{x^2-a^2}\right|$ or $\dfrac{x}{2}\sqrt{x^2-a^2} - \dfrac{a^2}{2}\cosh^{-1}\left(\dfrac{x}{a}\right)$

(ix) $\int \sqrt{a^2-x^2} \, dx = \dfrac{x}{2}\sqrt{a^2-x^2} + \dfrac{a^2}{2}\sin^{-1}\dfrac{x}{a}$

Example: – (1) (a) $I = \int \dfrac{dx}{2x^2 + 3} = \int \dfrac{dx}{\left(\sqrt{2}\,x\right)^2 + \left(\sqrt{3}\right)^2}$ $\left[\text{use formula}, \int \dfrac{dx}{x^2 + a^2} = \dfrac{1}{a}\tan^{-1}\dfrac{x}{a}\right]$

$\therefore \ I = \int \dfrac{dx}{\left(\sqrt{2}\,x\right)^2 + \left(\sqrt{3}\right)^2} = \dfrac{1}{\sqrt{3}}\tan^{-1}\dfrac{\sqrt{2}\,x}{\sqrt{3}} \cdot \dfrac{1}{\sqrt{2}} = \dfrac{1}{\sqrt{6}}\tan^{-1}\left(\dfrac{\sqrt{2}\,x}{\sqrt{3}}\right) + c$ Ans.

(b) $I = \int \dfrac{dx}{(3)^2 + (5-x)^2}$ (solve same as above question) $\ \therefore \ I = \dfrac{1}{3}\tan^{-1}\left(\dfrac{5-x}{3}\right)\cdot\dfrac{1}{-1} \ \therefore \ I = -\dfrac{1}{3}\tan^{-1}\left(\dfrac{5-x}{3}\right) + c$ Ans.

(c) $I = \int \dfrac{dx}{\sqrt{(3)^2 + (4-x)^2}}$ $\left[\text{formula}, \int \dfrac{dx}{\sqrt{x^2 + a^2}} = \log\left|x + \sqrt{x^2 + a^2}\right|\right]$

$\therefore \ I = \int \dfrac{dx}{\sqrt{(3)^2 + (4-x)^2}} = \log\left|(4-x) + \sqrt{(3)^2 + (4-x)^2}\right| \cdot \dfrac{1}{-1} = -\log\left|(4-x) + \sqrt{(3)^2 + (4-x)^2}\right|$ Ans.

(d) $I = \int \dfrac{dx}{\sqrt{(x+3)^2 - 4}} = \int \dfrac{dx}{\sqrt{(x+3)^2 - (2)^2}}$ $\left[\text{formula}, \int \dfrac{dx}{\sqrt{x^2 - a^2}} = \log\left|x + \sqrt{x^2 - a^2}\right|\right]$

$\therefore \ I = \int \dfrac{dx}{\sqrt{(x+3)^2 - (2)^2}} = \log\left|(x+3) + \sqrt{(x+3)^2 - (2)^2}\right| \cdot \dfrac{1}{1} = \log\left|(x+3) + \sqrt{(x+3)^2 - 4}\right|$ Ans.

(e) $I = \int \sqrt{x^2 - 16}\,dx = \int \sqrt{x^2 - 4^2}\,dx$ $\left[\text{formula}, \int \sqrt{x^2 - a^2}\,dx = \dfrac{x}{2}\sqrt{x^2 - a^2} - \dfrac{a^2}{2}\log\left|x + \sqrt{x^2 - a^2}\right|\right]$

$\therefore \ I = \int \sqrt{x^2 - 4^2}\,dx = \dfrac{x}{2}\sqrt{x^2 - 4^2} - \dfrac{4^2}{2}\log\left|x + \sqrt{x^2 - 4^2}\right| + c = \dfrac{x}{2}\sqrt{x^2 - 16} - \dfrac{16}{2}\log\left|x + \sqrt{x^2 - 16}\right| + c$

$\qquad = \dfrac{x}{2}\sqrt{x^2 - 16} - 8\log\left|x + \sqrt{x^2 - 16}\right| + c$ Ans.

(f) $I = \int \dfrac{dx}{16 - 3x^2} = \int \dfrac{dx}{(4)^2 - \left(\sqrt{3}\,x\right)^2}$ $\left[\text{formula}, \int \dfrac{dx}{a^2 - x^2} = \dfrac{1}{2a}\log\left|\dfrac{a+x}{a-x}\right|\right]$

$\therefore \ I = \int \dfrac{dx}{(4)^2 - \left(\sqrt{3}\,x\right)^2} = \dfrac{1}{2.4}\log\left|\dfrac{4 + \sqrt{3}x}{4 - \sqrt{3}x}\right| \cdot \dfrac{1}{\sqrt{3}} + c = \dfrac{1}{8\sqrt{3}}\log\left|\dfrac{4 + \sqrt{3}x}{4 - \sqrt{3}x}\right| + c$ Ans.

(g) $I = \int \sqrt{9 - x^2}\,dx = \int \sqrt{(3)^2 - x^2}\,dx$ $\left[\text{formula}, \int \sqrt{a^2 - x^2}\,dx = \dfrac{x}{2}\sqrt{a^2 - x^2} + \dfrac{a^2}{2}\sin^{-1}\dfrac{x}{a}\right]$

or $I = \int \sqrt{(3)^2 - x^2}\,dx = \dfrac{x}{2}\sqrt{9 - x^2} + \dfrac{9}{2}\sin^{-1}\dfrac{x}{3}$ Ans.

$I = \int \dfrac{f'(x)}{f(x)}\,dx = \log|f(x)|$

Proof: – Let $f(x) = z$ $\ \therefore \ f'(x)\,dx = dz$ then $\int \dfrac{f'(x)}{f(x)}\,dx = \int \dfrac{dz}{z} = \log|z| = \log|f(x)|$ Proved.

Example: – (a) $I = \int \dfrac{(4x+9)}{2x^2 + 9x + 14}\,dx$, Let $f(x) = 2x^2 + 9x + 14$, $f'(x) = 4x + 9$

or $I = \int \dfrac{f'(x)}{f(x)}\,dx = \log|f(x)| + c = \log|2x^2 + 9x + 14| + c$ Ans.

(b) $I = \int \cot x\,dx = \int \dfrac{\cos x}{\sin x}\,dx$, Let $f(x) = \sin x$ $\ \therefore \ f'(x) = \cos x$ $\ \therefore \ I = \int \dfrac{\cos x}{\sin x}\,dx = \int \dfrac{f'(x)}{f(x)}\,dx = \log|f(x)| + c = \log|\sin x| + c$ Ans.

(c) $I = \int \dfrac{x-2}{x^2 - 4x + 8}\,dx$, Let $f(x) = x^2 - 4x + 8$ $\ \therefore \ f'(x) = 2x - 4 = 2(x-2)$

$I = \int \dfrac{2x-4}{2(x^2-4x+8)}\,dx = \dfrac{1}{2}\int \dfrac{2x-4}{x^2-4x+8}\,dx = \dfrac{1}{2}\int \dfrac{f'(x)}{f(x)}\,dx = \dfrac{1}{2}\log|f(x)| + c = \dfrac{1}{2}\log|x^2-4x+8| + c$ Ans.

(d) $I = \int \dfrac{(\cos x - \sin x)}{\sin\left(x+\frac{\pi}{4}\right)}\,dx = \int \dfrac{\cos x - \sin x}{\sin x.\cos\frac{\pi}{4} + \cos x.\sin\frac{\pi}{4}}\,dx = \int \dfrac{\cos x - \sin x}{\frac{1}{\sqrt{2}}(\sin x + \cos x)}\,dx = \sqrt{2}\int \dfrac{\cos x - \sin x}{\sin x + \cos x}\,dx$

Let $f(x) = \sin x + \cos x$, $f'(x) = \cos x - \sin x$ then $\therefore\ I = \sqrt{2}\int \dfrac{f'(x)}{f(x)}\,dx = \sqrt{2}\log|f(x)| + c$ $\therefore\ I = \sqrt{2}\log|\sin x + \cos x| + c$ Ans.

\# $\quad I = \int \{f(x)\}^n.f'(x)\,dx = \dfrac{\{f(x)\}^{n+1}}{n+1}$, $n \neq -1$

Proof: − Let $f(x) = z$ $\therefore\ f'(x)\,dx = dz$ then $I = \int \{f(x)\}^n.f'(x)\,dx = \int z^n.dz = \dfrac{z^{n+1}}{n+1}$ $\therefore\ I = \dfrac{z^{n+1}}{n+1} = \dfrac{\{f(x)\}^{n+1}}{n+1}$ Proved

Example: − (a) $I = \int \tan x.\sec^2 x\,dx$ $\quad$ Let $f(x) = \tan x$, $f'(x) = \sec^2 x$

or $I = \int \{f(x)\}^n.f'(x)\,dx = \dfrac{\{f(x)\}^{n+1}}{n+1}$ $\quad$ or $I = \int \tan x.\sec^2 x\,dx = \int (\tan x)^1.\sec^2 x\,dx = \dfrac{(\tan x)^{1+1}}{1+1} + c = \dfrac{(\tan x)^2}{2} + c = \dfrac{\tan^2 x}{2} + c$ Ans.

IInd Method: − $I = \int \tan x.\sec^2 x\,dx = \int \dfrac{\sin x}{\cos x}.\dfrac{1}{\cos^2 x}\,dx = \int \dfrac{\sin x}{\cos^3 x}\,dx$, Let $f(x) = \cos x$, $f'(x) = -\sin x$

$\therefore\ I = \int \cos^{-3} x.\sin x\,dx = -\int [f(x)]^{-3}.(-f'(x))\,dx = -\dfrac{\{f(x)\}^{n+1}}{n+1} = -\dfrac{\{\cos x\}^{-3+1}}{-3+1} = \dfrac{\cos^{-2} x}{2} = \dfrac{1}{2\cos^2 x} + c = \dfrac{1}{2}\sec^2 x + c$

$= \dfrac{1}{2}(1 + \tan^2 x) + c = \dfrac{\tan^2 x}{2} + \dfrac{1}{2} + c$ Ans.

IIIrd Method: − $I = \int \tan x.\sec^2 x\,dx = \int \dfrac{\sin x}{\cos x}.\dfrac{1}{\cos^2 x}\,dx = \int \dfrac{\sin x}{\cos^3 x}\,dx$ (solve above Method) $\therefore\ I = \dfrac{1}{2}\sec^2 x = \dfrac{\tan^2 x}{2} + \dfrac{1}{2}$ Ans.

IVth Method: − $I = \int \tan x.\sec^2 x\,dx$ $\quad$ Put $z = \tan x$ $\therefore\ dz = \sec^2 x\,dx$ $\therefore\ I = \int z.dz = \dfrac{z^2}{2} + c = \dfrac{\tan^2 x}{2} + c$ Ans.

(b) $I = \int \dfrac{x^2\,dx}{(2x^3-5)^4} = \int (2x^3-5)^{-4}.x^2\,dx$ $\quad$ Let $f(x) = 2x^3 - 5$, $f'(x) = 6x^2$

$\therefore\ I = \dfrac{1}{6}\int (2x^3-5)^{-4}.6x^2\,dx$ $\quad \left[\text{formula, } I = \int \{f(x)\}^n.f'(x)\,dx = \dfrac{\{f(x)\}^{n+1}}{n+1}\right]$

$\therefore\ I = \dfrac{1}{6}.\dfrac{\{f(x)\}^{-4+1}}{-4+1} = \dfrac{1}{6}.\dfrac{(2x^3-5)^{-3}}{-3} = -\dfrac{1}{18}(2x^3-5)^{-3} = -\dfrac{1}{18(2x^3-5)^3}$ Ans.

(c) $I = \int \sin^4 x.\cos x\,dx$, Let $f(x) = \sin x$ $\therefore\ f'(x) = \cos x$ $\quad$ and $n = 4$ $\therefore\ I = \dfrac{[f(x)]^{4+1}}{4+1} = \dfrac{(\sin x)^5}{5} + c = \dfrac{\sin^5 x}{5} + c$ Ans.

IInd Method: − $I = \int \sin^4 x.\cos x\,dx$, Put $\sin x = z$ $\therefore\ \cos x\,dx = dz$ $\therefore\ I = \int z^4.dz = \dfrac{z^{4+1}}{4+1} = \dfrac{z^5}{5} = \dfrac{(\sin x)^5}{5} + c = \dfrac{\sin^5 x}{5} + c$ Ans.

\# $\quad I = \int \dfrac{f'(x)}{\sqrt{f(x)}}\,dx = 2\sqrt{f(x)}$

Proof: − Let $f(x) = z$ $\therefore\ f'(x)\,dx = dz$ then $I = \int \dfrac{dz}{\sqrt{z}} = \int z^{-\frac{1}{2}}.dz = \dfrac{z^{-\frac{1}{2}+1}}{-\frac{1}{2}+1} = \dfrac{z^{\frac{1}{2}}}{\frac{1}{2}} = 2\sqrt{z} = 2\sqrt{f(x)}$ Proved.

Example: − (a) $I = \int \dfrac{2ax+b}{\sqrt{ax^2+bx+c}}\,dx$ $\quad$ Let $f(x) = ax^2 + bx + c$ $\therefore\ f'(x) = 2ax + b$

$\qquad\qquad$ or $I = \int \dfrac{f'(x)}{\sqrt{f(x)}}\,dx = 2\sqrt{f(x)} = 2\sqrt{ax^2+bx+c} + d$ Ans.

(b) $I = \int \dfrac{\cos x}{\sqrt{\sin x}}\, dx$ Let $f(x) = \sin x$ $\therefore$ $f'(x) = \cos x$ or $I = \int \dfrac{f(x)}{\sqrt{f(x)}}\, dx = 2\sqrt{f(x)}$ $\therefore$ $I = \int \dfrac{\cos x}{\sqrt{\sin x}}\, dx = 2\sqrt{\sin x} + c$ Ans.

(c) $I = \int \dfrac{\sec^2 x}{\sqrt{3 - \tan x}}\, dx$ Let $f(x) = 3 - \tan x$ $\therefore$ $f'(x) = -\sec^2 x$

or $I = \int \dfrac{f(x)}{\sqrt{f(x)}}\, dx = 2\sqrt{f(x)}$ $\therefore$ $I = \int \dfrac{\sec^2 x}{\sqrt{3 - \tan x}}\, dx = -\int \dfrac{\sec^2 x}{\sqrt{3 - \tan x}}\, dx = -2\sqrt{3 - \tan x} + c$ Ans.

\# $\dfrac{d}{dx} \int f(x)\, dx = f(x)$ or $\int \dfrac{d}{dx}[f(x)]\, dx = f(x)$

(i) $\dfrac{d}{dx} \int \dfrac{\log(2 + x)}{3x}\, dx = \dfrac{\log(2 + x)}{3x}$ Ans. (ii) $\int \dfrac{d}{dx} \left(\dfrac{x^2 + 3x + 5}{2x - 3} \right) dx = \dfrac{x^2 + 3x + 5}{2x - 3}$ Ans.

Form of the integrals: –

$\int \dfrac{dx}{ax^2 + bx + c}$ or $\int \dfrac{dx}{\sqrt{ax^2 + bx + c}}$ or $\int \sqrt{ax^2 + bx + c}\, dx$ where $a \ne 0$

write $ax^2 + bx + c$ in the form $a\left\{ \left(x + \dfrac{b}{2a}\right)^2 - \dfrac{D}{4a^2} \right\}$ and then use standard result of integration.

Example: – (a) $I = \int \dfrac{dx}{x^2 + 2x + 3} = \int \dfrac{dx}{ax^2 + bx + c}$ Here $a = 1, b = 2$ and $c = 3$ then $D = b^2 - 4ac = 4 - 12 = -8$

or $I = \int \dfrac{dx}{a\left\{ \left(x + \frac{b}{2a}\right)^2 - \frac{D}{4a^2} \right\}} = \int \dfrac{dx}{1\left\{ \left(x + \frac{2}{2}\right)^2 - \frac{-8)}{4.1} \right\}} = \int \dfrac{dx}{(x + 1)^2 + 2} = \int \dfrac{dx}{(x + 1)^2 + \left(\sqrt{2}\right)^2}$

$\therefore$ $I = \int \dfrac{dx}{x^2 + a^2} = \dfrac{1}{a}\tan^{-1}\dfrac{x}{a}$ $\therefore$ $I = \int \dfrac{dx}{(x + 1)^2 + \left(\sqrt{2}\right)^2} = \dfrac{1}{\sqrt{2}}\tan^{-1}\left(\dfrac{x + 1}{\sqrt{2}}\right) + c$ Ans.

(b) $I = \int \dfrac{dx}{\sqrt{5 + 4x - 3x^2}} = \int \dfrac{dx}{\sqrt{ax^2 + bx + c}}$, here $a = -3, b = 4$ and $c = 5$ then $D = 16 + 60 = 76$

or $I = \int \dfrac{dx}{\sqrt{a\left\{ \left(x + \frac{b}{2a}\right)^2 - \frac{D}{4a^2} \right\}}} = \int \dfrac{dx}{\sqrt{-3\left\{ \left(x + \frac{4}{-6}\right)^2 - \frac{76}{36} \right\}}} = \int \dfrac{dx}{\sqrt{3\left\{ \frac{19}{9} - \left(x - \frac{2}{3}\right)^2 \right\}}} = \dfrac{1}{\sqrt{3}} \int \dfrac{dx}{\sqrt{\left(\frac{\sqrt{19}}{3}\right)^2 - \left(x - \frac{2}{3}\right)^2}}$

$\therefore$ $I = \dfrac{1}{\sqrt{3}} \int \dfrac{dx}{\sqrt{a^2 - x^2}} = \dfrac{1}{\sqrt{3}}\sin^{-1}\dfrac{x}{a} = \dfrac{1}{\sqrt{3}}\sin^{-1}\left(\dfrac{x - \frac{2}{3}}{\frac{\sqrt{19}}{3}}\right) = \dfrac{1}{\sqrt{3}}\sin^{-1}\left(\dfrac{3x - 2}{\sqrt{19}}\right)$ Ans.

(c) $I = \int \sqrt{-x^2 + 3x + 8}\, dx = \int \sqrt{ax^2 + bx + c}\, dx$, here $a = -1, b = 3$ and $c = 8$ then $D = 9 + 32 = 41$

or $I = \int \sqrt{a\left\{ \left(x + \frac{b}{2a}\right)^2 - \frac{D}{4a^2} \right\}}\, dx = \int \sqrt{-1\left\{ \left(x + \frac{3}{-2}\right)^2 - \frac{41}{4} \right\}}\, dx = \int \sqrt{\frac{41}{4} - \left(x - \frac{3}{2}\right)^2}\, dx$

$I = \int \sqrt{\left(\frac{\sqrt{41}}{2}\right)^2 - \left(x - \frac{3}{2}\right)^2}\, dx$ formula, $\int \sqrt{a^2 - x^2}\, dx = \dfrac{x}{2}\sqrt{a^2 - x^2} + \dfrac{a^2}{2}\sin^{-1}\dfrac{x}{a}$

or $I = \int \sqrt{\left(\frac{\sqrt{41}}{2}\right)^2 - \left(x - \frac{3}{2}\right)^2}\, dx$ or $dx = \dfrac{\left(x - \frac{3}{2}\right)}{2}\sqrt{\dfrac{41}{4} - \left(x - \dfrac{3}{2}\right)^2} + \dfrac{41}{8}\sin^{-1}\left(\dfrac{x - \frac{3}{2}}{\frac{\sqrt{41}}{2}}\right) + c$

$= \dfrac{(2x - 3)}{4}\sqrt{\dfrac{41}{4} - \left(x - \dfrac{3}{2}\right)^2} + \dfrac{41}{8}\sin^{-1}\left(\dfrac{2x - 3}{\sqrt{41}}\right) + c$ Ans.

Form of the integrals: $-\int \dfrac{dx}{ax^2 + bx + c}$ or $\int \dfrac{dx}{\sqrt{ax^2 + bx + c}}$ or $\int \sqrt{ax^2 + bx + c}\ dx$

Express $ax^2 + bx + c$ as sum of difference of two square and apply formulae.

Example:– (1) (a) $I = \int \dfrac{dx}{x^2 + 2x + 1}$ (b) $I = \int \dfrac{dx}{\sqrt{x^2 + 2x + 1}}$ (c) $I = \int \sqrt{x^2 + 2x + 1}\ dx$

Solution:– (a) $I = \int \dfrac{dx}{x^2 + 2x + 1} = \int \dfrac{dx}{(x+1)^2}$ Let $x + 1 = t$ or $x = t - 1$ $\therefore\ dx = dt$

$\therefore\ I = \int \dfrac{dt}{t^2} = \int t^{-2}\ dt = \dfrac{t^{-2+1}}{-2+1} + c = -t^{-1} + c = -\dfrac{1}{t} + c = -\dfrac{1}{x+1} + c$ Ans.

(b) $I = \int \dfrac{dx}{\sqrt{x^2 + 2x + 1}} = \int \dfrac{dx}{\sqrt{(x+1)^2}} = \int \dfrac{dx}{x+1} = \log(x + 1) + c$ Ans.

(c) $I = \int \sqrt{x^2 + 2x + 1}\ dx = \int \sqrt{(x+1)^2}\ dx = \int (x+1)\ dx = \int x\ dx + \int dx = \dfrac{x^2}{2} + x + c$ Ans.

Example:– (2) (a) $I = \int \dfrac{dx}{4x^2 + 4x + 5}$ (b) $I = \int \dfrac{dx}{\sqrt{4x^2 + 4x + 5}}$ (c) $I = \int \sqrt{4x^2 + 4x + 5}\ dx$

Solution:– (a) $I = \int \dfrac{dx}{4x^2 + 4x + 5} = \int \dfrac{dx}{ax^2 + bx + c}$, here $a = 4, b = 4$ and $c = 5$ then $D = 16 - 80 = -64$

$\therefore\ I = \int \dfrac{dx}{a\left\{\left(x + \frac{b}{2a}\right)^2 - \frac{D}{4a^2}\right\}} = \int \dfrac{dx}{4\left\{\left(x + \frac{4}{8}\right)^2 - \frac{-64}{64}\right\}} = \int \dfrac{dx}{(2x+1)^2 + 4} = \int \dfrac{dx}{(2x+1)^2 + (2)^2}$

use formula, $\therefore\ I = \int \dfrac{dx}{x^2 + a^2} = \dfrac{1}{a}\tan^{-1}\dfrac{x}{a}$ $\therefore\ I = \int \dfrac{dx}{(2x+1)^2 + (2)^2}$

Let $t = 2x + 1$, $dt = 2\ dx$ $\therefore\ I = \dfrac{1}{2}\int \dfrac{dt}{(t)^2 + (2)^2} = \dfrac{1}{2}\cdot\dfrac{1}{2}\tan^{-1}\left(\dfrac{t}{2}\right) + c = \dfrac{1}{4}\tan^{-1}\left(\dfrac{2x+1}{2}\right) + c$ Ans.

(b) $I = \int \dfrac{dx}{\sqrt{4x^2 + 4x + 5}} = \int \dfrac{dx}{\sqrt{ax^2 + bx + c}} = \int \dfrac{dx}{\sqrt{(2x+1)^2 + (2)^2}}$ formula, $\int \dfrac{dx}{\sqrt{x^2 + a^2}} = \log\left|x + \sqrt{x^2 + a^2}\right|$

$\therefore\ I = \int \dfrac{dx}{\sqrt{(2x+1)^2 + (2)^2}} = \log\left[(2x+1) + \sqrt{4x^2 + 4x + 5}\right]\cdot\dfrac{1}{\frac{d}{dx}(2x+1)} + c = \dfrac{1}{2}\log\left[(2x+1) + \sqrt{4x^2 + 4x + 5}\right] + c$ Ans.

(c) $I = \int \sqrt{4x^2 + 4x + 5}\ dx = \int \sqrt{ax^2 + bx + c}\ dx = \int \sqrt{(2x+1)^2 + (2)^2}\ dx$

Let $t = 2x + 1$ $\therefore\ dt = 2\ dx$ use formula, $\int \sqrt{x^2 + a^2}\ dx = \dfrac{x}{2}\sqrt{x^2 + a^2} + \dfrac{a^2}{2}\log\left|x + \sqrt{x^2 + a^2}\right|$

$\therefore\ I = \dfrac{1}{2}\int \sqrt{(t)^2 + (2)^2}\ dt = \dfrac{1}{2}\left[\dfrac{t}{2}\sqrt{t^2 + 4} + 2\log\left|t + \sqrt{t^2 + 4}\right|\right]$ Put $t = 2x + 1$

$\therefore\ I = \dfrac{(2x+1)}{4}\sqrt{(2x+1)^2 + (2)^2} + \log\left|(2x+1) + \sqrt{(2x+1)^2 + (2)^2}\right| + c$ Ans.

 # Form of the integrals: $-\int \dfrac{Ax + B}{ax^2 + bx + c}\ dx$, $\int \dfrac{Ax + B}{\sqrt{ax^2 + bx + c}}\ dx$, $\int (Ax + B)\sqrt{ax^2 + bx + c}\ dx$

Express $Ax + B = l[\ d.c\ of\ (ax^2 + bx + c)] + m \ldots\ldots\ldots\ldots\ldots\ldots\ldots (i)$

find the value of l, m by comparing the coefficient of x and constant term on both sides of (i)

$$\int \frac{f'(x)}{f(x)} \, dx = \log[f(x)] \text{ by putting } f(x) = t \quad \text{or} \quad \int \frac{f'(x)}{\sqrt{f(x)}} \, dx = 2\sqrt{f(x)} \text{ by putting } f(x) = t$$

$$\int f'(x) . \sqrt{f(x)} \, dx = \frac{2}{3}[f(x)]^{\frac{3}{2}} \text{ by putting } f(x) = t$$

Example: − (1) (a) $I = \int \dfrac{2x + 3}{4x^2 + 4x + 5} \, dx$ [use form of the integral $Ax + B = l(d.c \text{ of } ax^2 + bx + c) + m$]

$$\therefore \ 2x + 3 = l(d.c \text{ of } 4x^2 + 4x + 5) + m \quad \text{or} \quad 2x + 3 = l(8x + 4) + m$$

find the value of l, m by comparing the coefficient of x and constant term on both of sides.

$2x + 3 = l(8x + 4) + m \ \ldots\ldots\ldots\ldots (i) \quad \therefore \ 8l = 2 \ \text{and} \ 4l + m = 3 \quad \therefore \ l = \dfrac{1}{4} \ \text{and} \ 4.\dfrac{1}{4} + m = 3 \ \text{or} \ m = 2$

Divide $4x^2 + 4x + 5$ both of sides in equation (i), we have $\dfrac{2x + 3}{4x^2 + 4x + 5} = \dfrac{l(8x + 4)}{4x^2 + 4x + 5} + \dfrac{m}{4x^2 + 4x + 5}$

Integrating both of sides and put value of l and m, we have

$$\int \frac{2x + 3}{4x^2 + 4x + 5} \, dx = \frac{1}{4} \int \frac{8x + 4}{4x^2 + 4x + 5} \, dx + \int \frac{2}{4x^2 + 4x + 5} \, dx$$

or $I = I_1 + I_2$ where $I_1 = \dfrac{1}{4} \int \dfrac{8x + 4}{4x^2 + 4x + 5} \, dx$ and $I_2 = \int \dfrac{2}{4x^2 + 4x + 5} \, dx$

$I_1 = \dfrac{1}{4} \int \dfrac{8x + 4}{4x^2 + 4x + 5} \, dx$ Let $f(x) = 4x^2 + 4x + 5$ $\therefore \ f'(x) = 8x + 4$ $\left[\text{formula, } \int \dfrac{f'(x)}{f(x)} \, dx = \log[f(x)]\right]$

$I_1 = \dfrac{1}{4} \int \dfrac{f'(x)}{f(x)} \, dx = \dfrac{1}{4} \int \dfrac{8x + 4}{4x^2 + 4x + 5} \, dx = \dfrac{1}{4} \log[4x^2 + 4x + 5] + c$

$I_2 = \int \dfrac{2}{4x^2 + 4x + 5} \, dx = 2 \int \dfrac{dx}{(2x + 1)^2 + (2)^2} = 2 \int \dfrac{dx}{x^2 + a^2}$ $\left[\text{formula, } \int \dfrac{dx}{x^2 + a^2} = \dfrac{1}{a}\tan^{-1}\dfrac{x}{a}\right]$

$I_2 = 2.\dfrac{1}{2}\tan^{-1}\left(\dfrac{2x + 1}{2}\right).\dfrac{1}{d.c \text{ of } (2x + 1)} + c = \tan^{-1}\left(\dfrac{2x + 1}{2}\right).\dfrac{1}{2} + c = \dfrac{1}{2}\tan^{-1}\left(\dfrac{2x + 1}{2}\right) + c$

or $I = I_1 + I_2 = \dfrac{1}{4}\log[4x^2 + 4x + 5] + c + \dfrac{1}{2}\tan^{-1}\left(\dfrac{2x + 1}{2}\right) + c = \dfrac{1}{4}\log[4x^2 + 4x + 5] + \dfrac{1}{2}\tan^{-1}\left(\dfrac{2x + 1}{2}\right) + 2c$ Ans.

(b) $I = \int \dfrac{2x + 3}{\sqrt{4x^2 + 4x + 5}} \, dx$ (see above question) $l = \dfrac{1}{4}$ and $m = 2$

$$\int \frac{2x + 3}{\sqrt{4x^2 + 4x + 5}} \, dx = l \int \frac{8x + 4}{\sqrt{4x^2 + 4x + 5}} \, dx + m \int \frac{dx}{\sqrt{4x^2 + 4x + 5}}$$

$\left[\text{formula, } \int \dfrac{f'(x)}{\sqrt{f(x)}} \, dx = 2\sqrt{f(x)} \ \text{and} \ \int \dfrac{dx}{\sqrt{x^2 + a^2}} = \log\left|x + \sqrt{x^2 + a^2}\right|\right]$

or $I = \int \dfrac{2x + 3}{\sqrt{4x^2 + 4x + 5}} \, dx = \dfrac{1}{4} \int \dfrac{8x + 4}{\sqrt{4x^2 + 4x + 5}} \, dx + 2 \int \dfrac{dx}{\sqrt{(2x + 1)^2 + (2)^2}}$

$= \dfrac{1}{4}.2\sqrt{4x^2 + 4x + 5} + 2.\dfrac{1}{2}\log\left|(2x + 1) + \sqrt{4x^2 + 4x + 5}\right| + c$

$= \dfrac{1}{2}\sqrt{4x^2 + 4x + 5} + \log\left|(2x + 1) + \sqrt{4x^2 + 4x + 5}\right| + c$ Ans.

(c) $I = \int (2x + 3)\sqrt{4x^2 + 4x + 5} \, dx$ (see above question) $l = \dfrac{1}{4}$ and $m = 2$

$$\int (2x + 3)\sqrt{4x^2 + 4x + 5} \, dx = l \int (8x + 4)\sqrt{4x^2 + 4x + 5} \, dx + m \int \frac{dx}{\sqrt{4x^2 + 4x + 5}}$$

$$\therefore \; I = \int (2x+3)\sqrt{4x^2+4x+5}\; dx = \frac{1}{4}\int (8x+4)\sqrt{4x^2+4x+5}\; dx + 2\int \frac{dx}{\sqrt{4x^2+4x+5}}$$

$$\left[\text{formula} \; \int \sqrt{x^2+a^2}\; dx = \frac{x}{2}\sqrt{x^2+a^2} + \frac{a^2}{2}\log\left|x+\sqrt{x^2+a^2}\right| \; \text{and} \; \int f'(x)\cdot\sqrt{f(x)}\; dx = \frac{2}{3}[f(x)]^{\frac{3}{2}} \right]$$

$$I = \frac{1}{4}\cdot\frac{2}{3}[4x^2+4x+5]^{\frac{3}{2}} + 2\int \frac{dx}{\sqrt{(2x+1)^2+(2)^2}}$$

$$I = \frac{1}{6}[4x^2+4x+5]^{\frac{3}{2}} + \frac{1}{2}\left\{ 2\cdot\frac{(2x+1)}{2}\sqrt{4x^2+4x+5} + 2\cdot\frac{4}{2}\log\left[(2x+1)+\sqrt{4x^2+4x+5}\right] \right\} + c$$

$$\therefore \; I = \frac{1}{6}[4x^2+4x+5]^{\frac{3}{2}} + \frac{(2x+1)}{2}\sqrt{4x^2+4x+5} + 2\log\left[(2x+1)+\sqrt{4x^2+4x+5}\right] + c \quad \text{Ans.}$$

(2) (a) $I = \displaystyle\int \frac{1+\sin 4x}{\sin 4x + \cos 4x + 1}\; dx = \int \frac{\sin^2 2x + \cos^2 2x + 2\sin 2x\cos 2x}{\cos^2 2x - \sin^2 2x + 2\sin 2x\cos 2x + 1}\; dx$

$$I = \int \frac{(\sin 2x + \cos 2x)^2}{\cos^2 2x + 2\sin 2x\cos 2x + (1-\sin^2 2x)}\; dx = \int \frac{(\sin 2x + \cos 2x)^2}{2\cos^2 2x + 2\sin 2x\cos 2x}\; dx = \int \frac{(\sin 2x + \cos 2x)^2}{2\cos 2x(\cos 2x + \sin 2x)}\; dx$$

$$= \int \frac{\sin 2x + \cos 2x}{2\cos 2x}\; dx = \frac{1}{2}\int \left(\frac{\sin 2x}{\cos 2x} + 1\right) dx$$

$$I = \frac{1}{2}\int \frac{\sin 2x}{\cos 2x}\; dx + \frac{1}{2}\int dx = I_1 + I_2 \quad \text{where} \; I_1 = \frac{1}{2}\int \frac{\sin 2x}{\cos 2x}\; dx \quad \text{and} \quad I_2 = \frac{1}{2}\int dx$$

$$I_1 = \frac{1}{2}\int \frac{\sin 2x}{\cos 2x}\; dx \qquad \text{Let} \; \cos 2x = t \quad \therefore \; -\sin 2x \cdot 2\; dx = dt \quad \therefore \; \sin 2x\; dx = -\frac{dt}{2}$$

$$I_1 = \frac{1}{2}\int \frac{\sin 2x}{\cos 2x}\; dx = -\frac{1}{4}\int \frac{dt}{t} = -\frac{1}{4}\log t = -\frac{1}{4}\log(\cos 2x) + c$$

$$I_2 = \frac{1}{2}\int dx = \frac{1}{2}x + c$$

$$\therefore \; I = I_1 + I_2 = -\frac{1}{4}\log(\cos 2x) + c + \frac{1}{2}x + c = -\frac{1}{4}\log(\cos 2x) + \frac{1}{2}x + k \quad \text{where} \; k = 2c \quad \text{Ans.}$$

(b) $I = \displaystyle\int \frac{1+\sin(4\log x)}{x(\log x + 1)}\; dx \qquad \text{Let} \; \log x = t \quad \therefore \; \frac{1}{x}dx = dt$

or $I = \displaystyle\int \frac{1+\sin(4\log x)}{(\log x + 1)}\cdot\frac{1}{x}\; dx = \int \frac{1+\sin 4t}{(t+1)}\; dt = \int \frac{1}{1+t}\; dt + \int \frac{\sin 4t}{1+t}\; dt = I_1 + I_2 \quad \text{(say)}$

where $I_1 = \displaystyle\int \frac{1}{1+t}\; dt = \log(1+t) + c = \log(1+\log x) + c$

$$\therefore \; I_2 = \int \frac{\sin 4t}{1+t}\; dt$$

Integrating by part formula, $\displaystyle\int u\cdot v\; dx = v\cdot\int u\; dx - \int \left[\frac{dv}{dx}\int u\; dx\right] dx$

or $I_2 = \displaystyle\int \frac{\sin 4t}{1+t}\; dt = \int \frac{1}{1+t}\cdot\sin 4t\; dt = \sin 4t\cdot\int \frac{1}{1+t}\; dt - \int \left[\frac{d}{dx}(\sin 4t)\int \frac{1}{1+t}\; dt\right] dt = \sin 4t\cdot\log(1+t) - \int \log(1+t)\cdot\cos 4t\cdot 4\; dt$

$$= \sin 4t\cdot\log(1+t) - 4\int \cos 4t\log(1+t)\; dt$$

Let $I_3 = \displaystyle\int \cos 4t\log(1+t)\; dt$, Again use integration by part formula

or $I_3 = \displaystyle\int \cos 4t\log(1+t)\; dt = \cos 4t\cdot\int \log(1+t)\; dt - \int \left[\frac{d}{dx}(\cos 4t)\int \log(1+t)\; dt\right] dt = \frac{\cos 4t}{1+t} + 4\int \sin 4t\cdot\frac{1}{1+t}\; dt = \frac{\cos 4t}{1+t} + 4I_2$

---

or $\quad I_2 = \sin 4t.\log(1+t) - 4\left[\dfrac{\cos 4t}{1+t} + 4I_2\right] = \sin 4t.\log(1+t) - 4.\dfrac{\cos 4t}{1+t} - 16I_2$

or $\quad 17I_2 = \sin 4t.\log(1+t) - \dfrac{4\cos 4t}{1+t} = \dfrac{(1+t).\sin 4t.\log(1+t) - 4\cos 4t}{1+t}$

or $\quad I_2 = \dfrac{(1+t).\sin 4t.\log(1+t) - 4\cos 4t}{17(1+t)} + c$

Now, $\quad I = I_1 + I_2 = \log(1+t) + c + \dfrac{(1+t).\sin 4t.\log(1+t) - 4\cos 4t}{17(1+t)} + c$

$$= \dfrac{17(1+t).\log(1+t) + (1+t).\sin 4t.\log(1+t) - 4\cos 4t}{17(1+t)} + k \quad \text{where } k = 2c$$

$\therefore \quad I = \dfrac{(1+t).\log(1+t)\,[17 + \sin 4t] - 4\cos 4t}{17(1+t)} + k \qquad \text{Ans.}$

(3) (a) $I = \displaystyle\int \dfrac{1 + \tan 2\theta}{\cot\theta - \tan\theta}\, d\theta = \int \dfrac{1 + \dfrac{\sin 2\theta}{\cos 2\theta}}{\dfrac{\cos\theta}{\sin\theta} - \dfrac{\sin\theta}{\cos\theta}}\, d\theta = \int \dfrac{\dfrac{\cos 2\theta + \sin 2\theta}{\cos 2\theta}}{\dfrac{\cos^2\theta - \sin^2\theta}{\sin\theta\cos\theta}}\, d\theta = \int \dfrac{(\cos 2\theta + \sin 2\theta)\sin\theta\cos\theta}{\cos 2\theta.\cos 2\theta}\, d\theta$

or $\quad I = \displaystyle\int \dfrac{(\cos 2\theta + \sin 2\theta)2\sin\theta\cos\theta}{2\cos^2 2\theta}\, d\theta = \int \dfrac{(\cos 2\theta + \sin 2\theta)\sin 2\theta}{2\cos^2 2\theta}\, d\theta = \int \dfrac{\sin 2\theta.\cos 2\theta + \sin^2 2\theta}{2\cos^2 2\theta}\, d\theta$

or $\quad I = \displaystyle\int \dfrac{\sin 2\theta.\cos 2\theta}{2\cos^2 2\theta}\, d\theta + \int \dfrac{\sin^2 2\theta}{2\cos^2 2\theta}\, d\theta = \dfrac{1}{2}\int \tan 2\theta\ d\theta + \dfrac{1}{2}\int \tan^2 2\theta\ d\theta = \dfrac{1}{2}I_1 + \dfrac{1}{2}I_2 \quad \text{(say)}$

where $I_1 = \displaystyle\int \tan 2\theta\ d\theta = \int \dfrac{\sin 2\theta}{\cos 2\theta}\ d\theta$ $\quad$ Let $\cos 2\theta = t$ $\quad \therefore\ -2\sin 2\theta\, d\theta = dt$ $\quad \therefore\ 2\sin 2\theta\, d\theta = -dt$

$\therefore\ I_1 = -\dfrac{1}{2}\displaystyle\int \dfrac{dt}{t} = -\dfrac{1}{2}\log t + c = -\dfrac{1}{2}\log(\cos 2\theta) + c$

where $I_2 = \displaystyle\int \tan^2 2\theta\ d\theta = \int (\sec^2 2\theta - 1)\, d\theta = \int \sec^2 2\theta\, d\theta - \int d\theta$

Let $2\theta = t$ $\quad \therefore\ 2\, d\theta = dt$ or $d\theta = \dfrac{dt}{2}$ $\quad \therefore\ I_2 = \dfrac{1}{2}\displaystyle\int \sec^2 t\, dt - \theta = \dfrac{1}{2}\tan t - \theta = \dfrac{1}{2}\tan 2\theta - \theta + c$

$\therefore\ I = \dfrac{1}{2}I_1 + \dfrac{1}{2}I_2 = \dfrac{-\log(\cos 2\theta)}{4} + \dfrac{c}{2} + \dfrac{1}{4}\tan 2\theta - \dfrac{\theta}{2} + \dfrac{c}{2} = -\dfrac{1}{4}\log(\cos 2\theta) + \dfrac{1}{4}\tan 2\theta - \dfrac{1}{2}\theta + c \quad \text{Ans.}$

(b) $I = \displaystyle\int \dfrac{\sec x + \tan x}{1 + \operatorname{cosec} x}\, dx = \int \dfrac{\dfrac{1}{\cos x} + \dfrac{\sin x}{\cos x}}{1 + \dfrac{1}{\sin x}}\, dx = \int \dfrac{\dfrac{1 + \sin x}{\cos x}}{\dfrac{\sin x + 1}{\sin x}}\, dx = \int \dfrac{1 + \sin x}{\cos x} \times \dfrac{\sin x}{1 + \sin x}\, dx = \int \dfrac{\sin x}{\cos x}\, dx$

Let $\cos x = t$ $\quad \therefore\ -\sin x\, dx = dt$ then $I = \displaystyle\int \dfrac{\sin x}{\cos x}\, dx = -\int \dfrac{1}{t}\, dt = -\log t + c = -\log(\cos x) + c$ $\quad$ Ans.

(c) $I = \displaystyle\int \dfrac{e^{\sqrt{x}}\left(\sin e^{\sqrt{x}} + \cos e^{\sqrt{x}}\right)}{\sqrt{x}\left(1 + e^{\sqrt{x}}\right)}\, dx$ $\quad$ Let $e^{\sqrt{x}} = t$ $\quad \therefore\ e^{\sqrt{x}}.\dfrac{1}{2\sqrt{x}}\, dx = dt$ or $\dfrac{e^{\sqrt{x}}}{\sqrt{x}}\, dx = 2dt$

$\therefore\ I = \displaystyle\int \dfrac{\left(\sin e^{\sqrt{x}} + \cos e^{\sqrt{x}}\right)}{\left(1 + e^{\sqrt{x}}\right)}.\dfrac{e^{\sqrt{x}}}{\sqrt{x}}\, dx = \int \dfrac{(\sin t + \cos t)}{(1+t)}.2\, dt = 2\int \dfrac{\sin t}{1+t}\, dt + 2\int \dfrac{\cos t}{1+t}\, dt = 2I_1 + 2I_2 \quad \text{(say)}$

Solve I_1 and I_2 seprately,

where $I_1 = \displaystyle\int \dfrac{\sin t}{1+t}\, dt$ and $I_2 = \displaystyle\int \dfrac{\cos t}{1+t}\, dt$ $\quad$ (using integration by part formula)

or $I_1 = \displaystyle\int \dfrac{\sin t}{1+t}\, dt = \sin t.\int \dfrac{dt}{1+t} - \int \left[\dfrac{d}{dt}(\sin t)\int \dfrac{dt}{1+t}\right] dt = \sin t.\log(1+t) - \int \log(1+t).\cos t\ dt$

or $I_1 = \sin t.\log(1+t) - \left\{\cos t.\int \log(1+t)\,dt - \int\left[\dfrac{d}{dt}(\cos t)\int \log(1+t)\,dt\right]dt\right\} = \sin t.\log(1+t) - \left[\dfrac{\cos t}{1+t} + \int \sin t.\dfrac{1}{1+t}\,dt\right]$

$\qquad = \sin t.\log(1+t) - \dfrac{\cos t}{1+t} - I_1$

or $2I_1 = \sin t.\log(1+t) - \dfrac{\cos t}{1+t} = \dfrac{(1+t)\sin t.\log(1+t) - \cos t}{1+t} + c \quad \therefore\ I_1 = \dfrac{(1+t)\sin t.\log(1+t) - \cos t}{2(1+t)} + c$

where $I_2 = \displaystyle\int \dfrac{\cos t}{1+t}\,dt \quad$ using integration by part formula

or $I_2 = \displaystyle\int \dfrac{\cos t}{1+t}\,dt = \cos t.\int \dfrac{dt}{1+t} - \int\left[\dfrac{d}{dt}(\cos t)\int \dfrac{dt}{1+t}\right]dt = \sin t.\log(1+t) + \int \sin t.\log(1+t)\,dt$

$\qquad = \sin t.\log(1+t) + \left\{\sin t.\dfrac{1}{1+t} - \int \cos t.\dfrac{1}{1+t}\,dt\right\} = \sin t.\log(1+t) + \dfrac{\sin t}{1+t} - I_2$

or $2I_2 = \sin t.\log(1+t) + \dfrac{\sin t}{(1+t)} + c = \dfrac{(1+t)\sin t.\log(1+t) + \sin t}{(1+t)} + c$

or $I = 2I_1 + 2I_2 = \dfrac{(1+t)\sin t.\log(1+t) - \cos t}{(1+t)} + c + \dfrac{(1+t)\sin t.\log(1+t) + \sin t}{(1+t)} + c$

$\qquad = \dfrac{(1+t)\sin t.\log(1+t) - \cos t + (1+t)\sin t.\log(1+t) + \sin t}{(1+t)} + k$

$\therefore\ I = \dfrac{(1+t)\log(1+t)\,[\sin t + \cos t] + (\sin t - \cos t)}{(1+t)} + k = \log(1+t)\,[\sin t + \cos t] + \dfrac{(\sin t - \cos t)}{(1+t)} + k \quad$ Ans.

Form of the integrals: $-\displaystyle\int \dfrac{dx}{a\cos^2 x + 2b\sin x\cos x + c\sin^2 x}$, $\quad\displaystyle\int \dfrac{dx}{a\cos^2 x + b}$, $\quad\displaystyle\int \dfrac{dx}{a + b\sin^2 x}$

In above type of question divide above and below by $\cos^2 x$.

Example: $-$ (a) $I = \displaystyle\int \dfrac{dx}{3\cos^2 x + 2\sin x\cos x + \sin^2 x}$, Divide above and below by $\cos^2 x$

or $I = \displaystyle\int \dfrac{\sec^2 x}{3 + 2\tan x + \tan^2 x}\,dx \quad$ Let $\tan x = t \quad \therefore\ \sec^2 x\,dx = dt$

or $I = \displaystyle\int \dfrac{dt}{3 + 2t + t^2} = \int \dfrac{dt}{(t+1)^2 + \left(\sqrt{2}\right)^2} \qquad \left[\text{formula, } \displaystyle\int \dfrac{dx}{x^2 + a^2} = \dfrac{1}{a}\tan^{-1}\dfrac{x}{a}\right]$

$\therefore\ I = \displaystyle\int \dfrac{dt}{(t+1)^2 + \left(\sqrt{2}\right)^2} = \dfrac{1}{\sqrt{2}}\tan^{-1}\left(\dfrac{t+1}{\sqrt{2}}\right) = \dfrac{1}{\sqrt{2}}\tan^{-1}\left(\dfrac{\tan x + 1}{\sqrt{2}}\right) + c \quad$ Ans.

(b) $I = \displaystyle\int \dfrac{dx}{2\cos^2 x + 3} \qquad$ Divide above and below by $\cos^2 x$, we have

or $I = \displaystyle\int \dfrac{\sec^2 x}{2 + 3\sec^2 x}\,dx = \int \dfrac{\sec^2 x}{2 + 3(1 + \tan^2 x)}\,dx = \int \dfrac{\sec^2 x}{5 + 3\tan^2 x}\,dx \quad$ Let $\tan x = t \quad \therefore\ \sec^2 x\,dx = dt$

or $I = \displaystyle\int \dfrac{dt}{5 + 3t^2} = \int \dfrac{dt}{\left(\sqrt{5}\right)^2 + \left(\sqrt{3}t\right)^2} \qquad \left[\text{formula, } \displaystyle\int \dfrac{dx}{x^2 + a^2} = \dfrac{1}{a}\tan^{-1}\dfrac{x}{a}\right]$

$\therefore\ I = \displaystyle\int \dfrac{dt}{\left(\sqrt{5}\right)^2 + \left(\sqrt{3}t\right)^2} = \dfrac{1}{\sqrt{5}}\tan^{-1}\left(\dfrac{\sqrt{3}t}{\sqrt{5}}\right).\dfrac{1}{\sqrt{3}} = \dfrac{1}{\sqrt{15}}\tan^{-1}\left(\dfrac{\sqrt{3}t}{\sqrt{5}}\right) + c = \dfrac{1}{\sqrt{15}}\tan^{-1}\left(\dfrac{\sqrt{3}\tan x}{\sqrt{5}}\right) + c \quad$ Ans.

(c) $I = \displaystyle\int \dfrac{dx}{3 - 5\sin^2 x} \qquad$ Divide above and below by $\cos^2 x$, we have

or $I = \displaystyle\int \dfrac{\sec^2 x}{3\sec^2 x - 5\tan^2 x}\,dx = \int \dfrac{\sec^2 x}{3(1 + \tan^2 x) - 5\tan^2 x}\,dx = \int \dfrac{\sec^2 x}{3 + 3\tan^2 x - 5\tan^2 x}\,dx = \int \dfrac{\sec^2 x}{3 - 2\tan^2 x}\,dx$

Let $\tan x = t$ $\therefore \sec^2 x \, dx = dt$ $\therefore I = \int \dfrac{\sec^2 x}{3 - 2\tan^2 x} dx = \int \dfrac{dt}{3 - 2t^2} = \int \dfrac{dt}{\left(\sqrt{3}\right)^2 - \left(\sqrt{2}t\right)^2}$

Let $\sqrt{2}t = z$ $\therefore \sqrt{2}\, dt = dz$ or $dt = \dfrac{dz}{\sqrt{2}}$ $\therefore I = \int \dfrac{dt}{\left(\sqrt{3}\right)^2 - \left(\sqrt{2}t\right)^2} = \dfrac{1}{\sqrt{2}} \int \dfrac{dz}{\left(\sqrt{3}\right)^2 - (z)^2}$

using formula, $\int \dfrac{dx}{a^2 - x^2} = \dfrac{1}{2a} \log \left|\dfrac{a + x}{a - x}\right|$, when $x < a$ and $put \ \tan x = t, z = \sqrt{2}t$

or $I = \dfrac{1}{\sqrt{2}} \int \dfrac{dz}{\left(\sqrt{3}\right)^2 - (z)^2} = \dfrac{1}{\sqrt{2}} \cdot \dfrac{1}{2.\sqrt{3}} \log \left|\dfrac{\sqrt{3} + z}{\sqrt{3} - z}\right| + c = \dfrac{1}{2\sqrt{6}} \log\left(\dfrac{\sqrt{3} + \sqrt{2}t}{\sqrt{3} - \sqrt{2}t}\right) + c = \dfrac{1}{2\sqrt{6}} \log\left(\dfrac{\sqrt{3} + \sqrt{2}\tan x}{\sqrt{3} - \sqrt{2}\tan x}\right) + c$ Ans.

Form of the integrals: $- \int \dfrac{dx}{a\cos x + b\sin x + c}$, $\int \dfrac{dx}{a + b\cos x}$ and $\int \dfrac{dx}{a + b\sin x}$

write $\cos x = \cos^2 \dfrac{x}{2} - \sin^2 \dfrac{x}{2}$, $\sin x = 2\sin\dfrac{x}{2}\cos\dfrac{x}{2}$ Then, Divide above and below by $\cos^2 \dfrac{x}{2}$.

Form of the integrals: $- \int \dfrac{p\cos x + q\sin x + r}{a\cos x + b\sin x + c} dx$

Express, $p\cos x + q\sin x + r = l(a\cos x + b\sin x + c) + m(\text{d.c of } a\cos x + b\sin x + c) + n$

Divide both of sides by $(a\cos x + b\sin x + c)$ and integrating,

$\int \dfrac{p\cos x + q\sin x + r}{a\cos x + b\sin x + c} dx = l\int \dfrac{a\cos x + b\sin x + c}{a\cos x + b\sin x + c} dx + m\int \dfrac{\text{d.c of } (a\cos x + b\sin x + c)}{a\cos x + b\sin x + c} dx + n\int \dfrac{dx}{a\cos x + b\sin x + c}$

$\int \dfrac{p\cos x + q\sin x + r}{a\cos x + b\sin x + c} dx = l\int dx + m\int \dfrac{\text{d.c of } (a\cos x + b\sin x + c)}{a\cos x + b\sin x + c} dx + n\int \dfrac{dx}{a\cos x + b\sin x + c}$

find l, m and n by comparing the coefficient of $\sin x, \cos x$ and constant term.

Form of the integrals: $- \int \dfrac{p\cos x + q\sin x}{a\cos x + b\sin x} dx$

Express, $p\cos x + q\sin x = l(a\cos x + b\sin x) + m(\text{d.c of } a\cos x + b\sin x)$ and find l and m by comparing the coefficient of $\sin x \cos x$.

Divide both of side by $a\cos x + b\sin x$ (D^r) and integrating.

$\int \dfrac{p\cos x + q\sin x}{a\cos x + b\sin x} dx = l\int \dfrac{a\cos x + b\sin x}{a\cos x + b\sin x} dx + m\int \dfrac{\text{d.c of } a\cos x + b\sin x}{a\cos x + b\sin x} dx = l\int dx + m\int \dfrac{\text{d.c of } a\cos x + b\sin x}{a\cos x + b\sin x} dx$

Example: $-$ (1) (a) $I = \int \dfrac{dx}{3\cos x + 5\sin x + 2} = \int \dfrac{dx}{3\left(\cos^2 \frac{x}{2} - \sin^2 \frac{x}{2}\right) + 5.2\sin\frac{x}{2}\cos\frac{x}{2} + 2} = \int \dfrac{dx}{3\cos^2 \frac{x}{2} - 3\sin^2 \frac{x}{2} + 10\sin\frac{x}{2}\cos\frac{x}{2} + 2}$

Divide above and below by $\cos^2 \dfrac{x}{2}$, we have $\therefore I = \int \dfrac{\sec^2 \frac{x}{2}}{3 - 3\tan^2 \frac{x}{2} + 10\tan\frac{x}{2} + 2\sec^2 \frac{x}{2}} dx$

or $I = \int \dfrac{\sec^2 \frac{x}{2} \, dx}{3 - 3\tan^2 \frac{x}{2} + 10\tan\frac{x}{2} + 2\left(1 + \tan^2 \frac{x}{2}\right)} = \int \dfrac{\sec^2 \frac{x}{2} \, dx}{5 + 10\tan\frac{x}{2} - \tan^2 \frac{x}{2}}$

Let $\tan \frac{x}{2} = t$ $\therefore \sec^2 \frac{x}{2} \cdot \dfrac{1}{2} dx = dt$ or $\sec^2 \frac{x}{2} \, dx = 2 \, dt$

or $I = \int \dfrac{\sec^2 \frac{x}{2} \, dx}{5 + 10\tan\frac{x}{2} - \tan^2 \frac{x}{2}} = \int \dfrac{2 \, dt}{5 + 10t - t^2} = -2\int \dfrac{dt}{t^2 - 10t - 5} = -2\int \dfrac{dt}{(t - 5)^2 - \left(\sqrt{30}\right)^2}$

using formula, $\int \dfrac{dx}{x^2 - a^2} = \dfrac{1}{2a} \log \left|\dfrac{x - a}{x + a}\right|$, when $x > a$

$$\therefore\ I = -2\int \frac{dt}{(t-5)^2 - \left(\sqrt{30}\right)^2} = -2.\frac{1}{2.\sqrt{30}}\log\left|\frac{(t-5)-\sqrt{30}}{(t-5)+\sqrt{30}}\right| + c = -\frac{1}{\sqrt{30}}\log\left[\frac{\left(\tan\frac{x}{2}-5\right)-\sqrt{30}}{\left(\tan\frac{x}{2}-5\right)+\sqrt{30}}\right] + c \quad \text{Ans.}$$

(b) $\ I = \int \dfrac{dx}{3 - 5\sin x} = \int \dfrac{dx}{3 - 5.2\sin\frac{x}{2}\cos\frac{x}{2}} = \int \dfrac{dx}{3 - 10\sin\frac{x}{2}\cos\frac{x}{2}}$ \quad Divide above and below by $\cos^2\frac{x}{2}$, we have

or $\ I = \int \dfrac{\sec^2\frac{x}{2}\,dx}{3\sec^2\frac{x}{2} - 10\tan\frac{x}{2}} = \int \dfrac{\sec^2\frac{x}{2}\,dx}{3\left(1+\tan^2\frac{x}{2}\right) - 10\tan\frac{x}{2}} = \int \dfrac{\sec^2\frac{x}{2}\,dx}{3 + 3\tan^2\frac{x}{2} - 10\tan\frac{x}{2}}$

Let $\ \tan\frac{x}{2} = t\ \ \therefore\ \sec^2\frac{x}{2}.\frac{1}{2}dx = dt\ $ or $\ \sec^2\frac{x}{2}\,dx = 2\,dt$

or $\ I = \int \dfrac{\sec^2\frac{x}{2}\,dx}{3 + 3\tan^2\frac{x}{2} - 10\tan\frac{x}{2}} = \int \dfrac{2\,dt}{3 + 3t^2 - 10t} = 2\int \dfrac{dt}{3t^2 - 10t + 3}$

using $\left[ax^2 + bx + c = 0\ \therefore\ a\left\{\left(x+\dfrac{b}{2a}\right)^2 - \dfrac{D}{4a^2}\right\} = 0\ \text{and}\ \int \dfrac{dx}{x^2 - a^2} = \dfrac{1}{2a}\log\left|\dfrac{x-a}{x+a}\right|,\ \text{when}\ x > a\right]$

or $\ I = 2\int \dfrac{dt}{3t^2 - 10t + 3} = 2\int \dfrac{dt}{\left(\sqrt{3}t - \frac{5}{\sqrt{3}}\right)^2 - \left(\frac{4}{\sqrt{3}}\right)^2} = 2.\dfrac{1}{2.\frac{4}{\sqrt{3}}}\log\left|\dfrac{\sqrt{3}t - \frac{5}{\sqrt{3}} - \frac{4}{\sqrt{3}}}{\sqrt{3}t - \frac{5}{\sqrt{3}} + \frac{4}{\sqrt{3}}}\right|.\dfrac{1}{\sqrt{3}} + c = \dfrac{1}{4}\log\left(\dfrac{3t - 5 - 4}{3t - 5 + 4}\right) + c$

$$\therefore\ I = \frac{1}{4}\log\left(\frac{3t - 9}{3t - 1}\right) + c = \frac{1}{4}\log\left(\frac{3\tan\frac{x}{2} - 9}{3\tan\frac{x}{2} - 1}\right) + c \quad \text{Ans.}$$

(c) $\ I = \int \dfrac{dx}{3 + \cos x} = \int \dfrac{dx}{3 + \cos^2\frac{x}{2} - \sin^2\frac{x}{2}}$, \quad Divide above and below by $\cos^2\frac{x}{2}$ we have

or $\ I = \int \dfrac{\sec^2\frac{x}{2}\,dx}{3\sec^2\frac{x}{2} + 1 - \tan^2\frac{x}{2}} = \int \dfrac{\sec^2\frac{x}{2}\,dx}{3\left(1+\tan^2\frac{x}{2}\right) + 1 - \tan^2\frac{x}{2}} = \int \dfrac{\sec^2\frac{x}{2}\,dx}{3 + 3\tan^2\frac{x}{2} + 1 - \tan^2\frac{x}{2}} = \int \dfrac{\sec^2\frac{x}{2}\,dx}{2\tan^2\frac{x}{2} + 4}$

$$= \int \dfrac{\sec^2\frac{x}{2}\,dx}{2\left(\tan^2\frac{x}{2} + 2\right)} = \dfrac{1}{2}\int \dfrac{\sec^2\frac{x}{2}\,dx}{\left(\tan^2\frac{x}{2} + 2\right)}$$

Let $\ \tan\frac{x}{2} = t\ \ \therefore\ \sec^2\frac{x}{2}.\frac{1}{2}dx = dt\ $ or $\ \sec^2\frac{x}{2}\,dx = 2\,dt$

or $\ I = 2.\dfrac{1}{2}\int \dfrac{dt}{2 + t^2} = \int \dfrac{dt}{\left(\sqrt{2}\right)^2 + t^2} = \dfrac{1}{\sqrt{2}}\tan^{-1}\left(\dfrac{t}{\sqrt{2}}\right) + c$ \quad $\left[\text{using formula,}\ \int \dfrac{dx}{x^2 + a^2} = \dfrac{1}{a}\tan^{-1}\dfrac{x}{a}\right]$

$$\therefore\ I = \frac{1}{\sqrt{2}}\tan^{-1}\left(\frac{t}{\sqrt{2}}\right) + c = \frac{1}{\sqrt{2}}\tan^{-1}\left(\frac{\tan\frac{x}{2}}{\sqrt{2}}\right) + c \quad \text{Ans.}$$

(d) $\ I = \int \dfrac{1 + \cos x}{\sin x + \cos x}dx = \int \dfrac{1 + \cos^2\frac{x}{2} - \sin^2\frac{x}{2}}{2\sin\frac{x}{2}\cos\frac{x}{2} + \cos^2\frac{x}{2} - \sin^2\frac{x}{2}}dx = \int \dfrac{2\cos^2\frac{x}{2}}{2\sin\frac{x}{2}\cos\frac{x}{2} + \cos^2\frac{x}{2} - \sin^2\frac{x}{2}}dx$

Divide above and below by $\cos^2\frac{x}{2}$, we have

or $\ I = \int \dfrac{2\cos^2\frac{x}{2}}{2\sin\frac{x}{2}\cos\frac{x}{2} + \cos^2\frac{x}{2} - \sin^2\frac{x}{2}}dx = \int \dfrac{2}{2\tan\frac{x}{2} + 1 - \tan^2\frac{x}{2}}dx = 2\int \dfrac{dx}{1 + 2\tan\frac{x}{2} - \tan^2\frac{x}{2}}$

Let $\ \tan\frac{x}{2} = t\ \ \therefore\ \sec^2\frac{x}{2}.\frac{1}{2}dx = dt\ $ or $\ \sec^2\frac{x}{2}\,dx = 2\,dt$

or $\ I = 4\int \dfrac{dt}{1 + 2t - t^2} = -4\int \dfrac{dt}{t^2 - 2t - 1} = -4\int \dfrac{dt}{(t-1)^2 - \left(\sqrt{2}\right)^2}$ \quad $\left[\int \dfrac{dx}{x^2 - a^2} = \dfrac{1}{2a}\log\left|\dfrac{x-a}{x+a}\right|,\ \text{when}\ x > a\right]$

$$\therefore\ I = -4.\frac{1}{2.\sqrt{2}}\log\left|\frac{(t-1)-\sqrt{2}}{(t-1)+\sqrt{2}}\right| + c = -\sqrt{2}\log\left[\frac{\left(\tan\frac{x}{2}-1\right)-\sqrt{2}}{\left(\tan\frac{x}{2}-1\right)+\sqrt{2}}\right] + c \quad \text{Ans.}$$

---

(2) (a) $I = \int \dfrac{3\cos x + 5\sin x + 7}{2\cos x + 3\sin x + 8}\,dx = \int \dfrac{p\cos x + q\sin x + r}{a\cos x + b\sin x + c}\,dx$ (integral of the form)

using $\int \dfrac{p\cos x + q\sin x + r}{a\cos x + b\sin x + c}\,dx$ Express, $p\cos x + q\sin x + r = l(a\cos x + b\sin x + c) + m(d.c\ of\ a\cos x + b\sin x + c) + n$

Divide both of sides by $(a\cos x + b\sin x + c)$ and integrating,

Solution: $-$ $I = \int \dfrac{3\cos x + 5\sin x + 7}{2\cos x + 3\sin x + 8}\,dx$

$3\cos x + 5\sin x + 7 = l(2\cos x + 3\sin x + 8) + m(d.c\ of\ 2\cos x + 3\sin x + 8) + n$

find l, m and n by comparing the coefficient of $\sin x, \cos x$ and constant.

$3\cos x + 5\sin x + 7 = l(2\cos x + 3\sin x + 8) + m(3\cos x - 2\sin x) + n \ldots\ldots\ldots\ldots (A)$

$\therefore\ 2l + 3m = 3 \ldots\ldots\ldots (i),\ \ 3l - 2m = 5 \ldots\ldots\ldots\ldots (ii)\ \ \text{and}\ \ 8l + n = 7 \ldots\ldots\ldots\ldots (iii)$

solve (i) & (ii), we get $2l + 3m = 3$ and $3l - 2m = 5$ or $2m = 3l - 5$ or $m = \dfrac{3l - 5}{2}$

Put value of m in equation (i), we get $2l + 3\left(\dfrac{3l - 5}{2}\right) = 3$ or $4l + 9l - 15 = 6$ or $13l = 21$ $\therefore l = \dfrac{21}{13}$

Put value of l in equation (i) and (iii) then find value of m and n, we get

or $2l + 3m = 3$ or $2 \cdot \dfrac{21}{13} + 3m = 3$ or $3m = 3 - \dfrac{42}{13} = -\dfrac{3}{13}$ $\therefore\ m = -\dfrac{1}{13}$

and $8l + n = 7$ or $n = 7 - 8 \cdot \dfrac{21}{13} = \dfrac{91 - 168}{13} = -\dfrac{77}{13}$

Put value of l, m and n in equation (A), we get

$3\cos x + 5\sin x + 7 = l(2\cos x + 3\sin x + 8) + m(3\cos x - 2\sin x) + n = \dfrac{21}{13}(2\cos x + 3\sin x + 8) - \dfrac{1}{13}(3\cos x - 2\sin x) - \dfrac{77}{13}$

Divide both of sides by $(2\cos x + 3\sin x + 8)$ and integrate

$\int \dfrac{3\cos x + 5\sin x + 7}{2\cos x + 3\sin x + 8}\,dx = \dfrac{21}{13}\int \dfrac{2\cos x + 3\sin x + 8}{2\cos x + 3\sin x + 8}\,dx - \dfrac{1}{13}\int \dfrac{3\cos x - 2\sin x}{2\cos x + 3\sin x + 8}\,dx - \dfrac{77}{13}\int \dfrac{dx}{2\cos x + 3\sin x + 8}$

or $I = \int \dfrac{3\cos x + 5\sin x + 7}{2\cos x + 3\sin x + 8}\,dx = \dfrac{21}{13}\int dx - \dfrac{1}{13}\int \dfrac{3\cos x - 2\sin x}{2\cos x + 3\sin x + 8}\,dx - \dfrac{77}{13}\int \dfrac{dx}{2\cos x + 3\sin x + 8}$

Let $I_1 = \int \dfrac{3\cos x - 2\sin x}{2\cos x + 3\sin x + 8}\,dx$ and $I_2 = \int \dfrac{dx}{2\cos x + 3\sin x + 8}$

$I_1 = \int \dfrac{3\cos x - 2\sin x}{2\cos x + 3\sin x + 8}\,dx$

Let $2\cos x + 3\sin x + 8 = t$ $\therefore (-2\sin x + 3\cos x)dx = dt$ $\therefore (3\cos x - 2\sin x)dx = dt$ $\left[\text{formula,}\ \int \dfrac{f'(x)}{f(x)}\,dx = \log[f(x)]\right]$

$\therefore\ I_1 = \int \dfrac{dt}{t} = \log t + c = \log(2\cos x + 3\sin x + 8) + c$

Now, $I_2 = \int \dfrac{dx}{2\cos x + 3\sin x + 8} = \int \dfrac{dx}{2\left(\cos^2 {}^x\!/_2 - \sin^2 {}^x\!/_2\right) + 3 \cdot 2\sin {}^x\!/_2 \cos {}^x\!/_2 + 8}$

Divide above and below by $\cos^2 {}^x\!/_2$

$I_2 = \int \dfrac{\sec^2 {}^x\!/_2\,dx}{2 - 2\tan^2 {}^x\!/_2 + 6\tan {}^x\!/_2 + 8\sec^2 {}^x\!/_2} = \int \dfrac{\sec^2 {}^x\!/_2\,dx}{2 - 2\tan^2 {}^x\!/_2 + 6\tan {}^x\!/_2 + 8\left(1 + \tan^2 {}^x\!/_2\right)} = \int \dfrac{\sec^2 {}^x\!/_2\,dx}{6\tan^2 {}^x\!/_2 + 6\tan {}^x\!/_2 + 10}$

Let $\tan {}^{x}\!/_{2} = t$ ∴ $\sec^2 {}^{x}\!/_2 \cdot \dfrac{1}{2} dx = dt$ or $\sec^2 {}^{x}\!/_2 \, dx = 2 \, dt$

$$I_2 = \int \dfrac{2 \, dt}{6t^2 + 6t + 10} = \int \dfrac{2 \, dt}{2(3t^2 + 3t + 5)} = \int \dfrac{dt}{3t^2 + 3t + 5} = \int \dfrac{dt}{\left(\sqrt{3}t + \dfrac{\sqrt{3}}{2}\right)^2 + \left(\dfrac{\sqrt{17}}{2}\right)^2} \quad \left[\text{use formula,} \quad \int \dfrac{dx}{x^2 + a^2} = \dfrac{1}{a} \tan^{-1} {}^{x}\!/_a \cdot \dfrac{1}{\text{d.c of } x} \right]$$

$$I_2 = \int \dfrac{dt}{\left(\sqrt{3}t + \dfrac{\sqrt{3}}{2}\right)^2 + \left(\dfrac{\sqrt{17}}{2}\right)^2} = \dfrac{1}{\dfrac{\sqrt{17}}{2}} \tan^{-1}\left(\sqrt{3}t \Big/ \dfrac{\sqrt{17}}{2}\right) \cdot \dfrac{1}{\sqrt{3}} = \dfrac{2}{\sqrt{51}} \tan^{-1}\left(\dfrac{2\sqrt{3}t}{\sqrt{17}}\right) + c = \dfrac{2}{\sqrt{51}} \tan^{-1}\left(\dfrac{2\sqrt{3} \tan {}^{x}\!/_2}{\sqrt{17}}\right) + c$$

or $I = \dfrac{21}{13} \int dx - \dfrac{1}{13} I_1 - \dfrac{77}{13} I_2$

or $I = \dfrac{21}{13} x + c - \dfrac{1}{13} [\log(2\cos x + 3\sin x + 8) + c] - \dfrac{77}{13} \left[\dfrac{2}{\sqrt{51}} \tan^{-1}\left(\dfrac{2\sqrt{3}\tan {}^{x}\!/_2}{\sqrt{17}}\right) + c \right]$

∴ $I = \dfrac{21}{13} x - \dfrac{1}{13}[\log(2\cos x + 3\sin x + 8)] - \dfrac{77}{13}\left[\dfrac{2}{\sqrt{51}}\tan^{-1}\left(\dfrac{2\sqrt{3}\tan {}^{x}\!/_2}{\sqrt{17}}\right)\right] + k$ Ans. where $k = \left(c - \dfrac{c}{13} - \dfrac{77c}{13}\right)$

(b) $I = \int \dfrac{7\cos x + 4\sin x}{2\cos x + 3\sin x} dx$, use form of integral $\int \dfrac{p\cos x + q\sin x}{a\cos x + b\sin x} dx$

Express, $p\cos x + q\sin x = l(a\cos x + b\sin x) + m(\text{d. c of } a\cos x + b\sin x)$ and find l and m by comparing the coefficient of $\sin x \cos x$.

Divide both of side by $a\cos x + b\sin x$ (D^r) and integrating.

Solution: — $I = \int \dfrac{7\cos x + 4\sin x}{2\cos x + 3\sin x} dx$

$7\cos x + 4\sin x = l(2\cos x + 3\sin x) + m(\text{d. c of } 2\cos x + 3\sin x) = l(2\cos x + 3\sin x) + m(3\cos x - 2\sin x) \dots \dots \dots \dots \dots \text{(A)}$

find l and m by comparing the coefficient of $\sin x, \cos x$.

or $2l + 3m = 7 \dots \dots \dots \dots \dots \text{(i)}$ and $3l - 2m = 4 \dots \dots \dots \dots \dots \dots \text{(ii)}$

solve equation (i) and (ii) and find the value of l and m, we have ∴ $l = 2$ and $m = 1$

Now, Put value of l and m in equation (A), we have $7\cos x + 4\sin x = 2(2\cos x + 3\sin x) + 1(3\cos x - 2\sin x)$

Divide both of sides by $(2\cos x + 3\sin x)$ and integrating

$$\int \dfrac{7\cos x + 4\sin x}{2\cos x + 3\sin x} dx = 2\int \dfrac{2\cos x + 3\sin x}{2\cos x + 3\sin x} dx + \int \dfrac{3\cos x - 2\sin x}{2\cos x + 3\sin x} dx = 2\int dx + \int \dfrac{3\cos x - 2\sin x}{2\cos x + 3\sin x} dx = 2\int dx + I_1$$

$I_1 = \int \dfrac{3\cos x - 2\sin x}{2\cos x + 3\sin x} dx$ Let $2\cos x + 3\sin x = t$ ∴ $(-2\sin x + 3\cos x)dx = dt$ ∴ $(3\cos x - 2\sin x)dx = dt$

$I_1 = \int \dfrac{dt}{t} = \log t + c = \log(2\cos x + 3\sin x) + c$

∴ $I = 2\int dx + I_1 = 2x + c + \log(2\cos x + 3\sin x) + c = 2x + \log(2\cos x + 3\sin x) + k$ Ans. where $2c = k$

(c) $I = \int \dfrac{\tan x + \cot x}{\sec x - \cosec x} dx = \int \dfrac{\dfrac{\sin x}{\cos x} + \dfrac{\cos x}{\sin x}}{\dfrac{1}{\cos x} - \dfrac{1}{\sin x}} dx = \int \dfrac{\dfrac{\sin^2 x + \cos^2 x}{\sin x \cos x}}{\dfrac{\sin x - \cos x}{\sin x \cos x}} dx = \int \dfrac{1}{\sin x - \cos x} dx$

$= \int \dfrac{1}{2\sin {}^{x}\!/_2 \cos {}^{x}\!/_2 - (\cos^2 {}^{x}\!/_2 - \sin^2 {}^{x}\!/_2)} dx = \int \dfrac{1}{2\sin {}^{x}\!/_2 \cos {}^{x}\!/_2 - \cos^2 {}^{x}\!/_2 + \sin^2 {}^{x}\!/_2} dx$

Divide above and below by $\cos^2 {}^{x}\!/_2$

$I = \int \dfrac{\sec^2 x/2 \, dx}{2 \tan x/2 - 1 + \tan^2 x/2}$ $\qquad$ Let $\tan x/2 = t$ $\quad \therefore \sec^2 x/2 \cdot \dfrac{1}{2} dx = dt$ $\quad$ or $\quad \sec^2 x/2 \, dx = 2 \, dt$

$I = \int \dfrac{2 \, dt}{2t - 1 + t^2} = 2 \int \dfrac{dt}{t^2 + 2t - 1} = 2 \int \dfrac{dt}{(t+1)^2 - (\sqrt{2})^2}$ $\quad \left[\text{use formula,} \quad \int \dfrac{dx}{x^2 - a^2} = \dfrac{1}{2a} \log \left|\dfrac{x-a}{x+a}\right| \right]$

$I = 2 \cdot \dfrac{1}{2\sqrt{2}} \log \left[\dfrac{(t+1) - \sqrt{2}}{(t+1) + \sqrt{2}}\right] + c = \dfrac{1}{\sqrt{2}} \log \left[\dfrac{(\tan x/2 + 1) - \sqrt{2}}{(\tan x/2 + 1) + \sqrt{2}}\right] + c$ $\quad$ Ans.

(d) $I = \int \dfrac{\cos 2\theta}{\sin^4 \theta - \cos^4 \theta} d\theta = \int \dfrac{\cos^2 \theta - \sin^2 \theta}{\sin^4 \theta - \cos^4 \theta} d\theta$ $\quad$ Divide above and below by $\cos^4 \theta$

$I = \int \dfrac{\dfrac{\cos^2 \theta - \sin^2 \theta}{\cos^4 \theta}}{\dfrac{\sin^4 \theta - \cos^4 \theta}{\cos^4 \theta}} d\theta = \int \dfrac{\sec^2 \theta - \tan^2 \theta \sec^2 \theta}{\tan^4 \theta - 1} d\theta = \int \dfrac{\sec^2 \theta (1 - \tan^2 \theta)}{\tan^4 \theta - 1} d\theta$

Let $\tan \theta = t$ $\quad \therefore \sec^2 \theta \, d\theta = dt$

$I = \int \dfrac{(1 - t^2)}{t^4 - 1} dt = \int \dfrac{-(t^2 - 1)}{(t^2)^2 - 1} dt = -\int \dfrac{(t^2 - 1)}{(t^2 - 1)(t^2 + 1)} dt = -\int \dfrac{dt}{(t^2 + 1)}$ $\quad \left[\text{formula} \quad \int \dfrac{dx}{x^2 + a^2} = \dfrac{1}{a} \tan^{-1} x/a \right]$

$I = -\int \dfrac{dt}{1 + t^2} = -\dfrac{1}{1} \tan^{-1} t/1 + c = -\tan^{-1} t + c = -\tan^{-1}(\tan \theta) + c = -\theta + c$ $\quad$ Ans.

Form of the integrals: – (a) $\int \sqrt{\sec^2 x \pm a^2} \, dx$ $\quad$ (b) $\int \sqrt{\cosec^2 x \pm a^2} \, dx$ $\quad$ (c) $\int \sqrt{\tan^2 x \pm a^2} \, dx$ $\quad$ (d) $\int \sqrt{\cot^2 x \pm a^2} \, dx$

Solution: – $\quad$ (a) Let $I = \int \sqrt{\sec^2 x \pm a^2} \, dx = \int \dfrac{\sqrt{\sec^2 x \pm a^2}}{\sqrt{\sec^2 x \pm a^2}} \times \sqrt{\sec^2 x \pm a^2} \, dx = \int \dfrac{\sec^2 x \pm a^2}{\sqrt{\sec^2 x \pm a^2}} \, dx$

$I = \int \dfrac{\sec^2 x \pm a^2}{\sqrt{\sec^2 x \pm a^2}} \, dx = \int \dfrac{\sec^2 x \, dx}{\sqrt{\sec^2 x \pm a^2}} \pm \int \dfrac{a^2}{\sqrt{\sec^2 x \pm a^2}} \, dx = \int \dfrac{\sec^2 x \, dx}{\sqrt{(1 + \tan^2 x) \pm a^2}} \pm a^2 \int \dfrac{dx}{\sqrt{\dfrac{1}{\cos^2 x} \pm a^2}}$

$= \int \dfrac{\sec^2 x \, dx}{\sqrt{(1 \pm a^2) + \tan^2 x}} \pm a^2 \int \dfrac{\cos x \, dx}{\sqrt{1 \pm a^2 \cos^2 x}} = \int \dfrac{\sec^2 x \, dx}{\sqrt{(1 \pm a^2) + \tan^2 x}} \pm a^2 \int \dfrac{\cos x \, dx}{\sqrt{1 \pm a^2 (1 - \sin^2 x)}}$

In first part putting $t = \tan x$ and in second part putting $z = \sin x$

$\therefore t = \tan x \quad \therefore dt = \sec^2 x \, dx \quad$ and $\quad z = \sin x \quad \therefore dz = \cos x \, dx$

$I = \int \dfrac{dt}{\sqrt{(1 \pm a^2) + t^2}} \pm a^2 \int \dfrac{dz}{\sqrt{1 \pm a^2 (1 - z^2)}} = \int \dfrac{dt}{\sqrt{\left(\sqrt{(1 \pm a^2)}\right)^2 + t^2}} \pm a^2 \int \dfrac{dz}{\sqrt{\left(\sqrt{(1 \pm a^2)}\right)^2 - (az)^2}}$

Using formula, $\quad \int \dfrac{dx}{\sqrt{a^2 + x^2}} = \log\left(x + \sqrt{a^2 + x^2}\right)$ or $\sinh^{-1}\left(\dfrac{x}{a}\right)$ and $\int \dfrac{dx}{\sqrt{a^2 - x^2}} = \sin^{-1}\left(\dfrac{x}{a}\right)$

$I = \int \dfrac{dt}{\sqrt{\left(\sqrt{(1 \pm a^2)}\right)^2 + t^2}} \pm a^2 \int \dfrac{dz}{\sqrt{\left(\sqrt{(1 \pm a^2)}\right)^2 - (az)^2}} = \log\left(t + \sqrt{\left(\sqrt{(1 \pm a^2)}\right)^2 + t^2}\right) \pm \dfrac{a^2}{a} \sin^{-1}\left(\dfrac{az}{\sqrt{(1 \pm a^2)}}\right)$

$I = \log\left(\tan x + \sqrt{\left(\sqrt{(1 \pm a^2)}\right)^2 + (\tan x)^2}\right) \pm a \sin^{-1}\left(\dfrac{a \sin x}{\sqrt{(1 \pm a^2)}}\right) + c$ $\quad$ Ans.

(b), (c) and (d) Do yourself. $\quad$ (solve same as above question)

Integral of the form: $-\quad \int \dfrac{dx}{P\sqrt{Q}}\quad$ and $\quad \int \dfrac{f(x)}{P\sqrt{Q}}\,dx$

Linear	Quadratic	Substitution
P, Q	– – – – –	$Q = z^2$
Q	P	$Q = z^2$
P	Q	$P = {}^1\!/_z$
– – – – – –	P, Q	$z^2 = {}^Q\!/_P$

Example: $-$ (a) $\displaystyle\int \dfrac{dx}{(x+2)\sqrt{x-3}}$ (b) $\displaystyle\int \dfrac{dx}{(x^2+2x+3)\sqrt{2-x}}$ (c) $\displaystyle\int \dfrac{dx}{(1+x)\sqrt{2x^2+2x+1}}$ (d) $\displaystyle\int \dfrac{dx}{x^2\sqrt{x^2+1}}$ (e) $\displaystyle\int \dfrac{dx}{(x^2+1)\sqrt{x^2-1}}$

Solution: $-$ (a) Let $I = \displaystyle\int \dfrac{dx}{(x+2)\sqrt{x-3}}$ $\left[\,(x+2)\text{ and }(x-3)\text{ are linear then use formula }\displaystyle\int \dfrac{dx}{P\sqrt{Q}}\,,\ \text{put }Q = z^2\right]$

Putting $Q = z^2$ $\therefore$ $x - 3 = z^2$ or $x = z^2 + 3$ or $z = \sqrt{x-3}$ $\therefore$ $dx = 2z\,dz$

$I = \displaystyle\int \dfrac{dx}{(x+2)\sqrt{x-3}} = \int \dfrac{2z\,dz}{(z^2+3+2)\sqrt{z^2}} = \int \dfrac{2z\,dz}{(z^2+5)z} = 2\int \dfrac{dz}{z^2+5}$ $\left[\,\text{using formula, }\displaystyle\int \dfrac{dx}{x^2+a^2} = \dfrac{1}{a}\tan^{-1}\left(\dfrac{x}{a}\right)\right]$

$I = 2\displaystyle\int \dfrac{dz}{z^2+\left(\sqrt{5}\right)^2} = 2.\dfrac{1}{\sqrt{5}}\tan^{-1}\left(\dfrac{z}{\sqrt{5}}\right) + c = \dfrac{2}{\sqrt{5}}\tan^{-1}\left(\dfrac{\sqrt{x-3}}{\sqrt{5}}\right) + c$ Ans. put $z = \sqrt{x-3}$

(b) Let $I = \displaystyle\int \dfrac{dx}{(x^2+2x+3)\sqrt{2-x}} = \int \dfrac{dx}{P\sqrt{Q}}$ Here P is quadratic and Q is linear then putting $Q = z^2$

Putting $Q = z^2$ or $2 - x = z^2$ or $z = \sqrt{2-x}$ or $x = 2 - z^2$ $\therefore$ $dx = -2z\,dz$

$I = -\displaystyle\int \dfrac{2z\,dz}{[(2-z^2)^2+2(2-z^2)+3]\sqrt{z^2}} = -\int \dfrac{2z\,dz}{z[4-4z^2+z^4+4-2z^2+3]} = -2\int \dfrac{dz}{z^4-6z^2+11}$

$I = -2\displaystyle\int \dfrac{dz}{(z^2-3)^2+2} = -2\int \dfrac{dz}{(z^2-3)^2+\left(\sqrt{2}\right)^2}$ $\left[\,\text{using formula, }\displaystyle\int \dfrac{dx}{x^2+a^2} = \dfrac{1}{a}\tan^{-1}\left(\dfrac{x}{a}\right).\dfrac{1}{\text{d.c of }x}\right]$

$I = -2.\dfrac{1}{\sqrt{2}}\tan^{-1}\left[\dfrac{(z^2-3)}{\sqrt{2}}\right].\dfrac{1}{\text{d.c of }(z^2-3)} = -2.\dfrac{1}{\sqrt{2}}\tan^{-1}\left[\dfrac{(z^2-3)}{\sqrt{2}}\right].\dfrac{1}{2z} = -\dfrac{1}{\sqrt{2}}\tan^{-1}\left[\dfrac{2-x-3}{\sqrt{2}}\right].\dfrac{1}{\sqrt{2-x}} + c$

$I = -\dfrac{1}{\sqrt{2}.\sqrt{2-x}}\tan^{-1}\left[\dfrac{-(x+1)}{\sqrt{2}}\right] + c$ Ans. $\left[\text{put } 2-x = z^2 \ \therefore\ z = \sqrt{2-x}\right]$

(c) Let $I = \displaystyle\int \dfrac{dx}{(1+x)\sqrt{2x^2+2x+1}} = \int \dfrac{dx}{P\sqrt{Q}}$ Here P is linear and Q is quadratic then putting $P = {}^1\!/_z$

Put $1+x = \dfrac{1}{z}$ $\therefore$ $dx = -\dfrac{1}{z^2}dz$ or $x = \dfrac{1}{z} - 1 = \dfrac{1-z}{z}$ or $z = \dfrac{1}{1+x}$

$I = \displaystyle\int \dfrac{-\dfrac{1}{z^2}dz}{\dfrac{1}{z}.\sqrt{2\left(\dfrac{1-z}{z}\right)^2+2\left(\dfrac{1-z}{z}\right)+1}} = -\int \dfrac{dz}{z.\sqrt{\dfrac{2(1-2z+z^2)}{z^2}+\dfrac{2-2z}{z}+1}} = -\int \dfrac{dz}{z.\sqrt{\dfrac{2-4z+2z^2+2z-2z^2+z^2}{z^2}}} = -\int \dfrac{dz}{\sqrt{z^2-2z+2}}$

$= -\displaystyle\int \dfrac{dz}{\sqrt{(z-1)^2+1}}$ $\left[\,\text{using formula, }\displaystyle\int \dfrac{dx}{x^2+a^2} = \dfrac{1}{a}\tan^{-1}\left(\dfrac{x}{a}\right)\right]$

$I = -\displaystyle\int \dfrac{dz}{\sqrt{(z-1)^2+1^2}} = -\dfrac{1}{1}\tan^{-1}\left(\dfrac{z-1}{1}\right) = -\tan^{-1}\left(\dfrac{1}{1+x}-1\right) + c = -\tan^{-1}\left(\dfrac{1-1-x}{1+x}\right) + c = -\tan^{-1}\left(\dfrac{-x}{1+x}\right) + c$ Ans.

(d) Let $I = \int \dfrac{dx}{x^2\sqrt{x^2+1}}$

Particular Cases:— $\int \dfrac{dx}{(ax^2+b)\sqrt{cx^2+d}}$ Putting $x = \dfrac{1}{t}$ and simplify and then putting expression under radical sign $= z^2$.

Solution:— $I = \int \dfrac{dx}{x^2\sqrt{x^2+1}}$ Put $x = \dfrac{1}{t}$ $\therefore$ $dx = -\dfrac{1}{t^2}dt$ and $t = \dfrac{1}{x}$

$I = \int \dfrac{-\frac{1}{t^2}dt}{\left(\frac{1}{t^2}\right)\sqrt{\left(\frac{1}{t}\right)^2+1}} = -\int \dfrac{dt}{\sqrt{\frac{1+t^2}{t^2}}} = -\int \dfrac{t\,dt}{\sqrt{1+t^2}}$ Again put $1+t^2 = z^2$ $\therefore$ $2t\,dt = 2z\,dz$ $\therefore$ $t\,dt = z\,dz$

$I = -\int \dfrac{z\,dz}{\sqrt{z^2}} = -\int \dfrac{z\,dz}{z} = -\int dz = -z + c = -\sqrt{1+t^2} + c = -\sqrt{1+\dfrac{1}{x^2}} + c = -\dfrac{\sqrt{1+x^2}}{x} + c$ Ans.

(e) Let $I = \int \dfrac{dx}{(x^2+1)\sqrt{x^2-1}}$ Put $x = \dfrac{1}{t}$ $\therefore$ $dx = -\dfrac{1}{t^2}dt$ and $t = \dfrac{1}{x}$

$I = \int \dfrac{-\frac{1}{t^2}dt}{\left(\frac{1}{t^2}+1\right)\sqrt{\frac{1}{t^2}-1}} = -\int \dfrac{dt}{t^2.\left(\frac{1+t^2}{t^2}\right)\sqrt{\frac{1-t^2}{t^2}}} = -\int \dfrac{t\,dt}{(1+t^2)\sqrt{1-t^2}}$

Again put $1-t^2 = z^2$ $\therefore$ $-2t\,dt = 2z\,dz$ $\therefore$ $-t\,dt = z\,dz$ and $t^2 = 1-z^2$

$I = \int \dfrac{-t\,dt}{(1+t^2)\sqrt{1-t^2}} = \int \dfrac{z\,dz}{(1+1-z^2)\sqrt{z^2}} = \int \dfrac{z\,dz}{(2-z^2).z} = \int \dfrac{dz}{2-z^2}$ $\left[\text{use formula, } \int \dfrac{dx}{a^2-x^2} = \dfrac{1}{2a}\log\left|\dfrac{a+x}{a-x}\right|\right]$

$I = \int \dfrac{dz}{2-z^2} = \int \dfrac{dz}{\left(\sqrt{2}\right)^2-z^2} = \dfrac{1}{2.\sqrt{2}}\log\left[\dfrac{\sqrt{2}+z}{\sqrt{2}-z}\right] + c = \dfrac{1}{2\sqrt{2}}\log\left[\dfrac{\sqrt{2}+\sqrt{1-t^2}}{\sqrt{2}-\sqrt{1-t^2}}\right] + c = \dfrac{1}{2\sqrt{2}}\log\left[\dfrac{\sqrt{2}+\sqrt{1-\frac{1}{x^2}}}{\sqrt{2}-\sqrt{1-\frac{1}{x^2}}}\right] + c$

$I = \dfrac{1}{2\sqrt{2}}\log\left[\dfrac{\sqrt{2}+\sqrt{\frac{x^2-1}{x^2}}}{\sqrt{2}-\sqrt{\frac{x^2-1}{x^2}}}\right] + c = \dfrac{1}{2\sqrt{2}}\log\left[\dfrac{\frac{\sqrt{2}x+\sqrt{x^2-1}}{x}}{\frac{\sqrt{2}x-\sqrt{x^2-1}}{x}}\right] + c = \dfrac{1}{2\sqrt{2}}\log\left[\dfrac{\sqrt{2}x+\sqrt{x^2-1}}{\sqrt{2}x-\sqrt{x^2-1}}\right] + c$ Ans.

Integral of the form:— $\int \dfrac{dx}{1 \pm \sin ax}$, $\int \dfrac{dx}{\sqrt{1 \pm \sin ax}}$ and $\int \sqrt{1 \pm \sin ax}\ dx$

Working rule:— change $\sin ax$ into $\cos\left(\dfrac{\pi}{2} \pm ax\right)$ and then use result of integration.

Example:— (a) $I = \int \dfrac{dx}{1+\sin x} = \int \dfrac{dx}{1+\cos\left(\frac{\pi}{2}-x\right)} = \int \dfrac{dx}{2\cos^2\left(\frac{\frac{\pi}{2}-x}{2}\right)} = \dfrac{1}{2}\int \dfrac{dx}{\cos^2\left(\frac{\pi}{4}-\frac{x}{2}\right)}$

$$I = \dfrac{1}{2}\int \sec^2\left(\dfrac{\pi}{4}-\dfrac{x}{2}\right)dx = -\dfrac{1}{2}.\dfrac{\tan\left(\frac{\pi}{4}-\frac{x}{2}\right)}{-\frac{1}{2}} + c = \tan\left(\dfrac{\pi}{4}-\dfrac{x}{2}\right) + c \qquad \text{Ans.}$$

Remember:— $\sin\left(\dfrac{\pi}{2}-x\right) = \sin\dfrac{\pi}{2}.\cos x - \cos\dfrac{\pi}{2}.\sin x = \cos x - 0 = \cos x$

$\sin\left(\dfrac{\pi}{2}+x\right) = \sin\dfrac{\pi}{2}.\cos x + \cos\dfrac{\pi}{2}.\sin x = \cos x + 0 = \cos x$

$\cos\left(\dfrac{\pi}{2}-x\right) = \cos\dfrac{\pi}{2}.\cos x + \sin\dfrac{\pi}{2}.\sin x = 0 + \sin x = \sin x$

$\cos\left(\dfrac{\pi}{2}+x\right) = \cos\dfrac{\pi}{2}.\cos x - \sin\dfrac{\pi}{2}.\sin x = 0 - \sin x = -\sin x$

$$\tan\left(\frac{\pi}{2}-x\right)=\frac{\sin\left(\frac{\pi}{2}-x\right)}{\cos\left(\frac{\pi}{2}-x\right)}=\frac{\cos x}{\sin x}=\cot x \;, \quad \cot\left(\frac{\pi}{2}-x\right)=\frac{\cos\left(\frac{\pi}{2}-x\right)}{\sin\left(\frac{\pi}{2}-x\right)}=\frac{\sin x}{\cos x}=\tan x$$

$$\tan\left(\frac{\pi}{2}+x\right)=\frac{\sin\left(\frac{\pi}{2}+x\right)}{\cos\left(\frac{\pi}{2}+x\right)}=\frac{\cos x}{-\sin x}=-\cot x \;, \quad \cot\left(\frac{\pi}{2}+x\right)=\frac{\cos\left(\frac{\pi}{2}+x\right)}{\sin\left(\frac{\pi}{2}+x\right)}=\frac{-\sin x}{\cos x}=-\tan x$$

$$\sec\left(\frac{\pi}{2}-x\right)=\frac{1}{\cos\left(\frac{\pi}{2}-x\right)}=\frac{1}{\sin x}=\operatorname{cosec} x \;, \quad \sec\left(\frac{\pi}{2}+x\right)=\frac{1}{\cos\left(\frac{\pi}{2}+x\right)}=\frac{1}{-\sin x}=-\operatorname{cosec} x$$

$$\operatorname{cosec}\left(\frac{\pi}{2}-x\right)=\frac{1}{\sin\left(\frac{\pi}{2}-x\right)}=\frac{1}{\cos x}=\sec x \;, \quad \operatorname{cosec}\left(\frac{\pi}{2}+x\right)=\frac{1}{\sin\left(\frac{\pi}{2}+x\right)}=\frac{1}{\cos x}=\sec x$$

$$\sin(\pi-x)=\sin\pi.\cos x-\cos\pi.\sin x=0-(-\sin x)=\sin x \;, \quad \sin(\pi+x)=-\sin x$$

$$\cos(\pi-x)=\cos\pi.\cos x+\sin\pi.\sin x=-\cos x \;, \quad \cos(\pi+x)=\cos\pi.\cos x-\sin\pi.\sin x=-\cos x$$

IInd Method: $-\;\; I=\int\dfrac{dx}{1+\sin x}=\int\dfrac{dx}{\sin^2{}^{x}\!/_2+\cos^2{}^{x}\!/_2+2\sin^{x}\!/_2.\cos^{x}\!/_2}$ Divide above and below by $\cos^2{}^{x}\!/_2$

$$I=\int\frac{\sec^2{}^{x}\!/_2\;dx}{\tan^2{}^{x}\!/_2+2\tan^{x}\!/_2+1} \qquad \text{Put } \tan^{x}\!/_2=t \quad \therefore \frac{1}{2}.\sec^2{}^{x}\!/_2\;dx=dt \quad \therefore \sec^2{}^{x}\!/_2\;dx=2\;dt$$

$$I=\int\frac{2\;dt}{t^2+2t+1}=2\int\frac{dt}{(t+1)^2} \qquad \text{Put } t+1=z \; \therefore dt=dz$$

$$\text{then } I=2\int\frac{dz}{z^2}=2.\frac{(z)^{-2+1}}{-2+1}=-2\frac{1}{z}=-\frac{2}{z}+c=-\frac{2}{t+1}+c=-\frac{2}{\tan^{x}\!/_2+1}+c \qquad \text{Ans.}$$

IIIrd Method: $-\;\; I=\int\dfrac{dx}{1+\sin x}=\int\left(\dfrac{1}{1+\sin x}\times\dfrac{1-\sin x}{1-\sin x}\right)dx=\int\dfrac{1-\sin x}{1-\sin^2 x}dx=\int\dfrac{1-\sin x}{\cos^2 x}dx=\int\dfrac{dx}{\cos^2 x}-\int\dfrac{\sin x}{\cos^2 x}dx$

$$=\int\sec^2 x\;dx-\int\frac{\sin x}{\cos^2 x}dx=\tan x-I_1$$

or $I_1=\int\dfrac{\sin x}{\cos^2 x}dx \qquad \text{Put } \cos x=t \quad \therefore -\sin x\,dx=dt$

$$\therefore \; I_1=\int\frac{\sin x}{\cos^2 x}dx=-\int\frac{dt}{t^2}=-\frac{t^{-2+1}}{-2+1}=\frac{1}{t}=\frac{1}{\cos x}+c=\sec x+c=\tan x-\sec x+c \qquad \text{Ans.}$$

(b) $I=\int\dfrac{dx}{\sqrt{1+\sin x}}=\int\dfrac{dx}{\sqrt{\sin^2{}^{x}\!/_2+\cos^2{}^{x}\!/_2+2\sin^{x}\!/_2.\cos^{x}\!/_2}}$ Divide above and below by $\cos^4{}^{x}\!/_2$

$$I=\int\frac{\sec^4{}^{x}\!/_2}{\sqrt{\tan^2{}^{x}\!/_2+2\tan^{x}\!/_2+1}}dx=\int\frac{\sec^2{}^{x}\!/_2.\sec^2{}^{x}\!/_2}{\sqrt{\tan^2{}^{x}\!/_2+2\tan^{x}\!/_2+1}}dx=\int\frac{(1+\tan^2{}^{x}\!/_2).\sec^2{}^{x}\!/_2}{\sqrt{\tan^2{}^{x}\!/_2+2\tan^{x}\!/_2+1}}dx$$

Put $\tan^{x}\!/_2=t \quad \therefore \dfrac{1}{2}.\sec^2{}^{x}\!/_2\;dx=dt \quad \therefore \sec^2{}^{x}\!/_2\;dx=2\;dt$

$$I=\int\frac{1+t^2}{\sqrt{t^2+2t+1}}.2\;dt=2\int\frac{1+t^2}{\sqrt{(1+t)^2}}dt=2\int\frac{1+t^2}{1+t}dt=2\int\frac{dt}{1+t}+2\int\frac{t^2}{1+t}dt=2\int\frac{dt}{1+t}+2\int\left(t-\frac{t}{1+t}\right)dt$$

$$=2\log(1+t)+2\int t\,dt-2\int\frac{t}{1+t}dt=2\log(1+t)+2.\frac{t^2}{2}-2\int\left(1-\frac{1}{1+t}\right)dt$$

$$=2\log(1+t)+t^2-2\int dt+2\int\frac{1}{1+t}dt$$

$$I=2\log(1+t)+t^2-2t+2\log(1+t)+c=4\log(1+t)+t^2-2t+c \qquad \text{Ans.} \quad \text{where } t=\tan^{x}\!/_2$$

IInd Method: $-$ $\quad I = \int \dfrac{dx}{\sqrt{1+\sin x}} = \int \dfrac{dx}{\sqrt{1+\cos\left(\frac{\pi}{2}-x\right)}} = \int \dfrac{dx}{\sqrt{2\cos^2\left(\frac{\frac{\pi}{2}-x}{2}\right)}} = \dfrac{1}{\sqrt{2}}\int \dfrac{dx}{\cos\left(\frac{\pi}{4}-\frac{x}{2}\right)}$

Put $\dfrac{\pi}{4}-\dfrac{x}{2} = t \quad \therefore -\dfrac{1}{2}dx = dt \quad \therefore -dx = 2dt$

$I = -\dfrac{1}{\sqrt{2}}\int \dfrac{2\,dt}{\cos t} = -\sqrt{2}\int \dfrac{\sin^2 t + \cos^2 t}{\cos t}dt = -\sqrt{2}\int \dfrac{\sin^2 t}{\cos t}dt - \sqrt{2}\int \dfrac{\cos^2 t}{\cos t}dt = -\sqrt{2}\int \dfrac{\sqrt{1-\cos^2 t}.\sin t}{\cos t}dt - \sqrt{2}\int \cos t\,dt$

$\qquad\qquad = -\sqrt{2}I_1 - \sqrt{2}\sin t$

$I_1 = \int \dfrac{\sqrt{1-\cos^2 t}.\sin t}{\cos t}dt = \int \dfrac{\sqrt{1-\cos^2 t}.\sin t.\cos t}{\cos^2 t}dt \quad$ Put $1-\cos^2 t = z^2 \quad \therefore -2\cos t \sin t\,dt = 2z\,dz$

$I_1 = -\int \dfrac{z.z\,dz}{1-z^2} = -\int \dfrac{z^2}{1-z^2}dz = -\int \left(\dfrac{1}{1-z^2}-1\right)dz = -\int \dfrac{1}{1-z^2}dz + \int dz \quad \left[\text{using formula, } \int \dfrac{dx}{a^2-x^2} = \dfrac{1}{2a}\log\left|\dfrac{a+x}{a-x}\right|\right]$

$I_1 = -\dfrac{1}{2.1}\log\left|\dfrac{1+z}{1-z}\right| + z = -\dfrac{1}{2}\log\left[\dfrac{1+\sqrt{1-\cos^2 t}}{1-\sqrt{1-\cos^2 t}}\right] + \sqrt{1-\cos^2 t} + c$

$I = -\sqrt{2}I_1 - \sqrt{2}\sin t = \sqrt{2}.\dfrac{1}{2}\log\left[\dfrac{1+\sqrt{1-\cos^2 t}}{1-\sqrt{1-\cos^2 t}}\right] - \sqrt{2}.\sqrt{1-\cos^2 t} - \sqrt{2}\sin t + c$

$I = \dfrac{1}{\sqrt{2}}\log\left[\dfrac{1+\sin t}{1-\sin t}\right] - 2\sqrt{2}\sin t + c \quad$ Ans. $\quad$ where $t = \left(\dfrac{\pi}{4}-\dfrac{x}{2}\right)$

(c) $I = \int \sqrt{1+\sin x}\,dx = \int \sqrt{1+\cos\left(\frac{\pi}{2}-x\right)}\,dx = \int \sqrt{2\cos^2\left(\frac{\frac{\pi}{2}-x}{2}\right)}\,dx = \sqrt{2}\int \cos\left(\frac{\pi}{4}-\frac{x}{2}\right)dx = \sqrt{2}.\sin\left(\frac{\pi}{4}-\frac{x}{2}\right).\dfrac{1}{-\frac{1}{2}}$

$\qquad\qquad = -2\sqrt{2}\sin\left(\dfrac{\pi}{4}-\dfrac{x}{2}\right) + c = -2\sqrt{2}\left(\sin\dfrac{\pi}{4}.\cos\dfrac{x}{2} - \cos\dfrac{\pi}{4}.\sin\dfrac{x}{2}\right) + c$

$I = -2\sqrt{2}\left(\dfrac{1}{\sqrt{2}}.\cos\dfrac{x}{2} - \dfrac{1}{\sqrt{2}}.\sin\dfrac{x}{2}\right) + c = -2\sqrt{2}.\left(\dfrac{\cos\frac{x}{2}-\sin\frac{x}{2}}{\sqrt{2}}\right) + c = -2\left(\cos\dfrac{x}{2}-\sin\dfrac{x}{2}\right) + c \quad$ Ans.

IInd Method: $-$ $\quad I = \int \sqrt{1+\sin x}\,dx = I = \int \sqrt{1+\sin x \times \dfrac{1-\sin x}{1-\sin x}}\,dx = \int \sqrt{\dfrac{1-\sin^2 x}{1-\sin x}}\,dx = \int \dfrac{\cos x}{\sqrt{1-\sin x}}\,dx$

$\qquad\qquad$ Put $1-\sin x = t^2 \quad \therefore -\cos x\,dx = 2t\,dt$

$\qquad I = \int \dfrac{\cos x}{\sqrt{1-\sin x}}\,dx = -\int \dfrac{2t\,dt}{\sqrt{t^2}} = -2\int \dfrac{t\,dt}{t} = -2\int dt = -2t + c = -2\sqrt{1-\sin x} + c \quad$ Ans.

Integral of improper rational function: $-$ $\quad$ Improper rational function: $-$

A rational function in which degree of $N^r \geq$ degree of D^r is called improper rational function.

e.g. (a) $\dfrac{x^2+1}{2x+3}$ $\quad$ (b) $\dfrac{x^2+1}{x^2-1}$ $\quad$ (c) $\dfrac{3x^3-3x^2+1}{2+3x}$

Remarks: $-$ In improper rational we always divide N^r by D^r and then use result of integration.

Example: $-$ $\quad$ (a) $I = \int \dfrac{x^6}{x-1}\,dx$, $\qquad$ Divide N^r by D^r and then use result of integration.

Divide

$\qquad\qquad \therefore \quad x^6 = (x-1)(x^5+x^4+x^3+x^2+x+1) + 1$

Divide both of sides by $(x-1)$ and integrate

$$\int \frac{x^6}{x-1}dx = \int \frac{(x-1)(x^5+x^4+x^3+x^2+x+1)+1}{(x-1)}dx = \int \frac{(x-1)(x^5+x^4+x^3+x^2+x+1)}{(x-1)}dx + \int \frac{1}{x-1}dx$$

$$= \int (x^5+x^4+x^3+x^2+x+1)\,dx + \int \frac{1}{x-1}dx$$

$$I = \int \frac{x^6}{x-1}dx = \int x^5\,dx + \int x^4\,dx + \int x^3\,dx + \int x^2\,dx + \int x\,dx + \int dx + \int \frac{1}{x-1}dx$$

$$= \frac{x^6}{6} + \frac{x^5}{5} + \frac{x^4}{4} + \frac{x^3}{3} + \frac{x^2}{2} + x + \log|x-1| + c \qquad \text{Ans.}$$

IInd Method: $-\quad \int \frac{x^6}{x-1}dx = \int \frac{(x-1)x^5 + (x-1)x^4 + (x-1)x^3 + (x-1)x^2 + (x-1)x + (x-1) + 1}{x-1}dx$

$$I = \int \frac{(x-1)(x^5+x^4+x^3+x^2+x+1)+1}{(x-1)}dx = \int (x^5+x^4+x^3+x^2+x+1)\,dx + \int \frac{1}{x-1}dx$$

$$= \int x^5\,dx + \int x^4\,dx + \int x^3\,dx + \int x^2\,dx + \int x\,dx + \int dx + \int \frac{1}{x-1}dx$$

$$= \frac{x^6}{6} + \frac{x^5}{5} + \frac{x^4}{4} + \frac{x^3}{3} + \frac{x^2}{2} + x + \log|x-1| + c \qquad \text{Ans.}$$

(b) $I = \int \frac{x^2+1}{x^2-1}dx = \int \frac{(x^2-1)+2(x^2-1)+6}{3(x^2-1)}dx = \frac{1}{3}\int \frac{x^2-1}{x^2-1}dx + \frac{2}{3}\int \frac{x^2-1}{x^2-1}dx + 2\int \frac{1}{x^2-1}dx$

$$I = \frac{1}{3}\int dx + \frac{2}{3}\int dx + 2\int \frac{1}{x^2-1}dx \qquad \left[\text{use formula,}\ \int \frac{dx}{x^2-a^2} = \frac{1}{2a}\log\left|\frac{x-a}{x+a}\right|\right]$$

$$I = \frac{1}{3}x + \frac{2}{3}x + 2.\frac{1}{2.1}\log\left|\frac{x-1}{x+1}\right| + c = x + \log\left[\frac{x-1}{x+1}\right] + c \qquad \text{Ans.}$$

(c) $I = \int \frac{6x^3 + 7x^2 - 4x - 4}{(3x+2)}dx$, Divide N^r by D^r and then use result of integration.

Divide

$\therefore\ 6x^3 + 7x^2 - 4x - 4 = (3x+2)(2x^2 + x - 2) + 0$ \qquad Divide $(3x+2)$ both of sides and integrate

$$\int \frac{6x^3 + 7x^2 - 4x - 4}{(3x+2)}dx = \int \frac{(3x+2)(2x^2+x-2)}{(3x+2)}dx = \int (2x^2 + x - 2)\,dx = \int 2x^2\,dx + \int x\,dx - \int 2\,dx$$

$$\therefore\ I = \int \frac{6x^3 + 7x^2 - 4x - 4}{(3x+2)}dx = 2.\frac{x^3}{3} + \frac{x^2}{2} - 2x + c = \frac{2}{3}x^3 + \frac{1}{2}x^2 - 2x + c \qquad \text{Ans.}$$

\# **Integration by substitution**: $-\quad \int f(g(x)).g'(x)\,dx$, Putting $g(x)=z \quad \therefore\ g'(x)\,dx = dz \quad \boxed{\therefore\ \int f(z)\,dz}$

After evaluating this integral we substitute the value of z in term of x, i.e $z = g(x)$.

Example: $-$ (a) $I = \int \frac{\cos x}{\cos(x-3)}dx$, \qquad Put $x-3 = z$ or $x = z+3 \quad \therefore\ dx = dz$

$$I = \int \frac{\cos(z+3)}{\cos z}dz = \int \frac{\cos z.\cos 3 - \sin z.\sin 3}{\cos z}dz = \int \frac{\cos z.\cos 3}{\cos z}dz - \int \frac{\sin z.\sin 3}{\cos z}dz\ [\text{formula, } \cos(A+B) = \cos A\cos B - \sin A\sin B]$$

$$I = \int \cos 3\,dz - \int \sin 3.\tan z\,dz = \cos 3.z - \sin 3.(-\log(\cos z)) + c \left[\int \tan x\,dx = -\log(\cos x) + c\right]$$

$$I = z\cos 3 + \sin 3.\log[\cos z] + c = (x-3).\cos 3 + \sin 3.\log[\cos(x-3)] + c \qquad \text{Ans.}$$

(b) $I = \int \frac{dx}{(\sin^{-1}x)^3.\sqrt{1-x^2}}$ \qquad Putting $\sin^{-1}x = z \quad \therefore\ \frac{1}{\sqrt{1-x^2}}dx = dz \qquad \left[\frac{d}{dx}(\sin^{-1}x) = \frac{1}{\sqrt{1-x^2}}\right]$

$$\therefore \ I = \int \frac{dz}{z^3} = \int z^{-3} \, dz = \frac{z^{-3+1}}{-3+1} + c = -\frac{1}{2}z^{-2} + c = -\frac{1}{2z^2} + c = -\frac{1}{2(\sin^{-1}x)^2} + c \qquad \text{Ans.}$$

(c) $I = \displaystyle\int \sin^3 x \cdot \sqrt{\cos x} \, dx = \int \sin^2 x \cdot \sqrt{\cos x} \cdot \sin x \, dx = \int (1 - \cos^2 x) \cdot \sqrt{\cos x} \cdot \sin x \, dx$

Putting $\cos x = z^2 \quad \therefore -\sin x \, dx = 2z \, dz$

$$I = -\int (1 - (z^2)^2) \cdot \sqrt{z^2} \cdot 2z \, dz = -2\int (1 - z^4) \cdot z^2 \, dz = -2\int (z^2 - z^6) \, dz = -2\int z^2 \, dz + 2\int z^6 \, dz = -2\frac{z^{2+1}}{2+1} + 2\frac{z^{6+1}}{6+1} + c$$

$$= \frac{2}{7}z^7 - \frac{2}{3}z^3 + c = \frac{2}{7}\left(\sqrt{\cos x}\right)^7 - \frac{2}{3}\left(\sqrt{\cos x}\right)^3 + c \qquad \text{Ans.}$$

Integral of the form: $-$ $\displaystyle\int e^{ax} \cos bx \, dx$ and $\displaystyle\int e^{ax} \sin bx \, dx$

Let $c = \displaystyle\int e^{ax} \cos bx \, dx \dots\dots\dots\dots (i)$ and $s = \displaystyle\int e^{ax} \sin bx \, dx \dots\dots\dots\dots (ii)$ use $\left[\cos\theta + i\sin\theta = e^{i\theta}\right]$

Multiplying equation (i) $\times 1$ and (ii) $\times i$ and adding equation (i) & (ii), we have

$$c + is = \int e^{ax} \cos bx \, dx + \int e^{ax} \cdot i\sin bx \, dx = \int e^{ax}(\cos bx + i\sin bx) \, dx = \int e^{ax} \cdot e^{ibx} \, dx$$

$$c + is = \int e^{x(a+ib)} \, dx = \frac{e^{x(a+ib)}}{(a+ib)} = \frac{e^{ax} \cdot e^{ibx}}{a+ib} \times \frac{a-ib}{a-ib} = \frac{e^{ax}}{a^2+b^2}(\cos bx + i\sin bx) \cdot (a-ib)$$

$$= \frac{e^{ax}}{a^2+b^2}[a\cos bx - ib\cos bx + ia\sin bx + b\sin bx] = \frac{e^{ax}}{a^2+b^2}[(a\cos bx + b\sin bx) + i(a\sin bx - b\cos bx)]$$

$$c + is = \frac{e^{ax}}{a^2+b^2}(a\cos bx + b\sin bx) + i\frac{e^{ax}}{a^2+b^2}(a\sin bx - b\cos bx)$$

$$c = \frac{e^{ax}}{a^2+b^2}(a\cos bx + b\sin bx) \quad \text{and} \quad s = \frac{e^{ax}}{a^2+b^2}(a\sin bx - b\cos bx) \qquad \text{Ans.}$$

Generalized rule for integration by part: $-$

$$\int u \cdot v \, dx = u \cdot v_1 - u' \cdot v_2 + u'' \cdot v_3 - u''' \cdot v_4 + \cdots \dots\dots\dots\dots..$$

where $u' = \dfrac{du}{dx}, \quad u'' = \dfrac{du'}{dx} = \dfrac{d^2y}{dx^2}, \quad u''' = \dfrac{du''}{dx} \dots\dots\dots..$

and $v_1 = \displaystyle\int v \, dx, \quad v_2 = \int v_1 \, dx, \quad v_3 = \int v_2 \, dx \dots\dots\dots\dots$

Example: $-$ (a) $I = \displaystyle\int x^2 \cdot \cos x \, dx = x^2 \cdot \int \cos x \, dx - \int \left[\frac{d(x^2)}{dx} \cdot \int \cos x \, dx\right] dx = x^2 \cdot \sin x - \int 2x \cdot \sin x \, dx$

$I = x^2 \cdot \sin x - 2\left\{x \cdot \displaystyle\int \sin x \, dx - \int\left[\frac{dx}{dx} \cdot \int \sin x \, dx\right] dx\right\} = x^2 \cdot \sin x - 2x \cdot (-\cos x) + 2\int -\cos x \, dx = x^2 \cdot \sin x + 2x \cdot \cos x - 2\sin x + c$ Ans.

IInd Method: $-$ $I = \displaystyle\int x^2 \cdot \cos x \, dx = x^2 \cdot \int \cos x \, dx - \frac{d(x^2)}{dx} \cdot \int dx \cdot \int \cos x \, dx + \cdots\dots\dots.$

$$= x^2 \cdot \sin x - 2x \cdot \int \sin x \, dx + \frac{d(2x)}{dx} \cdot \int \cos x \, dx - 0 + 0 \dots..$$

or $I = x^2 \cdot \sin x - 2x(-\cos x) + 2\sin x + c \quad \therefore \ I = x^2 \sin x + 2x\cos x + 2\sin x + c$ Ans.

(b) $I = \displaystyle\int x^3 \cdot e^x \, dx = x^3 \cdot \int e^x \, dx - \int\left[\frac{d(x^3)}{dx} \cdot \int e^x \, dx\right] dx = x^3 \cdot e^x - \int 3x^2 \cdot e^x \, dx = x^3 \cdot e^x - 3\left\{x^2 \cdot \int e^x \, dx - \int\left[\frac{d(x^2)}{dx} \cdot \int e^x \, dx\right] dx\right\}$

$$= x^3 \cdot e^x - 3x^2 \cdot e^x + 3\int 2x \cdot e^x \, dx$$

$$I = x^3 . e^x - 3x^2 . e^x + 6 \left\{ x . \int e^x \, dx - \int \left[\frac{dx}{dx} . \int e^x \, dx \right] dx \right\} = x^3 . e^x - 3x^2 . e^x + 6x . e^x - 6 \int e^x \, dx = x^3 . e^x - 3x^2 . e^x + 6x . e^x - 6e^x + c \quad \text{Ans.}$$

IInd Method: $- \quad I = \int x^3 . e^x \, dx \quad$ [use formula generalized rule for integration by part]

$$I = x^3 \int e^x \, dx - \frac{d(x^3)}{dx} . \int e^x \, dx + \frac{d(3x^2)}{dx} . \int e^x \, dx - \frac{d(6x)}{dx} . \int e^x \, dx + \frac{d(6)}{dx} . \int e^x \, dx - \cdots \cdots = x^3 . e^x - 3x^2 . e^x + 6x . e^x - 6e^x + 0 - \cdots . + c$$

$$= x^3 . e^x - 3x^2 . e^x + 6x . e^x - 6e^x + c \quad \text{Ans.}$$

here $u = x^3, \ u' = 3x^2, \ u'' = 6x, \ u''' = 6, \ u'''' = 0$ and $v = e^x, \ v_1 = \int v \, dx = e^x, \ v_2 = \int v_1 \, dx = e^x \cdots$

$$\left[\text{use formula,} \ \int u . v \, dx = u . v_1 - u' . v_2 + u'' . v_3 - u''' . v_4 + \cdots \cdots \cdots \cdots \cdots \right]$$

Integration of proper rational function by partial function: −

If the denominator of proper rational function $\frac{P(x)}{Q(x)}$ is at the form

$Q(x) = (x - a)^m (x - b)^n (x - c)^t (px^2 + qx + r)^s (ex^2 + f.x + d)^l$ where $q^2 - 4pr, \ f^2 - 4ed < 0$ *then,*

$$\frac{P(x)}{Q(x)} = \frac{A_1}{(x-a)} + \frac{A_2}{(x-a)^2} + \cdots \cdots \cdots + \frac{A_m}{(x-a)^m} + \frac{B_1}{(x-b)} + \frac{B_2}{(x-b)^2} + \cdots \cdots \cdots + \frac{B_n}{(x-b)^n} + \frac{C_1}{(x-c)} + \frac{C_2}{(x-c)^2} + \cdots \cdots \cdots + \frac{C_t}{(x-c)^t}$$
$$+ \frac{D_1 x + E_1}{(px^2 + qx + r)} + \frac{D_2 x + E_2}{(px^2 + qx + r)^2} + \cdots \cdots \cdots + \frac{D_s x + E_s}{(px^2 + qx + r)^s} + \frac{F_1 x + G_1}{(ex^2 + f.x + d)} + \frac{F_2 x + G_2}{(ex^2 + f.x + d)^2}$$
$$+ \cdots \cdots \cdots \cdots + \frac{F_l x + G_l}{(ex^2 + f.x + d)^l}$$

e.g. (a) $\dfrac{2x + 3}{(x+1)(x+2)^2} = \dfrac{A}{x+1} + \dfrac{B}{x+2} + \dfrac{C}{(x+2)^2}$

(b) $\dfrac{2 - x^2}{(x-2)(x+3)(x-5)} = \dfrac{A}{x-2} + \dfrac{B}{x+3} + \dfrac{C}{x-5}$ \quad (c) $\dfrac{1}{x^2(x^2 - 2x + 3)} = \dfrac{A}{x} + \dfrac{B}{x^2} + \dfrac{Cx + D}{(x^2 - 2x + 3)}$

(d) $\dfrac{2x^2 + 2x - 15}{(x+5)^2(x^2 + 2x + 3)^2} = \dfrac{A}{x+5} + \dfrac{B}{(x+5)^2} + \dfrac{Cx + D}{x^2 + 2x + 3} + \dfrac{Ex + F}{(x^2 + 2x + 3)^2}$

	Linear factors	Quadratic factors
Non repeated factor	$\dfrac{P(x)}{(x-a)(x-b)(x+c)} = \dfrac{A}{x-a} + \dfrac{B}{x-b} + \dfrac{C}{x+c}$	$\dfrac{P(x)}{(x^2 + x - 1)(x^2 + x - 3)(x - 2)}$
		$= \dfrac{A}{x-2} + \dfrac{Bx + C}{x^2 + x - 1} + \dfrac{Dx + E}{x^2 + x - 3}$
	find, $A = \left(\dfrac{P(x)}{(x-b)(x+c)} \right)_{x=a}$	
	$B = \left(\dfrac{P(x)}{(x-a)(x+c)} \right)_{x=b}$	A, B, C, D, E are obtained by multiplying by $(x^2 + x - 1)(x^2 + x - 3)(x - 2)$ on both of sides
	$C = \left(\dfrac{P(x)}{(x-a)(x-b)} \right)_{x=-c}$	and then equate coefficient of suitable power of x on both side.

Repeated factor	$\dfrac{P(x)}{(x+a)^2(x+b)^3} = \dfrac{A}{x+a} + \dfrac{B}{(x+a)^2} + \dfrac{C}{x+b}$ $+ \dfrac{D}{(x+b)^2} + \dfrac{E}{(x+b)^3}$ $B = \left(\dfrac{P(x)}{(x+b)^3}\right)_{x=-a}$ $E = \left(\dfrac{P(x)}{(x+a)^2}\right)_{x=-b}$ A, C, D are obtained by multiplying by $(x+a)^2(x+b)^3$ on both sides and then equate of coefficient of suitable power of x.	$\dfrac{P(x)}{(x^2-x+1)^2(x^2+x-3)}$ $= \dfrac{Ax+B}{x^2-x+1} + \dfrac{Cx+D}{(x^2-x+1)^2} + \dfrac{Ex+F}{x^2+x-3}$ A, B, C, D, E, F are obtained by multiplying by $(x^2-x+1)^2(x^2+x-3)$ on both sides and then equate coefficient of suitable power of x on both sides.

Example: − (a) $I = \displaystyle\int \dfrac{3x^2 - 5x + 6}{(2x-1)(x+1)}\, dx$, Let $\dfrac{3x^2 - 5x + 6}{(2x-1)(x+1)} = \dfrac{A}{2x-1} + \dfrac{B}{x+1}$ ………………. (A)

$A = \dfrac{3x^2 - 5x + 6}{(x+1)}$ $\therefore 2x - 1 = 0$ or $2x = 1$ $\therefore x = \dfrac{1}{2}$

$A = \left(\dfrac{3x^2 - 5x + 6}{(x+1)}\right)_{x=\frac{1}{2}} = \dfrac{3.\left(\frac{1}{2}\right)^2 - 5.\frac{1}{2} + 6}{\frac{1}{2} + 1} = \dfrac{\frac{3}{4} - \frac{5}{2} + 6}{\frac{1+2}{2}} = \dfrac{\frac{3-10+24}{4}}{\frac{3}{2}} = \dfrac{17}{4} \times \dfrac{2}{3} = \dfrac{17}{6}$

$B = \dfrac{3x^2 - 5x + 6}{(2x-1)}$ $\therefore x + 1 = 0$ $\therefore x = -1$

$B = \left(\dfrac{3x^2 - 5x + 6}{(2x-1)}\right)_{x=-1} = \dfrac{3.(-1)^2 - 5.(-1) + 6}{2.(-1) - 1} = \dfrac{3 + 5 + 6}{-2 - 1} = -\dfrac{14}{3}$

From equation (A), we have $\Rightarrow \dfrac{3x^2 - 5x + 6}{(2x-1)(x+1)} = \dfrac{17}{6}.\dfrac{1}{2x-1} - \dfrac{14}{3}.\dfrac{1}{x+1}$

Integrating, $\displaystyle\int \dfrac{3x^2 - 5x + 6}{(2x-1)(x+1)}\, dx = \dfrac{17}{6}\int \dfrac{1}{2x-1}\, dx - \dfrac{14}{3}\int \dfrac{1}{x+1}\, dx = \dfrac{17}{6}.\dfrac{\log(2x-1)}{2} - \dfrac{14}{3}.\log(x+1) + c$

$= \dfrac{17}{12}\log(2x-1) - \dfrac{14}{3}\log(x+1) + c$ Ans.

(b) $I = \displaystyle\int \dfrac{x^2 + 3x + 5}{(x-1)(x+3)(x-4)}\, dx$, Let $\dfrac{x^2 + 3x + 5}{(x-1)(x+3)(x-4)} = \dfrac{A}{x-1} + \dfrac{B}{x+3} + \dfrac{C}{x-4}$ ………… (A)

See above question, Solve equation (A) and find A, B, C $\therefore A = -\dfrac{3}{4}$, $B = \dfrac{5}{28}$ and $C = \dfrac{33}{21}$

$I = \displaystyle\int \dfrac{x^2 + 3x + 5}{(x-1)(x+3)(x-4)}\, dx = -\dfrac{3}{4}\log(x-1) + \dfrac{5}{28}\log(x+3) + \dfrac{33}{21}\log(x-4) + c$ Ans.

(c) $I = \displaystyle\int \dfrac{3x+2}{(x+1)(x-2)^2}\, dx$, Let $\dfrac{3x+2}{(x+1)(x-2)^2} = \dfrac{A}{x+1} + \dfrac{B}{x-2} + \dfrac{C}{(x-2)^2}$ …………………. (A)

$\therefore A = \dfrac{3x+2}{(x-2)^2}\Big|_{x=-1} = \dfrac{3.(-1) + 2}{(-3)^2} = \dfrac{-1}{9} = -\dfrac{1}{9}$

or $\dfrac{3x+2}{(x+1)(x-2)^2} = \dfrac{B}{x-2}$, Multiplying by $(x+1)(x-2)^2$ both of sides, we have

or $\dfrac{3x+2}{(x+1)(x-2)^2} \times (x+1)(x-2)^2 = \dfrac{B}{x-2} \times (x+1)(x-2)^2 = B(x+1)(x-2) = B(x^2-2x+x-2)$

$\Rightarrow \quad 3x+2 = B(x^2-x-2) = Bx^2 - Bx - 2B \quad \Rightarrow \quad 0.x^2 + 3x + 2 = Bx^2 - Bx - 2B \quad$ or $-Bx = 3x \quad \therefore \quad B = -3$

or $\dfrac{3x+2}{(x+1)(x-2)^2} = \dfrac{C}{(x-2)^2}$, Multiplying by $(x+1)(x-2)^2$ both of sides, we have $\Rightarrow 3x+2 = C(x+1) = Cx + C \quad \therefore \quad C = 3$

Put value of A, B and C in equation (A) and integrate, we have

or $\dfrac{3x+2}{(x+1)(x-2)^2} = -\dfrac{1}{9}.\dfrac{1}{x+1} + (-3).\dfrac{1}{x-2} + 3.\dfrac{1}{(x-2)^2}$

$\therefore \displaystyle\int \dfrac{3x+2}{(x+1)(x-2)^2} \, dx = -\dfrac{1}{9}\int \dfrac{dx}{x+1} - 3\int \dfrac{dx}{x-2} + 3\int \dfrac{dx}{(x-2)^2} = -\dfrac{1}{9}\log(x+1) - 3\log(x-2) + 3.\dfrac{(x-2)^{-2+1}}{-2+1}$

$\therefore I = \displaystyle\int \dfrac{3x+2}{(x+1)(x-2)^2} \, dx = -\dfrac{1}{9}\log(x+1) - 3\log(x-2) - 3.\dfrac{1}{x-2} + c \qquad$ Ans.

(d) $I = \displaystyle\int \dfrac{1-x^2}{x^2(x-3)^2} \, dx$, Let $\dfrac{1-x^2}{x^2(x-3)^2} = \dfrac{A}{x} + \dfrac{B}{x^2} + \dfrac{C}{x-3} + \dfrac{D}{(x-3)^2}$ ....(A)

$\therefore B = \dfrac{1-x^2}{(x-3)^2}\bigg|_{x=0} = \dfrac{1-0}{(0-3)^2} = \dfrac{1}{9}$ and $D = \dfrac{1-x^2}{x^2}\bigg|_{x=3} = \dfrac{1-(3)^2}{(3)^2} = \dfrac{1-9}{9} = -\dfrac{8}{9}$

To be A and C are obtained by multiplying by $x^2(x-3)^2$ on both of sides and then equate coefficient of suitable power of x on both sides.

or $\dfrac{1-x^2}{x^2(x-3)^2} \times x^2(x-3)^2 = \dfrac{A}{x} \times x^2(x-3)^2$

or $1-x^2 = Ax(x-3)^2 = Ax(x^2-6x+9) = Ax^3 - 6Ax^2 + 9Ax \quad \therefore \ -1 = -6A$ or $6A = 1 \quad \therefore \ A = \dfrac{1}{6}$

Again, $\dfrac{1-x^2}{x^2(x-3)^2} = \dfrac{C}{x-3}$ or $\dfrac{1-x^2}{x^2(x-3)^2} \times x^2(x-3)^2 = \dfrac{C}{x-3} \times x^2(x-3)^2$

or $1-x^2 = Cx^2(x-3)$ or $1-x^2 = Cx^3 - 3Cx^2$ or $-1 = -3C$ or $1 = 3C \quad \therefore \ C = \dfrac{1}{3}$

Put value of A, B, C and D in equation (A) and integrate, we have

or $\dfrac{1-x^2}{x^2(x-3)^2} = \dfrac{A}{x} + \dfrac{B}{x^2} + \dfrac{C}{x-3} + \dfrac{D}{(x-3)^2} = \dfrac{1}{6x} + \dfrac{1}{9x^2} + \dfrac{1}{3(x-3)} - \dfrac{8}{9(x-3)^2}$

or $\displaystyle\int \dfrac{1-x^2}{x^2(x-3)^2} \, dx = \dfrac{1}{6}\int \dfrac{1}{x} \, dx + \dfrac{1}{9}\int \dfrac{1}{x^2} \, dx + \dfrac{1}{3}\int \dfrac{1}{(x-3)} \, dx - \dfrac{8}{9}\int \dfrac{1}{(x-3)^2} \, dx$

$I = \dfrac{1}{6}\log x + \dfrac{1}{9}.\dfrac{x^{-2+1}}{-2+1} + \dfrac{1}{3}\log(x-3) - \dfrac{8}{9}.\dfrac{(x-3)^{-2+1}}{-2+1} + c = \dfrac{1}{6}\log x - \dfrac{1}{9}x^{-1} + \dfrac{1}{3}\log(x-3) + \dfrac{8}{9}(x-3)^{-1} + c$

$\qquad\qquad = \dfrac{1}{6}\log x - \dfrac{1}{9x} + \dfrac{1}{3}\log(x-3) + \dfrac{8}{9(x-3)} + c \qquad$ Ans.

Example

(1) Integrate:− (a) $I = \displaystyle\int \dfrac{\cos x}{\sin^4 x + \cos^4 x} \, dx$ (b) $I = \displaystyle\int \dfrac{dx}{4x^2+3}$ (c) $I = \displaystyle\int \dfrac{dx}{\sqrt{x^2-2x+3}}$

(d) $I = \displaystyle\int \dfrac{dx}{\sqrt{3+(1-2x)^2}}$ (e) $I = \displaystyle\int \dfrac{dx}{\sqrt{4x^2+4x-5}}$ (f) $I = \displaystyle\int \dfrac{dx}{\sqrt{3-(1-x)^2}}$ (g) $I = \displaystyle\int \dfrac{dx}{x^2-4x+3}$

(2) (a) $I = \int \dfrac{dx}{2 - (3 - x)^2}$ (b) $I = \int \dfrac{dx}{9 - 6x^2}$ (c) $I = \int \sqrt{x^2 - 8}\ dx$ (d) $I = \int \sqrt{x^2 + 3}\ dx$

(e) $I = \int \sqrt{2 - x^2}\ dx$ (f) $I = \int \sqrt{(5x + 2)^2 - 3}\ dx$ (g) $I = \int \sqrt{(2x - 1)^2 + 4}\ dx$ (h) $I = \int \sqrt{5 - (2 - x)^2}\ dx$

(3) (a) $I = \int \dfrac{x\ dx}{(2x^2 - 3)^4}$ (b) $I = \int \dfrac{2ax + b}{\sqrt{2ax^2 + 2bx + c}}\ dx$ (c) $I = \int \dfrac{x^2}{\sqrt{2 - 3x^3}}\ dx$ (d) $I = \int \dfrac{\cos x}{\sqrt{1 + \sin x}}\ dx$

(e) $I = \int \dfrac{dx}{e^x\sqrt{1 + e^{-x}}}$ (f) $I = \int \dfrac{x^x(1 + \log x) + \dfrac{1}{x}}{\sqrt{x^x + \log x}}\ dx$ (g) $I = \int \dfrac{dx}{(2x - 1)^{3/2}}$ (h) $I = \int \dfrac{(1 + \cos\theta)}{(\theta + \sin\theta)^{5/2}}\ d\theta$

(4) (a) $I = \int \dfrac{1 + \sin\theta}{\cos\theta}\ d\theta$ (b) $I = \int \dfrac{1 + \cos\theta}{\sin\theta}\ d\theta$ (c) $I = \int \dfrac{\sec^2\theta}{1 + \tan\theta}\ d\theta$ (d) $I = \int \dfrac{d}{dx}\left(\dfrac{\log(1 + x)}{x}\right)dx$

(e) $I = \dfrac{d}{dx}\left(\int \dfrac{\log x + 1}{2x}\ dx\right)$ (f) $I = \int \dfrac{1 + \log x}{x \log x}\ dx$ (g) $I = \int \dfrac{x \cos x}{x \sin x + \cos x}\ dx$ (h) $I = \int \dfrac{x}{2x^2 - 5}\ dx$

(5) (a) $I = \int \dfrac{(2x - 1)}{2x^2 - 2x + 5}\ dx$ (b) $I = \int \dfrac{\cos x}{\sin^3 x}\ dx$ (c) $I = \int \dfrac{\sec^2 x}{\tan^4 x}\ dx$ (d) $I = \int \dfrac{dx}{x^2 + x + 3}$

(e) $I = \int \dfrac{dx}{2x^2 + 4x - 5}$ (f) $I = \int \dfrac{dx}{\sqrt{1 + 4x - 2x^2}}$ (g) $I = \int \dfrac{dx}{\sqrt{x^2 + 2x + 3}}$ (h) $I = \int \dfrac{dx}{\sqrt{3 + 4x - 2x^2}}$

(6) (a) $I = \int \sqrt{-x^2 + 3x + 5}\ dx$ (b) $I = \int \dfrac{\sec^2 x}{(\sec x + \tan x)^{9/2}}\ dx$ (c) $I = \int \dfrac{\sin^2(\log x) + \cos(\log x)}{x \tan(\log x)}\ dx$

(d) $I = \int \dfrac{dx}{x \sin(\log x) . \cos(\log x)}$ (e) $I = \int \dfrac{\log x}{x^2}\ dx$ (f) $I = \int \dfrac{dx}{x\sqrt{1 + \log x}}$

(7) (a) $I = \int \dfrac{x^3}{(x - 1)^2(x^2 + 2x + 3)}\ dx$ (b) $I = \int \dfrac{x}{x^3 + 1}\ dx$ (c) $I = \int \dfrac{x^4}{(x - 1)(x^2 + 1)}\ dx$ (d) $I = \int \dfrac{x - 1}{x + 1} \cdot \dfrac{1}{\sqrt{x^3 + x^2 + x}}\ dx$

Solution

(1) (a) $I = \int \dfrac{\cos x}{\sin^4 x + \cos^4 x}\ dx$,   Put $\sin x = t$ $\therefore$ $\cos x\ dx = dt$ then $\sin^4 x = t^4$ and $\cos^4 x = (1 - \sin^2 x)^2 = (1 - t^2)^2$

$$I = \int \dfrac{dt}{t^4 + (1 - t^2)^2} = \int \dfrac{dt}{t^4 + 1 - 2t^2 + t^4} = \int \dfrac{dt}{2t^4 - 2t^2 + 1} = \int \dfrac{dt}{\left(\sqrt{2}t^2 - \dfrac{1}{\sqrt{2}}\right)^2 + \left(\dfrac{1}{\sqrt{2}}\right)^2} = \dfrac{1}{\dfrac{1}{\sqrt{2}}} \tan^{-1}\left(\dfrac{\sqrt{2}t^2 - \dfrac{1}{\sqrt{2}}}{\dfrac{1}{\sqrt{2}}}\right) . \dfrac{1}{\text{d. c of }\left(\sqrt{2}t^2 - \dfrac{1}{\sqrt{2}}\right)}$$

$$= \sqrt{2}\tan^{-1}(2t^2 - 1) . \dfrac{1}{2\sqrt{2}t} = \dfrac{1}{2t}\tan^{-1}(2t^2 - 1)$$

$I = \dfrac{1}{2\sin x}\tan^{-1}(2\sin^2 x - 1) + c = \dfrac{1}{2}\sec x . \tan^{-1}(2\sin^2 x - 1) + c$ Ans.

(b) $I = \int \dfrac{dx}{4x^2 + 3} = \int \dfrac{dx}{(2x)^2 + (\sqrt{3})^2}$ $\left[\text{use formula, }\ \int \dfrac{dx}{x^2 + a^2} = \dfrac{1}{a}\tan^{-1}\dfrac{x}{a}\right]$

$\therefore$ $I = \int \dfrac{dx}{(2x)^2 + (\sqrt{3})^2} = \dfrac{1}{\sqrt{3}}\tan^{-1}\left(\dfrac{2x}{\sqrt{3}}\right) . \dfrac{1}{2} + c = \dfrac{1}{2\sqrt{3}}\tan^{-1}\left(\dfrac{2x}{\sqrt{3}}\right) + c$ Ans.

(c) $I = \int \dfrac{dx}{\sqrt{x^2 - 2x + 3}} = \int \dfrac{dx}{\sqrt{(x - 1)^2 + 2}} = \int \dfrac{dx}{\sqrt{(x - 1)^2 + (\sqrt{2})^2}}$ formula, $\int \dfrac{dx}{\sqrt{x^2 + a^2}} = \log\left|x + \sqrt{x^2 + a^2}\right|$

$\therefore$ $I = \int \dfrac{dx}{\sqrt{(x - 1)^2 + (\sqrt{2})^2}} = \log\left[(x - 1) + \sqrt{(x - 1)^2 + (\sqrt{2})^2}\right] + c = \log\left[(x - 1) + \sqrt{x^2 - 2x + 3}\right] + c$ Ans.

(d) $I = \int \dfrac{dx}{\sqrt{3 + (1 - 2x)^2}} = \int \dfrac{dx}{\sqrt{\left(\sqrt{3}\right)^2 + (1 - 2x)^2}}$ $\left[\text{use formula,}\quad \int \dfrac{dx}{\sqrt{x^2 + a^2}} = \log\left|x + \sqrt{x^2 + a^2}\right|\right]$

$\therefore I = \int \dfrac{dx}{\sqrt{\left(\sqrt{3}\right)^2 + (1 - 2x)^2}} = \log\left|(1 - 2x) + \sqrt{\left(\sqrt{3}\right)^2 + (1 - 2x)^2}\right|.\dfrac{1}{\frac{d}{dx}(1 - 2x)} = \log\left[(1 - 2x) + \sqrt{3 + (1 - 2x)^2}\right].\dfrac{1}{-2} + c$

$\qquad\qquad = -\dfrac{1}{2}\log\left[(1 - 2x) + \sqrt{3 + (1 - 2x)^2}\right] + c$ Ans.

(e) $I = \int \dfrac{dx}{\sqrt{4x^2 + 4x - 5}} = \int \dfrac{dx}{\sqrt{(2x + 1)^2 - 6}} = \int \dfrac{dx}{\sqrt{(2x + 1)^2 - \left(\sqrt{6}\right)^2}}$ $\left[\text{use formula,}\quad \int \dfrac{dx}{\sqrt{x^2 - a^2}} = \log\left|x + \sqrt{x^2 - a^2}\right|\right]$

$\therefore I = \int \dfrac{dx}{\sqrt{(2x + 1)^2 - \left(\sqrt{6}\right)^2}} = \log\left|(2x + 1) + \sqrt{(2x + 1)^2 - \left(\sqrt{6}\right)^2}\right|.\dfrac{1}{2} + c = \dfrac{1}{2}\log\left|(2x + 1) + \sqrt{4x^2 + 4x - 5}\right| + c$ Ans.

(f) $I = \int \dfrac{dx}{\sqrt{3 - (1 - x)^2}} = \int \dfrac{dx}{\sqrt{\left(\sqrt{3}\right)^2 - (1 - x)^2}}$ $\left[\text{use formula,}\quad \int \dfrac{dx}{\sqrt{a^2 - x^2}} = \sin^{-1}\dfrac{x}{a}\right]$

$\therefore I = \int \dfrac{dx}{\sqrt{\left(\sqrt{3}\right)^2 - (1 - x)^2}} = \sin^{-1}\left[\dfrac{1 - x}{\sqrt{3}}\right].\dfrac{1}{-1} + c = -\sin^{-1}\left[\dfrac{1 - x}{\sqrt{3}}\right] + c$ Ans.

(g) $I = \int \dfrac{dx}{x^2 - 4x + 3} = \int \dfrac{dx}{(x - 2)^2 - 1}$ $\left[\text{use formula,}\quad \int \dfrac{dx}{x^2 - a^2} = \dfrac{1}{2a}\log\left|\dfrac{x - a}{x + a}\right|\right]$

$\therefore I = \int \dfrac{dx}{(x - 2)^2 - 1} = \dfrac{1}{2.1}\log\left|\dfrac{(x - 2) - 1}{(x - 2) + 1}\right| + c = \dfrac{1}{2}\log\left|\dfrac{x - 3}{x - 1}\right| + c$ Ans.

(2) (a) $I = \int \dfrac{dx}{2 - (3 - x)^2} = \int \dfrac{dx}{\left(\sqrt{2}\right)^2 - (3 - x)^2}$ $\left[\text{use formula,}\quad \int \dfrac{dx}{a^2 - x^2} = \dfrac{1}{2a}\log\left|\dfrac{a + x}{a - x}\right|\right]$

$\therefore I = \int \dfrac{dx}{\left(\sqrt{2}\right)^2 - (3 - x)^2} = \dfrac{1}{2.\sqrt{2}}\log\left|\dfrac{\sqrt{2} + (3 - x)}{\sqrt{2} - (3 - x)}\right|.\dfrac{1}{-1} + c = -\dfrac{1}{2\sqrt{2}}\log\left|\dfrac{\sqrt{2} + (3 - x)}{\sqrt{2} - (3 - x)}\right| + c$ Ans.

(b) $I = \int \dfrac{dx}{9 - 6x^2} = \int \dfrac{dx}{3^2 - \left(\sqrt{6}x\right)^2} = \dfrac{1}{6\sqrt{6}}\log\left|\dfrac{3 + \sqrt{6}x}{3 - \sqrt{6}x}\right| + c$ Ans.   (see above question)

(c) $I = \int \sqrt{x^2 - 8}\ dx = \int \sqrt{x^2 - \left(2\sqrt{2}\right)^2}\ dx$ $\left[\text{use formula,}\quad \int \sqrt{x^2 - a^2}\ dx = \dfrac{x}{2}\sqrt{x^2 - a^2} - \dfrac{a^2}{2}\log\left|x + \sqrt{x^2 - a^2}\right|\right]$

$\therefore I = \int \sqrt{x^2 - \left(2\sqrt{2}\right)^2}\ dx = \dfrac{x}{2}\sqrt{x^2 - 8} - 4\log\left|x + \sqrt{x^2 - 8}\right| + c$ Ans.

(d) $I = \int \sqrt{x^2 + 3}\ dx = \int \sqrt{x^2 + \left(\sqrt{3}\right)^2}\ dx$ $\left[\text{use formula,}\quad \int \sqrt{x^2 + a^2}\ dx = \dfrac{x}{2}\sqrt{x^2 + a^2} + \dfrac{a^2}{2}\log\left|x + \sqrt{x^2 + a^2}\right|\right]$

$\therefore I = \int \sqrt{x^2 + \left(\sqrt{3}\right)^2}\ dx = \dfrac{x}{2}\sqrt{x^2 + 3} + \dfrac{3}{2}\log\left|x + \sqrt{x^2 + 3}\right| + c$ Ans.

(e) $I = \int \sqrt{2 - x^2}\ dx = \int \sqrt{\left(\sqrt{2}\right)^2 - x^2}\ dx$ $\left[\text{use formula,}\quad \int \sqrt{a^2 - x^2}\ dx = \dfrac{x}{2}\sqrt{a^2 - x^2} + \dfrac{a^2}{2}\sin^{-1}\dfrac{x}{a}\right]$

$\therefore I = \int \sqrt{\left(\sqrt{2}\right)^2 - x^2}\ dx = \dfrac{x}{2}\sqrt{2 - x^2} + \sin^{-1}\left(\dfrac{x}{\sqrt{2}}\right) + c$ Ans.

(f) $I = \int \sqrt{(5x + 2)^2 - 3}\ dx = \int \sqrt{(5x + 2)^2 - \left(\sqrt{3}\right)^2}\ dx$ [see question no. −(2)(c)]

$\therefore\ I = \dfrac{(5x+2)}{10}\sqrt{(5x+2)^2-3} - \dfrac{3}{10}\log\left|(5x+2)+\sqrt{(5x+2)^2-3}\right| + c \qquad$ Ans.

(g) $I = \displaystyle\int \sqrt{(2x-1)^2+4}\ dx = \int \sqrt{(2x-1)^2+2^2}\ dx \qquad$ [see question no. –(2)(d)]

$\therefore\ I = \dfrac{1}{2}\left\{ \dfrac{(2x-1)}{2}\sqrt{(2x-1)^2+4} + 2\log\left|(2x-1)+\sqrt{(2x-1)^2+4}\right| + c\right\}$

$\qquad\qquad = \dfrac{(2x-1)}{4}\sqrt{(2x-1)^2+4} + \log\left|(2x-1)+\sqrt{(2x-1)^2+4}\right| + c \qquad$ Ans.

(h) $I = \displaystyle\int \sqrt{5-(2-x)^2}\ dx = \int \sqrt{\left(\sqrt5\right)^2-(2-x)^2}\ dx \qquad$ [see question no. –(2)(e)]

$\therefore\ I = \left[\dfrac{(2-x)}{2}\sqrt{5-(2-x)^2} + \dfrac{5}{2}\sin^{-1}\left(\dfrac{(2-x)}{\sqrt5}\right)\right].\dfrac{1}{-1} + c = -\left[\dfrac{(2-x)}{2}\sqrt{5-(2-x)^2} + \dfrac{5}{2}\sin^{-1}\left(\dfrac{(2-x)}{\sqrt5}\right)\right] + c$ Ans.

(3) (a) $I = \displaystyle\int \dfrac{x\ dx}{(2x^2-3)^4} = \int (2x^2-3)^{-4}.x\ dx$, Let $f(x) = 2x^2-3 \ \ \therefore f'(x) = 4x$

$\therefore\ I = \displaystyle\int \{f(x)\}^{-4}.\dfrac{f'(x)}{4}\ dx \qquad \left[\text{using formula, } I = \int \{f(x)\}^n.f'(x)\ dx = \dfrac{\{f(x)\}^{n+1}}{n+1}\ ,\ n\ne -1\right]$

$\qquad\qquad \therefore\ I = \dfrac{\{f(x)\}^{-4+1}}{-4+1}.\dfrac{1}{4} = -\dfrac{(2x^2-3)^{-3}}{12} + c = -\dfrac{1}{12}(2x^2-3)^{-3} + c \qquad$ Ans.

(b) $I = \displaystyle\int \dfrac{2ax+b}{\sqrt{2ax^2+2bx+c}}\ dx$, Let $f(x) = 2ax^2+2bx+c \ \ \therefore f'(x) = 4ax+2b = 2(2ax+b)$ or $2ax+b = \dfrac{f'(x)}{2}$

$\left[\text{use formula, } \displaystyle\int \dfrac{f'(x)}{\sqrt{f(x)}}\ dx = 2\sqrt{f(x)}\right] \quad \therefore\ I = \displaystyle\int \dfrac{f'(x)}{2.\sqrt{f(x)}}\ dx = \dfrac{2}{2}.\sqrt{f(x)} = \sqrt{2ax^2+2bx+c} + d \quad$ Ans.

(c) $I = \displaystyle\int \dfrac{x^2}{\sqrt{2-3x^3}}\ dx$, Let $f(x) = 2-3x^3 \ \ \therefore f'(x) = -9x^2$ or $x^2 = -\dfrac{f'(x)}{9}$

or $I = \displaystyle\int \dfrac{-\dfrac{f'(x)}{9}}{\sqrt{f(x)}}\ dx = -\dfrac{1}{9}\int \dfrac{f'(x)}{\sqrt{f'(x)}}\ dx$ (see above question) $\therefore\ I = -\dfrac{1}{9}.2\sqrt{f(x)} = -\dfrac{2}{9}\sqrt{2-3x^3} + c \quad$ Ans.

(d) $I = \displaystyle\int \dfrac{\cos x}{\sqrt{1+\sin x}}\ dx$, Let $1+\sin x = f(x) \qquad \therefore\ f'(x) = \cos x$

$\therefore\ I = \displaystyle\int \dfrac{f'(x)}{\sqrt{f(x)}}\ dx = 2\sqrt{f(x)} \ \therefore\ I = \displaystyle\int \dfrac{\cos x}{\sqrt{1+\sin x}}\ dx = 2\sqrt{1+\sin x} + c \ \ $ Ans. $\left[\text{formula, } \displaystyle\int \dfrac{f'(x)}{\sqrt{f(x)}}\ dx = 2\sqrt{f(x)}\right]$

(e) $I = \displaystyle\int \dfrac{dx}{e^x\sqrt{1+e^{-x}}}$, Let $f(x) = 1+e^{-x} \qquad \therefore\ f'(x) = -e^{-x} = -\dfrac{1}{e^x}$

or $I = -\displaystyle\int \dfrac{1}{\sqrt{1+e^{-x}}}.\left(-\dfrac{1}{e^x}\right)\ dx = -\int \dfrac{f'(x)}{\sqrt{f(x)}}\ dx \qquad \left[\text{formula, } \displaystyle\int \dfrac{f'(x)}{\sqrt{f(x)}}\ dx = 2\sqrt{f(x)}\right]$

$\qquad\qquad\qquad \therefore\ I = -\displaystyle\int \dfrac{f'(x)}{\sqrt{f(x)}}\ dx = -2\sqrt{f(x)} = -2\sqrt{1+e^{-x}} + c \qquad$ Ans.

(f) $I = \displaystyle\int \dfrac{x^x(1+\log x) + \dfrac{1}{x}}{\sqrt{x^x+\log x}}\ dx$, Let $f(x) = x^x+\log x$ put $y = x^x$ or $\log y = x\log x \ \ \therefore\ \dfrac{1}{y}.\dfrac{dy}{dx} = x.\dfrac{1}{x} + \log x.1$

or $\dfrac{dy}{dx} = y(1+\log x) = x^x(1+\log x)$ then $f'(x) = x^x(1+\log x) + \dfrac{1}{x}$

$\therefore\ I = \displaystyle\int \dfrac{x^x(1+\log x) + \dfrac{1}{x}}{\sqrt{x^x+\log x}}\ dx = \int \dfrac{f'(x)}{\sqrt{f(x)}}\ dx = 2\sqrt{f(x)} + c = 2\sqrt{x^x+\log x} + c \qquad$ Ans.

(g) $I = \int \dfrac{dx}{(2x-1)^{3/2}}$, Let $2x - 1 = z$ or $2\,dx = dz$ or $dx = \dfrac{dz}{2}$

$\therefore\ I = \int \dfrac{dz}{2.z^{3/2}} = \dfrac{1}{2}\int z^{-3/2}\,dz = \dfrac{1}{2}\cdot\dfrac{z^{-\frac{3}{2}+1}}{-\dfrac{3}{2}+1} + c = -z^{-\frac{1}{2}} + c = -\dfrac{1}{\sqrt{z}} + c = -\dfrac{1}{\sqrt{2x-1}} + c$ \quad Ans.

IInd Method:$-\ I = \int \dfrac{dx}{(2x-1)^{3/2}} = \int (2x-1)^{-3/2}\,dx$, \qquad Let $f(x) = 2x - 1$ \quad $\therefore f'(x) = 2$

or $I = \dfrac{1}{2}\int (2x-1)^{-3/2}.2\,dx = \dfrac{1}{2}\int [f(x)]^{-3/2}.f'(x)\,dx$ $\left[\text{using formula, } I = \int \{f(x)\}^n.f'(x)\,dx = \dfrac{\{f(x)\}^{n+1}}{n+1}\right]$

$\therefore\ I = \dfrac{1}{2}\int [f(x)]^{-3/2}.f'(x)\,dx = \dfrac{1}{2}\cdot\dfrac{[f(x)]^{-3/2+1}}{-\dfrac{3}{2}+1} + c = -[f(x)]^{-1/2} + c = -\dfrac{1}{\sqrt{f(x)}} + c = -\dfrac{1}{\sqrt{2x-1}} + c$ \quad Ans.

(h) $I = \int \dfrac{(1+\cos\theta)}{(\theta+\sin\theta)^{5/2}}\,d\theta = \int (\theta+\sin\theta)^{-5/2}.(1+\cos\theta)\,d\theta$, Let $f(\theta) = \theta + \sin\theta$ \quad $\therefore f'(\theta) = 1 + \cos\theta$

or $I = \int (\theta+\sin\theta)^{-5/2}.(1+\cos\theta)\,d\theta = \int [f(\theta)]^{-5/2}.f'(\theta)d\theta$ $\left[\text{formula, } \int \{f(x)\}^n.f'(x)\,dx = \dfrac{\{f(x)\}^{n+1}}{n+1}\right]$

or $I = \int [f(\theta)]^{-5/2}.f'(\theta)\,d\theta = \dfrac{[f(\theta)]^{-5/2+1}}{-\dfrac{5}{2}+1} + c = -\dfrac{2}{3}.[f(\theta)]^{-3/2} + c = -\dfrac{2}{3}\cdot\dfrac{1}{[f(\theta)]^{3/2}} + c = -\dfrac{2}{3}\cdot\dfrac{1}{[\theta+\sin\theta]^{3/2}} + c$

$\qquad\qquad = -\dfrac{2}{3}\cdot\dfrac{1}{(\theta+\sin\theta).\sqrt{\theta+\sin\theta}} + c$ \quad Ans.

(4) (a) $I = \int \dfrac{1+\sin\theta}{\cos\theta}\,d\theta = \int \dfrac{\sin^2\theta/2 + \cos^2\theta/2 + 2\sin\theta/2\cos\theta/2}{\cos^2\theta/2 - \sin^2\theta/2}\,d\theta$, Divide above and below by $\cos^2\theta/2$

$I = \int \dfrac{\tan^2\theta/2 + 1 + 2\tan\theta/2}{1 - \tan^2\theta/2}\,d\theta = \int \dfrac{\left(1+\tan\theta/2\right)^2}{\left(1+\tan\theta/2\right)\left(1-\tan\theta/2\right)}\,d\theta = \int \dfrac{\left(1+\tan\theta/2\right)}{\left(1-\tan\theta/2\right)}\,d\theta = \int \dfrac{\sec^2\theta/2}{\left(1-\tan\theta/2\right)}\,d\theta$

Put $\tan\theta/2 = t$ \quad $\therefore\ \dfrac{1}{2}\sec^2\theta/2\,d\theta = dt$ or $\sec^2\theta/2\,d\theta = 2\,dt$

$\qquad\qquad \therefore\ I = \int \dfrac{2\,dt}{1-t} = 2\log(1-t)\cdot\dfrac{1}{\text{d.c of }(1-t)} + c = -2\log\left(1-\tan\theta/2\right) + c$ \quad Ans.

(b) $I = \int \dfrac{1+\cos\theta}{\sin\theta}\,d\theta = \int \dfrac{\sin^2\theta/2 + \cos^2\theta/2 + \cos^2\theta/2 - \sin^2\theta/2}{2\sin\theta/2\cos\theta/2}\,d\theta = \int \dfrac{2\cos^2\theta/2}{2\sin\theta/2\cos\theta/2}\,d\theta = \int \dfrac{\cos\theta/2}{\sin\theta/2}\,d\theta$

$\therefore\ I = \int \dfrac{\cos\theta/2}{\sin\theta/2}\,d\theta$ \quad Put $\sin\theta/2 = t$ or $\dfrac{1}{2}\cos\theta/2\,d\theta = dt$ \quad or $\cos\theta/2\,d\theta = 2\,dt$

$\qquad\qquad \therefore\ I = \int \dfrac{2\,dt}{t} = 2\log t + c = 2\log\left(\sin\theta/2\right) + c$ \quad Ans.

(c) $I = \int \dfrac{\sec^2\theta}{1+\tan\theta}\,d\theta$ \quad Put $1 + \tan\theta = t$ \quad $\therefore \sec^2\theta\,d\theta = dt$ \quad $\therefore\ I = \int \dfrac{dt}{t} = \log t + c = \log(1+\tan\theta) + c$ \quad Ans.

(d) $I = \int \dfrac{d}{dx}\left(\dfrac{\log(1+x)}{x}\right)dx$ $\left[\text{use formula, } \dfrac{d}{dx}\int f(x)\,dx = f(x) \text{ or } \int \dfrac{d}{dx}[f(x)]\,dx = f(x)\right]$

$\qquad\qquad \therefore\ I = \int \dfrac{d}{dx}\left(\dfrac{\log(1+x)}{x}\right)dx = \dfrac{\log(1+x)}{x} + c$ \quad Ans.

(e) $I = \dfrac{d}{dx}\left(\int \dfrac{\log x + 1}{2x}\,dx\right) = \dfrac{\log x + 1}{2x} + c$ \quad Ans. \quad (see above formula and solve)

Integral & Differential Calculus

(f) $I = \int \dfrac{1 + \log x}{x \log x} \, dx$ Let $f(x) = x \log x$ $\therefore$ $f'(x) = x\dfrac{1}{x} + \log x . 1 = 1 + \log x$

or $I = \int \dfrac{f'(x)}{f(x)} \, dx = \log|f(x)| + c = \log|x \log x| + c = \log|x| + \log|\log x| + c$ Ans.

IInd Method: $-$ $I = \int \dfrac{1 + \log x}{x \log x} \, dx$,   Let $x \log x = t$ or $\left(x.\dfrac{1}{x} + \log x . 1\right) dx = dt$ or $(1 + \log x) \, dx = dt$

or $I = \int \dfrac{dt}{t} = \log t + c = \log(x \log x) + c = \log x + \log(\log x) + c$ Ans. $[\log(m.n) = \log m + \log n]$

(g) $I = \int \dfrac{x \cos x}{x \sin x + \cos x} \, dx$ Let $f(x) = x \sin x + \cos x$ or $f'(x) = x \cos x + \sin x - \sin x = x \cos x$

$$\therefore \quad I = \int \dfrac{f'(x)}{f(x)} \, dx = \log|f(x)| + c = \log|x \sin x + \cos x| + c \quad \text{Ans.}$$

(h) $I = \int \dfrac{x}{2x^2 - 5} \, dx = \dfrac{1}{4}\log|2x^2 - 5| + c$ Ans.   (solve, same as above question)

(5) (a) $I = \int \dfrac{(2x - 1)}{2x^2 - 2x + 5} \, dx$ Let $2x^2 - 2x + 5 = z$ $\therefore$ $(4x - 2) \, dx = dz$ or $(2x - 1) \, dx = \dfrac{dz}{2}$

or $I = \int \dfrac{(2x - 1)}{2x^2 - 2x + 5} \, dx = \int \dfrac{dz}{2z} = \dfrac{1}{2}\log z + c = \dfrac{1}{2}\log(2x^2 - 2x + 5) + c = \log\left(\sqrt{2x^2 - 2x + 5}\right) + c$ Ans.

(b) $I = \int \dfrac{\cos x}{\sin^3 x} \, dx = \int \sin^{-3} x . \cos x \, dx$ $\left[\text{formula,} \quad \int [f(x)]^n. f'(x) \, dx = \dfrac{[f(x)]^{n+1}}{n + 1}\right]$

or $I = \int \sin^{-3} x . \cos x \, dx$, Let $f(x) = \sin x$ $\therefore$ $f'(x) = \cos x$

or $I = \int [f(x)]^{-3}. f'(x) \, dx = \dfrac{[f(x)]^{-3+1}}{-3 + 1} = -\dfrac{1}{2}[f(x)]^{-2} + c$ where $f(x) = \sin x$ and $f'(x) = \cos x$

$$\therefore \quad I = -\dfrac{1}{2}[f(x)]^{-2} + c = -\dfrac{1}{2}(\sin x)^{-2} + c = -\dfrac{1}{2\sin^2 x} + c = -\dfrac{1}{2}\csc^2 x + c \quad \text{Ans.}$$

IInd Method: $-$ $I = \int \dfrac{\cos x}{\sin^3 x} \, dx$ Let $\sin x = z$ $\therefore$ $\cos x \, dx = dz$

or $I = \int \dfrac{\cos x}{\sin^3 x} \, dx = \int \dfrac{dz}{z^3} = \int z^{-3} \, dz = \dfrac{z^{-3+1}}{-3 + 1} + c = -\dfrac{1}{2}z^{-2} + c = -\dfrac{1}{2z^2} + c = -\dfrac{1}{2\sin^2 x} + c = -\dfrac{1}{2}\csc^2 x + c$ Ans.

(c) $I = \int \dfrac{\sec^2 x}{\tan^4 x} \, dx = \int \tan^{-4} x . \sec^2 x \, dx$ Let $f(x) = \tan x$ $\therefore$ $f'(x) = \sec^2 x$ $\left[\text{using formula,} \quad \int [f(x)]^n. f'(x) \, dx = \dfrac{[f(x)]^{n+1}}{n + 1}\right]$

or $I = \int [f(x)]^{-4}. f'(x) \, dx = \dfrac{[f(x)]^{-4+1}}{-4 + 1} + c = -\dfrac{1}{3}[f(x)]^{-3} + c = -\dfrac{1}{3[f(x)]^3} + c = -\dfrac{1}{3(\tan x)^3} + c$

$$= -\dfrac{1}{3}\cot^3 x + c \quad \text{Ans.} \quad \text{or} \quad -\dfrac{\cos^3 x}{3\sin^3 x} + c \quad \text{Ans.}$$

(d) $I = \int \dfrac{dx}{x^2 + x + 3}$ or $x^2 + x + 3 = a\left\{\left(x + \dfrac{b}{2a}\right)^2 - \dfrac{D}{4a^2}\right\}$ here $a = 1$, $b = 1$ and $c = 3$ then $D = b^2 - 4ac = 1 - 12 = -11$

or $x^2 + x + 3 = 1\left\{\left(x + \dfrac{1}{2}\right)^2 - \dfrac{-11}{4.1}\right\} = \left(x + \dfrac{1}{2}\right)^2 + \dfrac{11}{4} = \left(x + \dfrac{1}{2}\right)^2 + \left(\dfrac{\sqrt{11}}{2}\right)^2$

$I = \int \dfrac{dx}{x^2 + x + 3} = \int \dfrac{dx}{\left(x + \dfrac{1}{2}\right)^2 + \left(\dfrac{\sqrt{11}}{2}\right)^2}$ using formula, $\quad \int \dfrac{dx}{x^2 + a^2} = \dfrac{1}{a}\tan^{-1}\left(\dfrac{x}{a}\right)$

$$I = \int \frac{dx}{\left(x+\frac{1}{2}\right)^2 + \left(\frac{\sqrt{11}}{2}\right)^2} = \frac{1}{\frac{\sqrt{11}}{2}}\tan^{-1}\left(\frac{x+\frac{1}{2}}{\frac{\sqrt{11}}{2}}\right) + c = \frac{2}{\sqrt{11}}\tan^{-1}\left(\frac{2x+1}{\sqrt{11}}\right) + c \quad \text{Ans.}$$

(e) $I = \int \dfrac{dx}{2x^2 + 4x - 5} = \dfrac{1}{2}\int \dfrac{dx}{(x+1)^2 - \left(\frac{\sqrt{7}}{\sqrt{2}}\right)^2} = \dfrac{1}{2\sqrt{14}}\log\left|\dfrac{\sqrt{2}x + \sqrt{2} - \sqrt{7}}{\sqrt{2}x + \sqrt{2} + \sqrt{7}}\right| + c \quad \text{Ans.}$

(f) $I = \int \dfrac{dx}{\sqrt{1 + 4x - 2x^2}} \qquad \therefore \ 1 + 4x - 2x^2 = a\left\{\left(x + \dfrac{b}{2a}\right)^2 - \dfrac{D}{4a^2}\right\} \quad$ here $a = -2, b = 4, c = 1$ and $D = 24$

or $\ 1 + 4x - 2x^2 = -2\left\{\left(x + \dfrac{4}{-4}\right)^2 - \dfrac{24}{16}\right\} = -2\left\{(x-1)^2 - \dfrac{3}{2}\right\} = 2\left\{\dfrac{3}{2} - (x-1)^2\right\} = 2\left\{\left(\sqrt{\dfrac{3}{2}}\right)^2 - (x-1)^2\right\}$

or $I = \int \dfrac{dx}{\sqrt{2\left[\left(\frac{\sqrt{3}}{\sqrt{2}}\right)^2 - (x-1)^2\right]}} = \dfrac{1}{\sqrt{2}}\int \dfrac{dx}{\sqrt{\left(\frac{\sqrt{3}}{\sqrt{2}}\right)^2 - (x-1)^2}} \quad$ using formula, $\ \int \dfrac{dx}{\sqrt{a^2 - x^2}} = \sin^{-1}\dfrac{x}{a}$

$\therefore \ I = \dfrac{1}{\sqrt{2}}\int \dfrac{dx}{\sqrt{\left(\frac{\sqrt{3}}{\sqrt{2}}\right)^2 - (x-1)^2}} = \dfrac{1}{\sqrt{2}}\sin^{-1}\left(\dfrac{x-1}{\frac{\sqrt{3}}{\sqrt{2}}}\right) + c = \dfrac{1}{\sqrt{2}}\sin^{-1}\left(\dfrac{\sqrt{2}(x-1)}{\sqrt{3}}\right) + c \quad \text{Ans.}$

(g) $I = \int \dfrac{dx}{\sqrt{x^2 + 2x + 3}} = \int \dfrac{dx}{\sqrt{(x+1)^2 + 2}} = \int \dfrac{dx}{\sqrt{(x+1)^2 + \left(\sqrt{2}\right)^2}} \quad$ use formula, $\ \int \dfrac{dx}{\sqrt{x^2 + a^2}} = \log\left|x + \sqrt{x^2 + a^2}\right|$

or $\ I = \int \dfrac{dx}{\sqrt{(x+1)^2 + \left(\sqrt{2}\right)^2}} = \log\left|(x+1) + \sqrt{(x+1)^2 + \left(\sqrt{2}\right)^2}\right| + c = \log\left|(x+1) + \sqrt{x^2 + 2x + 3}\right| + c \quad \text{Ans.}$

(h) $I = \int \dfrac{dx}{\sqrt{3 + 4x - 2x^2}} = \int \dfrac{dx}{\sqrt{a\left\{\left(x + \frac{b}{2a}\right)^2 - \frac{D}{4a^2}\right\}}} = \int \dfrac{dx}{\sqrt{-2\left\{\left(x + \frac{4}{-4}\right)^2 - \frac{40}{16}\right\}}} = \int \dfrac{dx}{\sqrt{-2\left\{(x-1)^2 - \frac{5}{2}\right\}}} = \int \dfrac{dx}{\sqrt{2\left\{\frac{5}{2} - (x-1)^2\right\}}}$

$= \dfrac{1}{\sqrt{2}}\int \dfrac{dx}{\sqrt{\left(\frac{\sqrt{5}}{\sqrt{2}}\right)^2 - (x-1)^2}} \quad$ using formula, $\int \dfrac{dx}{\sqrt{a^2 - x^2}} = \sin^{-1}\dfrac{x}{a}$

or $I = \dfrac{1}{\sqrt{2}}\int \dfrac{dx}{\sqrt{\left(\frac{\sqrt{5}}{\sqrt{2}}\right)^2 - (x-1)^2}} = \dfrac{1}{\sqrt{2}}\sin^{-1}\left(\dfrac{x-1}{\frac{\sqrt{5}}{\sqrt{2}}}\right) + c = \dfrac{1}{\sqrt{2}}\sin^{-1}\left[\dfrac{\sqrt{2}(x-1)}{\sqrt{5}}\right] + c \quad \text{Ans.}$

(6) (a) $I = \int \sqrt{-x^2 + 3x + 5}\ dx = \int \sqrt{a\left\{\left(x + \frac{b}{2a}\right)^2 - \frac{D}{4a^2}\right\}}\ dx = \int \sqrt{-1\left\{\left(x + \frac{3}{-2}\right)^2 - \frac{29}{4}\right\}}\ dx = \int \sqrt{\frac{29}{4} - \left(x - \frac{3}{2}\right)^2}\ dx$

$= \int \sqrt{\left(\frac{\sqrt{29}}{2}\right)^2 - \left(x - \frac{3}{2}\right)^2}\ dx \quad$ using formula, $\ \int \sqrt{a^2 - x^2}\ dx = \dfrac{x}{2}\sqrt{a^2 - x^2} + \dfrac{a^2}{2}\sin^{-1}\dfrac{x}{a}$

or $I = \dfrac{x - \frac{3}{2}}{2}\sqrt{\left(\frac{\sqrt{29}}{2}\right)^2 - \left(x - \frac{3}{2}\right)^2} + \dfrac{29}{8}\sin^{-1}\left(\dfrac{x - \frac{3}{2}}{\frac{\sqrt{29}}{2}}\right) + c = \dfrac{2x - 3}{4}\sqrt{-x^2 + 3x + 5} + \dfrac{29}{8}\sin^{-1}\left(\dfrac{2x - 3}{\sqrt{29}}\right) + c \quad \text{Ans.}$

(b) $I = \int \dfrac{\sec^2 x}{(\sec x + \tan x)^{9/2}}\, dx = -\dfrac{1}{(\sec x + \tan x)^{11/2}} \left\{ \dfrac{1}{11} + \dfrac{1}{7}(\sec x + \tan x)^2 \right\} + k$ Ans.

(c) $I = \int \dfrac{\sin^2(\log x) + \cos(\log x)}{x \tan(\log x)}\, dx$ Let $\log x = t$ $\therefore$ $\dfrac{1}{x} dx = dt$

or $I = \int \dfrac{\sin^2(\log x) + \cos(\log x)}{\tan(\log x)} \cdot \dfrac{1}{x}\, dx = \int \dfrac{\sin^2 t + \cos t}{\tan t}\, dt = \int \dfrac{\sin^2 t + \cos t}{\dfrac{\sin t}{\cos t}}\, dt = \int \dfrac{\cos t\,(\sin^2 t + \cos t)}{\sin t}\, dt$

$\qquad = \int \dfrac{\sin^2 t . \cos t}{\sin t}\, dt + \int \dfrac{\cos^2 t}{\sin t}\, dt = \int \sin t . \cos t\, dt + \int \dfrac{1 - \sin^2 t}{\sin t}\, dt = I_1 + I_2 \text{ (say)}$

or $I_1 = \int \sin t . \cos t\, dt$ Let $\sin t = u$ $\therefore$ $\cos t\, dt = du$

$\therefore$ $I_1 = \int u.\, du = \dfrac{u^2}{2} + c = \dfrac{1}{2}. \sin^2 t + c = \dfrac{\sin^2(\log x)}{2} + c$

Now, $I_2 = \int \dfrac{1 - \sin^2 t}{\sin t}\, dt = \int \dfrac{1}{\sin t}\, dt - \int \dfrac{\sin^2 t}{\sin t}\, dt = \int \csc t\, dt - \int \sin t\, dt = \log\!\left(\tan \tfrac{t}{2}\right) + \cos t + c$

$\qquad = \log\!\left[\tan\!\left(\dfrac{\log x}{2}\right)\right] + \cos(\log x) + c$

or $I = I_1 + I_2 = \dfrac{\sin^2(\log x)}{2} + c + \log\!\left[\tan\!\left(\dfrac{\log x}{2}\right)\right] + \cos(\log x) + c = \dfrac{\sin^2(\log x)}{2} + \log\!\left[\tan\!\left(\dfrac{\log x}{2}\right)\right] + \cos(\log x) + k$ Ans.

(d) $I = \int \dfrac{dx}{x \sin(\log x) . \cos(\log x)}$ Let $\log x = t$ $\therefore$ $\dfrac{1}{x}\, dx = dt$

or $I = \int \dfrac{1}{\sin(\log x) . \cos(\log x)} \cdot \dfrac{1}{x}\, dx = \int \dfrac{dt}{\sin t . \cos t} = \int \dfrac{\sin^2 t + \cos^2 t}{\sin t . \cos t}\, dt = \int \dfrac{\sin^2 t}{\sin t . \cos t}\, dt + \int \dfrac{\cos^2 t}{\sin t . \cos t}\, dt$

$\therefore$ $I = \int \tan t\, dt + \int \cot t\, dt = -\log(\cos t) + \log(\sin t) + c = \log(\sin t) - \log(\cos t) + c = \log\!\left(\dfrac{\sin t}{\cos t}\right) + c = \log(\tan t) + c$

$\qquad = \log[\tan(\log x)] + c$ Ans.

(e) $I = \int \dfrac{\log x}{x^2}\, dx$,   use integration by part formula $\int f(x).\, g(x)\, dx = f(x). \int g(x)\, dx - \int \left[\dfrac{d(f(x))}{dx} . \int g(x)\, dx \right] dx$

or $I = \int x^{-2} . \log x\, dx = \log x . \int x^{-2}\, dx - \int \left[\dfrac{d(\log x)}{dx} . \int x^{-2}\, dx \right] dx = \log x . \dfrac{x^{-2+1}}{-2+1} - \int \dfrac{1}{x} . \dfrac{x^{-2+1}}{-2+1}\, dx = -\log x . \dfrac{1}{x} + \int \dfrac{1}{x^2}\, dx$

$\qquad = -\dfrac{\log x}{x} + \int x^{-2}\, dx = -\dfrac{\log x}{x} + \dfrac{x^{-2+1}}{-2+1} + c = -\dfrac{\log x}{x} - \dfrac{1}{x} + c$

$\therefore$ $I = \dfrac{-\log x - 1}{x} + c = \dfrac{-(\log x + 1)}{x} + c$ Ans.

(f) $I = \int \dfrac{dx}{x\sqrt{1 + \log x}}$ Let $\log x = t$ $\therefore$ $\dfrac{1}{x}\, dx = dt$

$\therefore$ $I = \int \dfrac{1}{\sqrt{1 + \log x}} . \dfrac{1}{x}\, dx = \int \dfrac{dt}{\sqrt{1 + t}} = \int (1 + t)^{-\frac{1}{2}}\, dt = \dfrac{(1 + t)^{-\frac{1}{2}+1}}{-\dfrac{1}{2}+1} + c = 2\sqrt{1 + t} + c = 2\sqrt{1 + \log x} + c$ Ans.

(7) (a) $I = \int \dfrac{x^3}{(x - 1)^2(x^2 + 2x + 3)}\, dx$,   Let $\dfrac{x^3}{(x - 1)^2(x^2 + 2x + 3)} = \dfrac{A}{x - 1} + \dfrac{B}{(x - 1)^2} + \dfrac{Cx + D}{x^2 + 2x + 3}$ .... (A)

$\therefore$ $(x - 1)^2 = 0$ or $x = 1$ then $B = \dfrac{x^3}{(x^2 + 2x + 3)}\bigg|_{x=1} = \dfrac{1}{1 + 2 + 3} = \dfrac{1}{6}$

To be find A, C and D are obtained by multiplying by $(x - 1)^2(x^2 + 2x + 3)$ on both of sides and then equate coefficient of suitable

power of x on both sides.

or $\dfrac{x^3}{(x-1)^2(x^2+2x+3)} \times (x-1)^2(x^2+2x+3) = \dfrac{A}{x-1} \times (x-1)^2(x^2+2x+3)$

or $x^3 = A(x-1)(x^2+2x+3) = A[x^3+2x^2+3x-x^2-2x-3] = A[x^3+x^2+x-3] = Ax^3+Ax^2+Ax-3A$

$$\therefore \ Ax^3 = x^3 \quad \text{or} \quad A = 1$$

similarly, $\ x^3 = A(x+1)(x^2+2x+3) + B(x^2+2x+3) + (Cx+D)(x-1)^2$

solve and find value of A, C & D *and put value of A, B, C and D in equation* (A) *and solve it.*

(b) $I = \displaystyle\int \dfrac{x}{x^3+1}\ dx$, Let $\dfrac{x}{x^3+1} = \dfrac{x}{(x+1)(x^2-x+1)} = \dfrac{A}{x+1} + \dfrac{Bx+C}{x^2-x+1}$ .... (A)

$\therefore \ x+1 = 0 \ $ or $ \ x = -1 \ $ then $ \ A = \dfrac{x}{(x^2-x+1)}\Big|_{x=-1} = \dfrac{-1}{1+1+1} = -\dfrac{1}{3}$

Now, $\dfrac{x}{(x+1)(x^2-x+1)} = \dfrac{Bx+C}{x^2-x+1}$ Multiplying both of side by $(x+1)(x^2-x+1)$

and then equate coefficient of suitable power of x on both sides.

or $\ x = (Bx+C)(x+1) = Bx^2+Bx+Cx+C = Bx^2+x(B+C)+C \quad \therefore \ B+C = 1 \ $ and $ \ C = 0 \ $ then $ \ B = 1$

Put value of A, B and C in equation (A) and integrate, we get

or $\dfrac{x}{(x+1)(x^2-x+1)} = \dfrac{A}{x+1} + \dfrac{Bx+C}{x^2-x+1} = -\dfrac{1}{3}\cdot\dfrac{1}{x+1} + \dfrac{x}{x^2-x+1}$

or $I = \displaystyle\int \dfrac{x}{(x+1)(x^2-x+1)}\ dx = -\dfrac{1}{3}\int \dfrac{dx}{x+1} + \int \dfrac{x}{x^2-x+1}\ dx = -\dfrac{1}{3}\log(x+1) + \int \dfrac{\frac{1}{2}[2x-1]+\frac{1}{2}}{x^2-x+1}\ dx$

$I = -\dfrac{1}{3}\log(x+1) + \displaystyle\int \dfrac{\frac{1}{2}[2x-1]}{x^2-x+1}\ dx + \dfrac{1}{2}\int \dfrac{dx}{x^2-x+1}\ dx = -\dfrac{1}{3}\log(x+1) + \dfrac{1}{2}\int \dfrac{f'(x)}{f(x)}\ dx + \dfrac{1}{2}\int \dfrac{dx}{\left(x-\frac{1}{2}\right)^2+\frac{3}{4}}$

use formula, $\displaystyle\int \dfrac{f'(x)}{f(x)}\ dx = \log[f(x)] \quad$ and $\quad \displaystyle\int \dfrac{dx}{x^2+a^2} = \dfrac{1}{a}\tan^{-1}\dfrac{x}{a}$

$I = -\dfrac{1}{3}\log(x+1) + \dfrac{1}{2}\displaystyle\int \dfrac{f'(x)}{f(x)}\ dx + \dfrac{1}{2}\int \dfrac{dx}{\left(x-\frac{1}{2}\right)^2+\left(\frac{\sqrt{3}}{2}\right)^2} = -\dfrac{1}{3}\log(x+1) + \dfrac{1}{2}\log f(x) + \dfrac{1}{2}\cdot\dfrac{1}{\frac{\sqrt{3}}{2}}\tan^{-1}\left(\dfrac{\left(x-\frac{1}{2}\right)}{\frac{\sqrt{3}}{2}}\right)$

$$= -\dfrac{1}{3}\log(x+1) + \log\left(\sqrt{x^2-x+1}\right) + \dfrac{1}{\sqrt{3}}\tan^{-1}\left(\dfrac{2x-1}{\sqrt{3}}\right) + c \quad \text{Ans.}$$

(c) $I = \displaystyle\int \dfrac{x^4}{(x-1)(x^2+1)}\ dx$, Let $\dfrac{x^4}{(x-1)(x^2+1)} = \dfrac{A}{x-1} + \dfrac{Bx+C}{x^2+1}$ .... (A)

$\therefore \ x-1 = 0 \ $ or $ \ x = 1 \ $ then $ \ A = \dfrac{x^4}{(x^2+1)}\Big|_{x=1} = \dfrac{1}{1+1} = \dfrac{1}{2}$

Now, $\dfrac{x^4}{(x-1)(x^2+1)} = \dfrac{Bx+C}{x^2+1}$

Multiplying both of side by $(x-1)(x^2+1)$ and then equate coefficient of suitable power of x on both sides.

or $x^4 = (Bx+C)(x-1) = Bx^2-Bx+Cx-C \quad \therefore \ C = 0 \ $ or $ \ x(C-B) = 0.x \ $ or $ \ B = 0$

(d) $I = \displaystyle\int \dfrac{x-1}{x+1}\cdot\dfrac{1}{\sqrt{x^3+x^2+x}}\ dx = \int \dfrac{(x-1)(x+1)}{(x+1)^2\cdot\sqrt{x^3+x^2+x}}\ dx = \int \dfrac{(x^2-1)\ dx}{(x^2+2x+1)\cdot x\cdot\sqrt{x+1+\frac{1}{x}}} = \int \dfrac{\left(1-\frac{1}{x^2}\right)}{\left(x+\frac{1}{x}+2\right)\cdot\sqrt{x+1+\frac{1}{x}}}\ dx$

Putting $x + \dfrac{1}{x} + 1 = z^2$ $\quad \therefore \left(1 - \dfrac{1}{x^2}\right) dx = 2z\, dz$ $\quad$ or $\quad x + \dfrac{1}{x} = z^2 - 1$

or $I = \displaystyle\int \dfrac{2z\, dz}{(z^2 - 1 + 2).\sqrt{z^2}} = \int \dfrac{2z\, dz}{(z^2 + 1).z} = 2\int \dfrac{dz}{z^2 + 1} = 2.\dfrac{1}{1}\tan^{-1}\left(\dfrac{z}{1}\right) = 2\tan^{-1}\left(\sqrt{x + \dfrac{1}{x} + 1}\right) + c$ $\quad$ Ans.

Integral of the form: $-\displaystyle\int x^m (a + bx^n)^p\, dx$ $\quad$ where m, n and p are rational numbers.

Case I: $-$ $\quad$ when $p \in I$

subcase: $-$ $\quad$ (a) when $p \in I_+$, $\;$ Expand $(a + bx^n)^p$ using binomal theorem.

(b) when $p \in I_-$, $\;$ Putting $x = t^\alpha$, $\;$ where $\alpha = $ L.C.M of D^r of fractions m and n.

Case II: $-$ $\quad$ when $p \notin I$

subcase: $-$ (a) when $\dfrac{m+1}{n} = $ integer, $\;$ then $\;$ putting $a + bx^n = t^\alpha$ $\;$ where α is the D^r of fractions p.

(b) when $\dfrac{m+1}{n} + p \neq $ integer, $\;$ then $\;$ putting $a + bx^n = x^n t^\alpha$ $\;$ where α is the D^r of fractions p.

Example: $-$ $\;$ (1) (a) $I = \displaystyle\int \sqrt{x}.\left(3 + \sqrt[3]{x}\right)^2 dx = \int x^{\frac{1}{2}}.\left(3 + x^{\frac{1}{3}}\right)^2 dx$

Here $m = \dfrac{1}{2}$, $\;$ $n = \dfrac{1}{3}$ $\;$ and $p = 2$ $\;$ where p is positive integer.

Putting $x = t^\alpha$ where $\alpha = $ L.C.M of D^r of fractions m and n. $\;$ $\therefore$ $\alpha = \{2,3\} = 6$

$\therefore$ $\;$ $x = t^6$ $\;$ or $\;$ $t = x^{\frac{1}{6}}$ $\;$ or $\;$ $dx = 6t^5 dt$ $\;$ and $\;$ $x^{\frac{1}{2}} = (t^6)^{\frac{1}{2}} = t^3$, $\;$ $x^{\frac{1}{3}} = (t^6)^{\frac{1}{3}} = t^2$

$I = \displaystyle\int t^3.(3 + t^2)^2.6t^5\, dt = 6\int t^8.(3 + t^2)^2\, dt = 6\int t^8.(9 + 6t^2 + t^4)\, dt = 6\int (9t^8 + 6t^{10} + t^{12})\, dt$

$\qquad = 54\displaystyle\int t^8\, dt + 36\int t^{10}\, dt + 6\int t^{12}\, dt = 54.\dfrac{t^{8+1}}{8+1} + 36.\dfrac{t^{10+1}}{10+1} + 6.\dfrac{t^{12+1}}{12+1} + c$

$I = 54.\dfrac{t^9}{9} + 36.\dfrac{t^{11}}{11} + 6.\dfrac{t^{13}}{13} + c = 6t^9 + \dfrac{36}{11}t^{11} + \dfrac{6}{13}t^{13} + c$

put $t = x^{\frac{1}{6}}$ then $t^9 = x^{\frac{3}{2}}$, $t^{11} = x^{\frac{11}{6}}$, $t^{13} = x^{\frac{13}{6}}$ $\quad \therefore$ $I = 6x^{\frac{3}{2}} + \dfrac{36}{11}x^{\frac{11}{6}} + \dfrac{6}{13}x^{\frac{13}{6}} + c$ $\quad$ Ans.

IInd Method: $-$ $I = \displaystyle\int \sqrt{x}.\left(3 + \sqrt[3]{x}\right)^2 dx = \int x^{\frac{1}{2}}.\left(3 + x^{\frac{1}{3}}\right)^2 dx = \int x^{\frac{1}{2}}.\left(9 + 6x^{\frac{1}{3}} + x^{\frac{2}{3}}\right) dx = \int \left(9x^{\frac{1}{2}} + 6x^{\frac{1}{2}}.x^{\frac{1}{3}} + x^{\frac{1}{2}}.x^{\frac{2}{3}}\right) dx$

$\qquad = 9\displaystyle\int x^{\frac{1}{2}}\, dx + 6\int x^{\frac{5}{6}}\, dx + \int x^{\frac{7}{6}}\, dx$

$I = 9.\dfrac{x^{\frac{1}{2}+1}}{\frac{1}{2}+1} + 6.\dfrac{x^{\frac{5}{6}+1}}{\frac{5}{6}+1} + \dfrac{x^{\frac{7}{6}+1}}{\frac{7}{6}+1} + c = 6x^{\frac{3}{2}} + \dfrac{36}{11}x^{\frac{11}{6}} + \dfrac{6}{13}x^{\frac{13}{6}} + c$ $\quad$ Ans.

(b) $I = \displaystyle\int x^{-\frac{1}{2}}.\left(2 + x^{\frac{1}{2}}\right)^{-1} dx$ $\quad \therefore$ here $m = -\dfrac{1}{2}$, $\;$ $n = \dfrac{1}{2}$ and $p = -1$ $\;$ (p is a negative integer)

Putting $x = t^\alpha$ $\;$ where $\alpha = $ L.C.M of D^r of fractions m and n. $\;$ $\therefore$ $\alpha = \{2,2\} = 2$ $\quad \therefore$ $x = t^2$ $\;$ or $\;$ $t = \sqrt{x}$ $\;$ $\therefore$ $dx = 2t\, dt$

$\therefore$ $\;$ $I = \displaystyle\int x^{-\frac{1}{2}}.\left(2 + x^{\frac{1}{2}}\right)^{-1} dx = \int (t^2)^{\frac{-1}{2}}.\left[2 + (t^2)^{\frac{1}{2}}\right]^{-1}.2t\, dt = 2\int t^{-1}.(2 + t)^{-1}.t\, dt = 2\int \dfrac{dt}{2+t} = 2\log(2 + t) + c = \log(2 + t)^2 + c$

$\qquad = \log\left(2 + \sqrt{x}\right)^2 + c = \log\left[2 + x^{\frac{1}{2}}\right]^2 + c$ $\quad$ Ans.

(c) $I = \int \dfrac{\sqrt{3 + \sqrt[3]{x}}}{\sqrt[3]{x^2}}\, dx = \int x^{-\frac{2}{3}}.\left(3 + x^{\frac{1}{3}}\right)^{\frac{1}{2}} dx$ Here $m = -\dfrac{2}{3}$, $n = \dfrac{1}{3}$ and $p = \dfrac{1}{2} \neq$ integer

when, $\dfrac{m+1}{n} = \dfrac{-\dfrac{2}{3}+1}{\dfrac{1}{3}} = 1 =$ integer.   Putting, $3 + x^{\frac{1}{3}} = t^{\alpha}$ where α is the D^r of fractions p $\therefore\ \alpha = 2$

$$\therefore\ 3 + x^{\frac{1}{3}} = t^2 \ \text{ or } \ \frac{1}{3}x^{\frac{1}{3}-1}dx = 2t\, dt \ \ \therefore\ x^{\frac{-2}{3}}dx = 6t\, dt$$

$I = \int x^{-\frac{2}{3}}.\left(3 + x^{\frac{1}{3}}\right)^{\frac{1}{2}} dx = \int (t^2)^{\frac{1}{2}}.6t\, dt = 6\int t^2\, dt = 6.\dfrac{t^{2+1}}{2+1} + c = 2t^3 + c$

put $3 + x^{\frac{1}{3}} = t^2$ $\therefore\ t = \left(3 + x^{\frac{1}{3}}\right)^{\frac{1}{2}}$ then $I = 2t^3 + c = 2\left[\left(3 + x^{\frac{1}{3}}\right)^{\frac{1}{2}}\right]^3 + c = 2\left(3 + x^{\frac{1}{3}}\right)^{\frac{3}{2}} + c$ Ans.

(d) $I = \int \dfrac{dx}{x^9(2 + x^3)^{\frac{1}{3}}} = \int x^{-9}.(2 + x^3)^{\frac{-1}{3}} dx$ Here $m = -9$, $n = 3$ and $p = -\dfrac{1}{3} \neq$ integer

when, $\dfrac{m+1}{n} = \dfrac{-9+1}{3} = -\dfrac{8}{3} \neq$ integer $\therefore\ \dfrac{m+1}{n} + p = -\dfrac{8}{3} - \dfrac{1}{3} = -\dfrac{9}{3} = -3 =$ negative integer

$\therefore$ Putting, $2 + x^3 = x^3 t^{\alpha}$ where α is D^r of the fractions p $\therefore\ \alpha = 3$

$\therefore\ 2 + x^3 = x^3 t^3$ or $x^3 t^3 - x^3 = 2$ or $x^3(t^3 - 1) = 2$ or $x^3 = \dfrac{2}{t^3 - 1}$ $\therefore\ x = \left(\dfrac{2}{t^3 - 1}\right)^{\frac{1}{3}}$

or $3x^2\, dx = \dfrac{(t^3 - 1).0 - 2.3t^2}{(t^3 - 1)^2}\, dt = \dfrac{-6t^2}{(t^3 - 1)^2}\, dt$ or $dx = \dfrac{-6t^2}{(t^3 - 1)^2 . 3x^2}\, dt = \dfrac{-2t^2}{(t^3 - 1)^2 . \left[\left(\frac{2}{t^3 - 1}\right)^{\frac{1}{3}}\right]^2}\, dt$

or $I = \int x^{-9}.(2 + x^3)^{\frac{-1}{3}} dx = \int \left(\dfrac{2}{t^3 - 1}\right)^{-3} . \left(2 + \dfrac{2}{t^3 - 1}\right)^{-\frac{1}{3}} . \dfrac{-2t^2}{(t^3 - 1)^2 . \left(\frac{2}{t^3 - 1}\right)^{\frac{2}{3}}}\, dt$

$= -\int \left(\dfrac{2}{t^3 - 1}\right)^{-3} \times \left(\dfrac{2t^3 - 2 + 2}{t^3 - 1}\right)^{-\frac{1}{3}} \times \dfrac{2t^2.(t^3 - 1)^{\frac{2}{3}}}{(t^3 - 1)^2 . 2^{\frac{2}{3}}}\, dt$

$I = -\int 2^{-3}.(t^3 - 1)^3 . \dfrac{(2t^3)^{-\frac{1}{3}}}{(t^3 - 1)^{\frac{-1}{3}}} . \dfrac{2t^2.(t^3 - 1)^{\frac{2}{3}}}{(t^3 - 1)^2 . 2^{\frac{2}{3}}}\, dt = -\int \dfrac{2^{-3}.2^{\frac{-2}{3}}.2^{\frac{-1}{3}}.2.t^2.t^{-1}}{(t^3 - 1)^{-1}.(t^3 - 1)^{\frac{-1}{3}}.(t^3 - 1)^{\frac{-2}{3}}}\, dt$

$I = -\int \dfrac{2^{\left(-3-\frac{1}{3}-\frac{2}{3}+1\right)}.t}{(t^3 - 1)^{\left(-1-\frac{1}{3}-\frac{2}{3}\right)}}\, dt = -\int \dfrac{2^{-3}.t}{(t^3 - 1)^{-2}}\, dt = -\dfrac{1}{8}\int t(t^3 - 1)^2\, dt = -\dfrac{1}{8}\int t(t^6 - 2t^3 + 1)\, dt = -\dfrac{1}{8}\int (t^7 - 2t^4 + t)\, dt$

$= -\dfrac{1}{8}\int t^7\, dt + \dfrac{1}{4}\int t^4\, dt - \dfrac{1}{8}\int t\, dt$

$I = -\dfrac{1}{8}.\dfrac{t^{7+1}}{7+1} + \dfrac{1}{4}.\dfrac{t^{4+1}}{4+1} - \dfrac{1}{8}.\dfrac{t^{1+1}}{1+1} + c = -\dfrac{1}{64}t^8 + \dfrac{1}{20}t^5 - \dfrac{1}{16}t^2 + c$

Putting, $2 + x^3 = x^3 t^3$ or $t^3 = \dfrac{2 + x^3}{x^3}$ $\therefore\ t = \left(\dfrac{2 + x^3}{x^3}\right)^{\frac{1}{3}}$

$\therefore\ I = -\dfrac{1}{64}t^8 + \dfrac{1}{20}t^5 - \dfrac{1}{16}t^2 + c = -\dfrac{1}{64}\left[\left(\dfrac{2 + x^3}{x^3}\right)^{\frac{1}{3}}\right]^8 + \dfrac{1}{20}\left[\left(\dfrac{2 + x^3}{x^3}\right)^{\frac{1}{3}}\right]^5 - \dfrac{1}{16}\left[\left(\dfrac{2 + x^3}{x^3}\right)^{\frac{1}{3}}\right]^2 + c$

$= -\dfrac{1}{64}\left(\dfrac{2 + x^3}{x^3}\right)^{\frac{8}{3}} + \dfrac{1}{20}\left(\dfrac{2 + x^3}{x^3}\right)^{\frac{5}{3}} - \dfrac{1}{16}\left(\dfrac{2 + x^3}{x^3}\right)^{\frac{2}{3}} + c$ Ans.

---

(2) Do yourself.   find the integral: $-\int x^m(a + bx^n)^p\ dx$

(a) $I = \int \sqrt[3]{x}\left(2 + \sqrt{x}\right)^2\ dx,$ Ans: $-\ 3x^{\frac{4}{3}} + \dfrac{24}{11}x^{\frac{11}{6}} + \dfrac{3}{7}x^{\frac{7}{3}} + c$

(b) $I = \int x^{-\frac{2}{3}}\left(1 + x^{\frac{2}{3}}\right)^{-1}\ dx,$ Ans: $-\ 3\tan^{-1}(x)^{\frac{1}{3}} + c$ (c) $I = \int \dfrac{\sqrt{1 + \sqrt[3]{x}}}{\sqrt[3]{x^2}}\ dx,$ Ans: $-\ 2\left(1 + x^{\frac{1}{3}}\right)^{\frac{3}{2}} + c$

(d) $I = \int \dfrac{dx}{x^{11}(1 + x^4)^{\frac{1}{2}}},$ Ans: $-\ -\dfrac{1}{10}\left(\dfrac{1 + x^4}{x^4}\right)^{\frac{5}{2}} + \dfrac{1}{3}\left(\dfrac{1 + x^4}{x^4}\right)^{\frac{3}{2}} - \dfrac{1}{2}\left(\dfrac{1 + x^4}{x^4}\right)^{\frac{1}{2}} + c$

Other method of integrating: $-\int R\left(x, \sqrt{ax^2 + bx + c}\right) dx$

Integral of the form: $-\int \dfrac{P_m(x)}{\sqrt{ax^2 + bx + c}}\ dx$, where P_m is polynomial in x of degree m and $x \neq 0$

Method of integration: $-$ Let $\int \dfrac{P_m(x)}{\sqrt{ax^2 + bx + c}}\ dx = P_{m-1}(x).\sqrt{ax^2 + bx + c} + k\int \dfrac{dx}{\sqrt{ax^2 + bx + c}}$ ... ... (A)

Differentiating both sides with respect to x and then equate coefficient of suitable power of x from this

find the value of constants and put in equation (A).

Example: $-$ $I = \int \dfrac{x^3 + 2}{\sqrt{x^2 + 2x + 3}}\ dx$ Let $\int \dfrac{x^3 + 2}{\sqrt{x^2 + 2x + 3}}\ dx = (ax^2 + bx + c).\sqrt{x^2 + 2x + 3} + k\int \dfrac{dx}{\sqrt{x^2 + 2x + 3}}$ ... ... (A)

Differentiating both of sides , we get

or $\dfrac{x^3 + 2}{\sqrt{x^2 + 2x + 3}} = (ax^2 + bx + c).\dfrac{1}{2\sqrt{x^2 + 2x + 3}}.(2x + 2) + \sqrt{x^2 + 2x + 3}.(2ax + b) + k.\dfrac{1}{\sqrt{x^2 + 2x + 3}}$

Multiplying both of sides by $\sqrt{x^2 + 2x + 3}$, we get

or $x^3 + 2 = (ax^2 + bx + c)(x + 1) + (x^2 + 2x + 3)(2ax + b) + k$

or $x^3 + 0.x^2 + 0.x + 2 = ax^3 + bx^2 + cx + ax^2 + bx + c + 2ax^3 + 4ax^2 + 6ax + bx^2 + 2bx + 3b + k$

∴ Equating coefficient of similar power of x on both sides, we get

$\therefore\ a + 2a = 1$ or $a = \dfrac{1}{3}$ and $a + b + 4a + b = 0$ or $5a + 2b = 0$ $\therefore\ b = -\dfrac{5}{6}$

or $c + b + 6a + 2b = 0$ or $6a + 3b + c = 0$ $\therefore\ c = -6.\dfrac{1}{3} - 3.\dfrac{-5}{6} = -2 + \dfrac{5}{2} = \dfrac{-4 + 5}{2} = \dfrac{1}{2}$

or $3b + c + k = 2$ or $3.\dfrac{-5}{6} + \dfrac{1}{2} + k = 2$ $\therefore\ k = 2 + \dfrac{5}{2} - \dfrac{1}{2} = \dfrac{4 + 5 - 1}{2} = \dfrac{8}{2} = 4$

Put value a, b, c and k in equation (A), we get

$\int \dfrac{x^3 + 2}{\sqrt{x^2 + 2x + 3}}\ dx = (ax^2 + bx + c).\sqrt{x^2 + 2x + 3} + k\int \dfrac{dx}{\sqrt{x^2 + 2x + 3}} = \left(\dfrac{1}{3}x^2 - \dfrac{5}{6}x + \dfrac{1}{2}\right).\sqrt{x^2 + 2x + 3} + 4\int \dfrac{dx}{\sqrt{(x + 1)^2 + \left(\sqrt{2}\right)^2}}$

$\int \dfrac{x^3 + 2}{\sqrt{x^2 + 2x + 3}}\ dx = \left(\dfrac{1}{3}x^2 - \dfrac{5}{6}x + \dfrac{1}{2}\right).\sqrt{x^2 + 2x + 3} + 4\log\left[(x + 1) + \sqrt{x^2 + 2x + 3}\right] + c$ Ans.

Integral of the form: $-\int \dfrac{dx}{(x \pm x_1)^m.\sqrt{ax^2 + bx + c}}$, Method of integral: $-$ Putting $x \pm x_1 = \dfrac{1}{t}$

Example:− (a) $I = \int \dfrac{dx}{(x-1)^2 \cdot \sqrt{x^2 - 2x + 2}}$, Putting $x - 1 = \dfrac{1}{t}$ or $x = \dfrac{1+t}{t}$ ∴ $dx = -\dfrac{1}{t^2} dt$

$I = \int -\dfrac{1}{t^2} \cdot \dfrac{dt}{\left(\frac{1}{t}\right)^2 \cdot \sqrt{\left(\frac{1+t}{t}\right)^2 - 2 \cdot \left(\frac{1+t}{t}\right) + 2}} = -\int \dfrac{dt}{\sqrt{\frac{(1+t)^2}{t^2} - 2 \cdot \frac{1+t}{t} + 2}} = -\int \dfrac{dt}{\sqrt{\frac{1 + 2t + t^2 - 2t - 2t^2 + 2t^2}{t^2}}} = -\int \dfrac{t\, dt}{\sqrt{t^2 + 1}}$

Let $1 + t^2 = z^2$ ∴ $2t\, dt = 2z\, dz$ or $t\, dt = z\, dz$

$I = -\int \dfrac{z\, dz}{z} = -\int dz = -z + c = -\sqrt{1 + t^2} + c = -\sqrt{1 + \left(\dfrac{1}{x-1}\right)^2} + c = -\sqrt{\dfrac{(x-1)^2 + 1}{(x-1)^2}} + c = -\dfrac{\sqrt{x^2 - 2x + 1 + 1}}{x - 1} + c$

$= -\dfrac{\sqrt{x^2 - 2x + 2}}{x - 1} + c$ Ans.

(b) $I = \int \dfrac{x\, dx}{(x^2 - 4x + 3) \cdot \sqrt{x^2 - 5x + 4}} = \int \dfrac{x\, dx}{(x-1)(x-3) \cdot \sqrt{x^2 - 5x + 4}}$

Let $\dfrac{x}{(x-1)(x-3)} = \dfrac{A}{x-1} + \dfrac{B}{x-3}$ .... (i)   then find A and B.

∴ $A = \dfrac{x}{(x-3)}\Big|_{x=1} = \dfrac{1}{1-3} = -\dfrac{1}{2}$ and $B = \dfrac{x}{(x-1)}\Big|_{x=3} = \dfrac{3}{3-1} = \dfrac{3}{2}$

∴ $\dfrac{x}{(x-1)(x-3)} = \dfrac{A}{x-1} + \dfrac{B}{x-3} = -\dfrac{1}{2(x-1)} + \dfrac{3}{2(x-3)}$

∴ $I = \int \dfrac{x\, dx}{(x-1)(x-3) \cdot \sqrt{x^2 - 5x + 4}} = \int \left\{ \dfrac{3}{2(x-3)} - \dfrac{1}{2(x-1)} \right\} \cdot \dfrac{dx}{\sqrt{x^2 - 5x + 4}} = \dfrac{3}{2} \int \dfrac{dx}{(x-3) \cdot \sqrt{x^2 - 5x + 4}} - \dfrac{1}{2} \int \dfrac{dx}{(x-1) \cdot \sqrt{x^2 - 5x + 4}}$

$= \dfrac{3}{2} I_1 - \dfrac{1}{2} I_2$ .... (ii)

solve I_1 and I_2 and put in equation (ii)

where $I_1 = \int \dfrac{dx}{(x-3) \cdot \sqrt{x^2 - 5x + 4}}$ Putting $x - 3 = \dfrac{1}{t}$ or $x = \dfrac{1 + 3t}{t}$ or $t = \dfrac{1}{x-3}$ ∴ $dx = -\dfrac{1}{t^2} dt$

or $I_1 = -\int \dfrac{1}{t^2} \cdot \dfrac{dt}{\frac{1}{t} \cdot \sqrt{\left(\frac{1+3t}{t}\right)^2 - 5 \cdot \left(\frac{1+3t}{t}\right) + 4}} = -\int \dfrac{dt}{t \cdot \sqrt{\frac{1 + 6t + 9t^2 - 5t - 15t^2 + 4t^2}{t^2}}} = -\int \dfrac{dt}{\sqrt{1 + t - 2t^2}} = -\int \dfrac{dt}{\sqrt{2\left[\frac{9}{16} - \left(t - \frac{1}{4}\right)^2\right]}}$

$= -\dfrac{1}{\sqrt{2}} \int \dfrac{dt}{\sqrt{\left(\frac{3}{4}\right)^2 - \left(t - \frac{1}{4}\right)^2}}$

use formula, $ax^2 + bx + c = a\left[\left(x + \dfrac{b}{2a}\right)^2 - \dfrac{D}{4a^2}\right]$ and $\int \dfrac{dx}{\sqrt{a^2 - x^2}} = \sin^{-1}\dfrac{x}{a}$

∴ $I_1 = -\dfrac{1}{\sqrt{2}} \cdot \sin^{-1}\left(\dfrac{t - \frac{1}{4}}{\frac{3}{4}}\right) + c = -\dfrac{1}{\sqrt{2}} \sin^{-1}\left(\dfrac{4t - 1}{3}\right) + c = -\dfrac{1}{\sqrt{2}} \sin^{-1}\left(\dfrac{4 \cdot \frac{1}{x-3} - 1}{3}\right) + c = -\dfrac{1}{\sqrt{2}} \sin^{-1}\left(\dfrac{4 - x + 3}{3x - 9}\right) + c$

$= -\dfrac{1}{\sqrt{2}} \sin^{-1}\left(\dfrac{7 - x}{3x - 9}\right) + c$

similarly, $I_2 = \int \dfrac{dx}{(x-1) \cdot \sqrt{x^2 - 5x + 4}}$, Putting $x - 1 = \dfrac{1}{t}$ or $x = \dfrac{1+t}{t}$ ∴ $dx = -\dfrac{1}{t^2} dt$

or $I_2 = -\int \dfrac{1}{t^2} \cdot \dfrac{dt}{\frac{1}{t} \cdot \sqrt{\left(\frac{1+t}{t}\right)^2 - 5 \cdot \left(\frac{1+t}{t}\right) + 4}} = -\int \dfrac{dt}{\sqrt{1 + 2t + t^2 - 5t - 5t^2 + 4t^2}} = -\int \dfrac{dt}{\sqrt{1 - 3t}} = \int \dfrac{-dt}{\sqrt{1 - 3t}}$

Put $1 - 3t = z^2$ ∴ $-3\, dt = 2z\, dz$ or $-dt = \dfrac{2}{3} z\, dz$

$$\therefore \ I_2 = \frac{2}{3}\int \frac{z\,dz}{z} = \frac{2}{3}\int dz = \frac{2}{3}z + c = \frac{2}{3}\sqrt{1-3t} + c = \frac{2}{3}\sqrt{1 - 3.\frac{1}{x-1}} + c = \frac{2}{3}\sqrt{\frac{x-1-3}{x-1}} + c = \frac{2}{3}\sqrt{\frac{x-4}{x-1}} + c$$

Put value of I_1 and I_2 in equation (ii), we get

$$\text{or } I = \frac{3}{2}I_1 - \frac{1}{2}I_2 = -\frac{3}{2}.\frac{1}{\sqrt{2}}\sin^{-1}\left(\frac{7-x}{3x-9}\right) - \frac{1}{2}.\frac{2}{3}\sqrt{\frac{x-4}{x-1}} + k = -\frac{3}{2\sqrt{2}}\sin^{-1}\left(\frac{7-x}{3x-9}\right) - \frac{1}{3}.\sqrt{\frac{x-4}{x-1}} + k \quad \text{Ans.}$$

Irrational function: – If in $y = f(x)$ operations of addition, subtraction, multiplication, division and

raising to a power with rational non integer are performed on the R. H. S , then $y = f(x)$ is called irrational function.

e.g. (i) $y = \dfrac{\sqrt[3]{x} + 3.\sqrt{x}}{2.\sqrt[5]{x} + 4} = \dfrac{x^{\frac{1}{3}} + 3.x^{\frac{1}{2}}}{2.x^{\frac{1}{5}} + 4} = R\left(x, x^{\frac{1}{3}}, x^{\frac{1}{2}}, x^{\frac{1}{5}}\right)$ (ii) $y = -\dfrac{1}{3}.x^{\frac{1}{4}} = R\left(x, x^{\frac{1}{4}}\right)$

(iii) $y = \dfrac{\sqrt{x+3} - \sqrt[3]{x-2}}{(x-5)^{\frac{1}{4}}} = R\left(x, (x+3)^{\frac{1}{2}}, (x-2)^{\frac{1}{3}}, (x-5)^{\frac{1}{4}}\right)$

Integral of the form: – $\int R\left(x, x^{\frac{p_1}{q_1}}, x^{\frac{p_2}{q_2}}, \dots\dots\dots x^{\frac{p_n}{q_n}}\right)dx$ Putting $x = t^{\alpha}$, where $\alpha = $ L.C.M of $\{q_1, q_2, q_3, \dots\dots\dots\dots\dots q_n\}$

Integral of the form: – $\int R\left[x, \left(\dfrac{ax+b}{cx+d}\right)^{\frac{1}{n}}\right]dx$, Putting $\dfrac{ax+b}{cx+d} = t^n$

Example: – find the integral $I = \int \dfrac{x - \sqrt[3]{x} + \sqrt[5]{x^2}}{x\left(1 + \sqrt[4]{x}\right)}\,dx$, Here integral is of the form $\int R\left(x, x^{\frac{1}{3}}, x^{\frac{2}{5}}, x^{\frac{1}{4}}\right)dx$

Putting $x = t^{\alpha}$, where $\alpha = $ L.C.M of $\{3,5,4\} = 60$ i.e $x = t^{60}$ $\therefore \ dx = 60t^{59}\,dt$

$$I = \int \frac{x - \sqrt[3]{x} + \sqrt[5]{x^2}}{x\left(1 + \sqrt[4]{x}\right)}\,dx = \int \frac{t^{60} - t^{\frac{60}{3}} + t^{\frac{2\times 60}{5}}}{t^{60}\left(1 + t^{\frac{60}{4}}\right)}.60t^{59}\,dt = 60\int \frac{t^{60} - t^{20} + t^{24}}{t(1 + t^{15})}\,dt = 60\int \frac{t^{59} - t^{19} + t^{23}}{(1 + t^{15})}\,dt$$

$$I = 60\int \frac{t^{59}}{(1 + t^{15})}\,dt - 60\int \frac{t^{19}}{(1 + t^{15})}\,dt + 60\int \frac{t^{23}}{(1 + t^{15})}\,dt = 60I_1 - 60I_2 + 60I_3 \quad \text{(say)}\dots\dots (A)$$

solve I_1, I_2 and I_3 and put equation (A),

where $I_1 = \int \dfrac{t^{59}}{(1 + t^{15})}\,dt$, Let $1 + t^{15} = z$ or $t^{15} = z - 1$ $\therefore \ 15t^{14}\,dt = dz$ or $t^{14}\,dt = \dfrac{dz}{15}$

where $I_1 = \int \dfrac{t^{45}.t^{14}}{(1 + t^{15})}\,dt = \int \dfrac{(z-1)^3}{z}.\dfrac{dz}{15} = \dfrac{1}{15}\int \dfrac{z^3 - 3z^2 + 3z - 1}{z}\,dz$ $[(a - b)^3 = a^3 - b^3 - 3a^2b + 3ab^2]$

or $I_1 = \dfrac{1}{15}\int \dfrac{z^3}{z}\,dz - \dfrac{3}{15}\int \dfrac{z^2}{z}\,dz + \dfrac{3}{15}\int \dfrac{z}{z}\,dz - \dfrac{1}{15}\int \dfrac{dz}{z} = \dfrac{1}{15}\int z^2\,dz - \dfrac{3}{15}\int z\,dz + \dfrac{3}{15}\int dz - \dfrac{1}{15}\int \dfrac{dz}{z}$

$$= \frac{1}{15}.\frac{z^{2+1}}{2+1} - \frac{3}{15}.\frac{z^{1+1}}{1+1} + \frac{3}{15}.z - \frac{1}{15}.\log z + c = \frac{z^3}{45} - \frac{z^2}{10} + \frac{3z}{15} - \frac{1}{15}.\log z + c$$

$$\therefore \ I_1 = \frac{(1 + t^{15})^3}{45} - \frac{(1 + t^{15})^2}{10} + \frac{3(1 + t^{15})}{15} - \frac{1}{15}.\log(1 + t^{15}) + c$$

similarly, $I_2 = \int \dfrac{t^{19}}{(1 + t^{15})}\,dt = \int \dfrac{t^5.t^{14}}{(1 + t^{15})}\,dt = \int \dfrac{(z-1)^{\frac{1}{3}}}{z}.\dfrac{dz}{15}$

Integral of the form: – Euler's substitution: – $\int R\left(x, \sqrt{ax^2 + bx + c}\right)dx$

Case I: – when root of equation $ax^2 + bx + c = 0$ are imaginary.

Subcase: – (a) when $a > 0$, *Putting* $\sqrt{ax^2 + bx + c} = t - x\sqrt{a}$

(b) when $c > 0$, *Putting* $\sqrt{ax^2 + bx + c} = tx - \sqrt{c}$

Case II: – when root of equation $ax^2 + bx + c = 0$ are real.

Putting $\sqrt{ax^2 + bx + c} = (x - \alpha)t$, where α is a root of equation $ax^2 + bx + c = 0$.

Example: – $I = \int \dfrac{dx}{x + \sqrt{x^2 + 4x + 8}}$ $\quad$ Integral of the form $\int R\left(x, \sqrt{x^2 + 4x + 8}\right) dx$

when $a = 1 > 0$, *Putting* $\sqrt{x^2 + 4x + 8} = t - x\sqrt{1} = t - x$ $\quad$ squaring both of sides,

or $x^2 + 4x + 8 = (t - x)^2 = t^2 - 2tx + x^2$ $\ $ or $\ 4x + 8 = t^2 - 2tx$ $\ $ or $\ 4x + 2tx = t^2 - 8$

or $2x(2 + t) = t^2 - 8$ $\quad \therefore \ 2x = \dfrac{t^2 - 8}{2 + t}$ $\ $ or $\ x = \dfrac{t^2 - 8}{4 + 2t}$

$\therefore \ dx = \dfrac{(4 + 2t).2t - (t^2 - 8).2}{[2(2 + t)]^2} dt = \dfrac{8t + 4t^2 - 2t^2 + 16}{4(2 + t)^2} dt = \dfrac{t^2 + 4t + 8}{2(2 + t)^2} dt$

$I = \int \dfrac{1}{x + t - x} \cdot \dfrac{t^2 + 4t + 8}{2(2 + t)^2} dt = \dfrac{1}{2} \int \dfrac{t^2 + 4t + 8}{t(2 + t)^2} dt = \dfrac{1}{2} \int \dfrac{(2 + t)^2 + 4}{t(2 + t)^2} dt = \dfrac{1}{2} \int \dfrac{(2 + t)^2}{t(2 + t)^2} dt + \int \dfrac{2}{t(2 + t)^2} dt$

$\qquad = \dfrac{1}{2} \int \dfrac{dt}{t} + 2 \int \left(\dfrac{1}{t} - \dfrac{t + 4}{(2 + t)^2}\right) dt = \dfrac{1}{2} \int \dfrac{dt}{t} + 2 \int \dfrac{dt}{t} - \int \dfrac{2t + 8}{(2 + t)^2} dt = \dfrac{5}{2} \int \dfrac{dt}{t} - \int \dfrac{2t + 8}{4 + 4t + t^2} dt$

$I = \dfrac{5}{2} \int \dfrac{dt}{t} - \int \dfrac{2(t + 2) + 4}{(2 + t)^2} dt = \dfrac{5}{2} \int \dfrac{dt}{t} - \int \dfrac{2(t + 2)}{(2 + t)^2} dt - \int \dfrac{4}{(2 + t)^2} dt = \dfrac{5}{2} . \log t - 2 \int \dfrac{dt}{t + 2} - 4 \int \dfrac{dt}{(2 + t)^2}$

$\qquad = \dfrac{5}{2} . \log t - 2 \log(t + 2) - 4. \dfrac{(2 + t)^{-2+1}}{-2 + 1} + c = \dfrac{5}{2} . \log t - 2 \log(t + 2) + \dfrac{4}{2 + t} + c$ $\quad$ Ans.

Put $\sqrt{x^2 + 4x + 8} = t - x$ $\ $ or $\ t = x + \sqrt{x^2 + 4x + 8}$

$I = \dfrac{5}{2} . \log\left(x + \sqrt{x^2 + 4x + 8}\right) - 2. \log\left[(x + 2) + \sqrt{x^2 + 4x + 8}\right] + \dfrac{4}{(x + 2) + \sqrt{x^2 + 4x + 8}} + c$ $\quad$ Ans.

Integral of the form: – $\int \sin^m x \cos^n x \ dx$

Case I: – when $m \in$ odd positive integer then putting $z = \cos x$

Case II: – when $n \in$ odd positive integer then putting $z = \sin x$

Case III: – when m and n both are odd positive integer then putting $z = \sin x$ or $\cos x$

Case IV: – when neither m nor n is odd integer,

(a) when $m + n =$ even negative integer then putting $z = \tan x$

(b) when $m + n =$ odd negative integer, multiplying suitable power of $\sin^2 x + \cos^2 x$

(c) when $m + n =$ even positive integer, Let $z = \cos x + i \sin x$ $\ $ or $\ \dfrac{1}{z} = \cos x - i \sin x$

or $\cos x = \dfrac{1}{2}\left(z + \dfrac{1}{z}\right)$ and $\sin x = \dfrac{1}{2i}\left(z - \dfrac{1}{z}\right)$ $\ $ or $\ z^n + \dfrac{1}{z^n} = 2 \cos nx$ and $z^n - \dfrac{1}{z^n} = 2i \sin nx$

Example: – (a) $I = \int \sin^5 x . \cos^2 x \ dx$ $\quad$ Here $m = 5$ and $n = 2$ $\ $ when m is odd positive integer

Then putting $z = \cos x$ $\quad \therefore \ dz = - \sin x \ dx$

$$I = \int \sin^5 x \cdot \cos^2 x \, dx = \int \cos^2 x \cdot \sin^4 x \cdot \sin x \, dx = \int \cos^2 x \cdot (1 - \cos^2 x)^2 \cdot \sin x \, dx = -\int z^2 \cdot (1 - z^2)^2 \, dz = -\int z^2 (1 - 2z^2 + z^4) \, dz$$

$$= -\int (z^2 - 2z^4 + z^6) \, dz = -\int z^2 \, dz + 2 \int z^4 \, dz - \int z^6 \, dz$$

$$I = -\frac{z^{2+1}}{2+1} + 2 \cdot \frac{z^{4+1}}{4+1} - \frac{z^{6+1}}{6+1} + c = -\frac{1}{3} z^3 + \frac{2}{5} z^5 - \frac{1}{7} z^7 + c = -\frac{1}{3} \cos^3 x + \frac{2}{5} \cos^5 x - \frac{1}{7} \cos^7 x + c \quad \text{Ans.}$$

(b) $I = \int \sin^4 x \cdot \cos^3 x \, dx$ Here $m = 4$ and $n = 3$ when n is odd positive integer

Putting, $z = \sin x$ $\therefore$ $dz = \cos x \, dx$

$$I = \int \sin^4 x \cdot \cos^2 x \cdot \cos x \, dx = \int \sin^4 x \, (1 - \sin^2 x) \cdot \cos x \, dx = \int z^4 (1 - z^2) \, dz = \int (z^4 - z^6) \, dz = \int z^4 \, dz - \int z^6 \, dz = \frac{z^{4+1}}{4+1} - \frac{z^{6+1}}{6+1} + c$$

$$= \frac{z^5}{5} - \frac{z^7}{7} + c = \frac{1}{5} \sin^5 x - \frac{1}{7} \sin^7 x + c \quad \text{Ans.}$$

(c) $I = \int \frac{\cos^3 x}{\sin^5 x} \, dx = \int \cos^3 x \cdot \sin^{-5} x \, dx$ Here $m = -5$ and $n = 3$ when n is odd positive integer

Putting $z = \sin x$ $\therefore$ $dz = \cos x \, dx$

$$I = \int \cos^2 x \cdot \sin^{-5} x \cdot \cos x \, dx = \int (1 - \sin^2 x) \cdot \sin^{-5} x \cdot \cos x \, dx = \int (1 - z^2) \cdot z^{-5} \, dz = \int (z^{-5} - z^{-3}) \, dz = \int z^{-5} \, dz - \int z^{-3} \, dz$$

$$= \frac{z^{-5+1}}{-5+1} - \frac{z^{-3+1}}{-3+1} + c = -\frac{z^{-4}}{4} + \frac{z^{-2}}{2} + c = \frac{1}{2z^2} - \frac{1}{4z^4} + c$$

$$I = \frac{1}{2 \cdot \sin^2 x} - \frac{1}{4 \cdot \sin^4 x} + c = \frac{1}{2 \sin^2 x} \left[1 - \frac{1}{2 \sin^2 x} \right] + c = \frac{1}{2 \sin^2 x} \left[\frac{2 \sin^2 x - 1}{2 \sin^2 x} \right] + c = \frac{1}{2 \sin^2 x} \left[\frac{-\cos 2x}{2 \sin^2 x} \right] + c$$

$$= -\frac{1}{4} \csc^4 x \cdot \cos 2x + c \quad \text{Ans.}$$

(d) $I = \int \frac{dx}{\sqrt[4]{\sin^3 x \cdot \cos^5 x}} = \int \sin^{-\frac{3}{4}} x \cdot \cos^{-\frac{5}{4}} x \, dx$ Here $m = -\frac{3}{4}$ and $n = -\frac{5}{4}$

when $m + n = -\frac{3}{4} - \frac{5}{4} = \frac{-3-5}{4} = -\frac{8}{4} = -2 = $ even negative integer

Putting $z = \tan x$ $\therefore$ $dz = \sec^2 x \, dx$ or $dx = \frac{dz}{\sec^2 x} = \frac{dz}{1 + \tan^2 x} = \frac{dz}{1 + z^2}$

or $z = \tan x$ $\therefore$ $\sin x = \frac{\tan x}{\sqrt{1 + \tan^2 x}} = \frac{z}{\sqrt{1 + z^2}}$ and $\cos x = \frac{1}{\sqrt{1 + \tan^2 x}} = \frac{1}{\sqrt{1 + z^2}}$

$$I = \int \sin^{-\frac{3}{4}} x \cdot \cos^{-\frac{5}{4}} x \, dx = \int \left(\frac{z}{\sqrt{1 + z^2}} \right)^{-\frac{3}{4}} \cdot \left(\frac{1}{\sqrt{1 + z^2}} \right)^{-\frac{5}{4}} \cdot \frac{dz}{1 + z^2} = \int \left(\frac{\sqrt{1 + z^2}}{z} \right)^{\frac{3}{4}} \cdot \left(\sqrt{1 + z^2} \right)^{\frac{5}{4}} \cdot \frac{dz}{1 + z^2}$$

$$I = \int \frac{(\sqrt{1 + z^2})^{\frac{3}{4}}}{z^{\frac{3}{4}}} \cdot \left(\sqrt{1 + z^2} \right)^{\frac{5}{4}} \cdot \frac{dz}{1 + z^2} = \int \frac{(\sqrt{1 + z^2})^{\frac{3}{4} + \frac{5}{4}}}{z^{\frac{3}{4}} (1 + z^2)} \, dz = \int \frac{(\sqrt{1 + z^2})^2}{z^{\frac{3}{4}} (1 + z^2)} \, dz = \int \frac{1 + z^2}{z^{\frac{3}{4}} (1 + z^2)} \, dz$$

$$I = \int z^{-\frac{3}{4}} \, dz = \frac{z^{-\frac{3}{4} + 1}}{-\frac{3}{4} + 1} + c = \frac{z^{\frac{1}{4}}}{\frac{1}{4}} + c = 4 \cdot z^{\frac{1}{4}} + c = 4 (\tan x)^{\frac{1}{4}} + c \quad \text{Ans.}$$

(e) Do yourself. $I = \int \frac{dx}{\sqrt{\sin^7 x \cdot \cos^5 x}} = \int \sin^{-\frac{7}{2}} x \cdot \cos^{-\frac{5}{2}} x \, dx$ Here $m = -\frac{7}{2}$ and $n = -\frac{5}{2}$

when $m + n = -\frac{7}{2} - \frac{5}{2} = -6 = $ even negative integer, putting $z = \tan x$ $\therefore$ $dz = \sec^2 x \, dx$

Ans: $- \, -\frac{2}{5} (\tan x)^{-\frac{5}{2}} + 2 (\tan x)^{\frac{5}{2}} + \frac{20}{3} (\tan x)^{\frac{3}{2}} + \frac{20}{7} (\tan x)^{\frac{7}{2}} + \frac{10}{11} (\tan x)^{\frac{11}{2}} + \frac{2}{15} (\tan x)^{\frac{15}{2}} + c$

(f) $I = \int \sin^4 x . \cos^6 x \, dx$, Here $m = 4$ and $n = 6$, when $m + n = 4 + 6 = 10$ = even positive integer

Let $z = \cos x + i \sin x$, $\dfrac{1}{z} = \cos x - i \sin x$ and $\cos x = \dfrac{1}{2}\left(z + \dfrac{1}{z}\right)$ and $\sin x = \dfrac{1}{2i}\left(z - \dfrac{1}{z}\right)$

Also, $z^n + \dfrac{1}{z^n} = 2 \cos nx$

Now, $\sin^4 x . \cos^6 x = \left\{\dfrac{1}{2i}\left(z - \dfrac{1}{z}\right)\right\}^4 . \left\{\dfrac{1}{2}\left(z + \dfrac{1}{z}\right)\right\}^6 = \dfrac{1}{1024}.\left(z - \dfrac{1}{z}\right)^4.\left(z + \dfrac{1}{z}\right)^6 = \dfrac{1}{1024}.\left(z - \dfrac{1}{z}\right)^2.\left(z - \dfrac{1}{z}\right)^2.\left(z + \dfrac{1}{z}\right)^2.\left(z + \dfrac{1}{z}\right)^2.\left(z + \dfrac{1}{z}\right)^2$

$\sin^4 x . \cos^6 x = \dfrac{1}{1024}.\left(z^2 - \dfrac{1}{z^2}\right)^2.\left(z^2 - \dfrac{1}{z^2}\right)^2.\left(z + \dfrac{1}{z}\right)^2 = \dfrac{1}{1024}.\left(z^4 - 2.z^2.\dfrac{1}{z^2} + \dfrac{1}{z^4}\right).\left(z^4 - 2.z^2.\dfrac{1}{z^2} + \dfrac{1}{z^4}\right).\left(z^2 - 2.z.\dfrac{1}{z} + \dfrac{1}{z^2}\right)$

$\sin^4 x . \cos^6 x = \dfrac{1}{1024}.\left(z^4 + \dfrac{1}{z^4} - 2\right).\left(z^4 + \dfrac{1}{z^4} - 2\right).\left(z^2 + \dfrac{1}{z^2} + 2\right)$

$\qquad = \dfrac{1}{1024}.\left[z^8 + z^4.\dfrac{1}{z^4} - 2z^4 + \dfrac{1}{z^4}.z^4 + \dfrac{1}{z^4}.\dfrac{1}{z^4} - \dfrac{2}{z^4} - 2z^4 - \dfrac{2}{z^4} + 4\right]\left(z^2 + \dfrac{1}{z^2} + 2\right)$

$\sin^4 x . \cos^6 x = \dfrac{1}{1024}.\left[\left(z^8 + \dfrac{1}{z^8}\right) - 2\left(z^4 + \dfrac{1}{z^4}\right) - 2\left(z^4 + \dfrac{1}{z^4}\right) + 6\right]\left(z^2 + \dfrac{1}{z^2} + 2\right) = \dfrac{1}{1024}.(2\cos 8x - 8\cos 4x + 6)(2\cos 2x + 2)$

$\qquad = \dfrac{4}{1024}.(\cos 8x - 4\cos 4x + 3)(\cos 2x + 1)$

$\sin^4 x . \cos^6 x = \dfrac{1}{256}.(\cos 8x - 4\cos 4x + 3 + \cos 8x.\cos 2x - 4\cos 4x.\cos 2x + 3\cos 2x)$

Integrating both of sides, we get

$\int \sin^4 x . \cos^6 x \, dx = \dfrac{1}{256}\int (\cos 8x - 4\cos 4x + 3 + \cos 8x.\cos 2x - 4\cos 4x.\cos 2x + 3\cos 2x)\, dx$

$I = \dfrac{1}{256}\int \cos 8x \, dx - \dfrac{4}{256}\int \cos 4x \, dx + \dfrac{3}{256}\int dx + \dfrac{1}{256}\int \cos 8x.\cos 2x \, dx - \dfrac{2}{256}\int 2\cos 4x.\cos 2x \, dx + \dfrac{3}{256}\int \cos 2x \, dx$

$I = \dfrac{1}{256}.\dfrac{\sin 8x}{8} - \dfrac{4}{256}.\dfrac{\sin 4x}{4} + \dfrac{3}{256}x + \dfrac{3}{256}.\dfrac{\sin 2x}{2} + \dfrac{1}{512}\int 2\cos 8x.\cos 2x \, dx - \dfrac{2}{256}\int 2\cos 4x.\cos 2x \, dx$

Let $I_1 = \int 2\cos 8x.\cos 2x \, dx$ [use formula, $2\cos A\cos B = \cos(A + B) + \cos(A - B)$]

$I_1 = \int (\cos 10x + \cos 6x) \, dx = \int \cos 10x \, dx + \int \cos 6x \, dx = \dfrac{\sin 10x}{10} + \dfrac{\sin 6x}{6} + c$

Again, $I_2 = \int 2\cos 4x.\cos 2x \, dx = \int \cos 6x \, dx + \int \cos 2x \, dx = \dfrac{\sin 6x}{6} + \dfrac{\sin 2x}{2} + c$

$I = \dfrac{1}{256}.\dfrac{\sin 8x}{8} - \dfrac{4}{256}.\dfrac{\sin 4x}{4} + \dfrac{3}{256}x + \dfrac{3}{256}.\dfrac{\sin 2x}{2} + \dfrac{1}{512}\left(\dfrac{\sin 10x}{10} + \dfrac{\sin 6x}{6}\right) - \dfrac{2}{256}\left(\dfrac{\sin 6x}{6} + \dfrac{\sin 2x}{2}\right) + c$

$I = \dfrac{1}{2048}\sin 8x - \dfrac{1}{256}\sin 4x + \dfrac{3}{256}x + \dfrac{3}{512}\sin 2x + \dfrac{1}{5120}\sin 10x + \dfrac{1}{3072}\sin 6x - \dfrac{1}{768}\sin 6x - \dfrac{1}{256}\sin 2x + c$

$I = \dfrac{1}{5120}\sin 10x + \dfrac{1}{2048}\sin 8x - \dfrac{1}{1024}\sin 6x - \dfrac{1}{256}\sin 4x + \dfrac{1}{512}\sin 2x + \dfrac{3}{256}x + c$ Ans.

Integral of the form: $- \int \sqrt{\dfrac{ax + b}{cx + d}} \, dx$, Multiplying, N^r and D^r by $\sqrt{ax + b}$

Example: $-$ (a) $I = \int \sqrt{\dfrac{x + 3}{x - 5}} \, dx$, Multiplying, N^r and D^r by $\sqrt{x + 3}$, we get

---

$$I = \int \frac{\sqrt{x+3}}{\sqrt{x-5}} \times \frac{\sqrt{x+3}}{\sqrt{x+3}}\, dx = \int \frac{x+3}{\sqrt{(x-5)(x+3)}}\, dx = \int \frac{x+3}{\sqrt{x^2+3x-5x-15}}\, dx = \int \frac{x+3}{\sqrt{x^2-2x-15}}\, dx = \int \frac{(x-1)+4}{\sqrt{x^2-2x-15}}\, dx$$

$$= \int \frac{x-1}{\sqrt{x^2-2x-15}}\, dx + \int \frac{4}{\sqrt{x^2-2x-15}}\, dx = I_1 + I_2 \quad \text{(say)}$$

Now, $I_1 = \int \frac{x-1}{\sqrt{x^2-2x-15}}\, dx$, Let $x^2-2x-15 = z^2$ $\quad\therefore\quad (2x-2)dx = 2z\, dz \quad$ or $\quad (x-1)dx = z\, dz$

or $I_1 = \int \frac{z\, dz}{\sqrt{z^2}} = \int dz = z + c = \sqrt{x^2-2x-15} + c$ $\quad \left[\text{Direct use formula,} \quad \int \frac{f'(x)}{f(x)}\, dx = 2\sqrt{f(x)} + c \right]$

Now, $I_2 = \int \frac{4}{\sqrt{x^2-2x-15}}\, dx = 4 \int \frac{dx}{\sqrt{(x-1)^2-16}}$ $\quad$ use formula, $\quad \int \frac{dx}{\sqrt{x^2-a^2}} = \log\left|x + \sqrt{x^2-a^2}\right|$

or $I_2 = 4 \int \frac{dx}{\sqrt{(x-1)^2-(4)^2}} = 4.\log\left[(x-1) + \sqrt{(x-1)^2-(4)^2}\right] + c = 4\log\left[(x-1) + \sqrt{x^2-2x-15}\right] + c$

or $I = I_1 + I_2 = \sqrt{x^2-2x-15} + c + 4\log\left[(x-1) + \sqrt{x^2-2x-15}\right] + c = \sqrt{x^2-2x-15} + 4\log\left[(x-1) + \sqrt{x^2-2x-15}\right] + k$ $\quad$ Ans.

(b) $I = \int \sqrt{\frac{3-x}{x+4}}\, dx$ $\quad$ Ans:$- \sqrt{-x^2-x+12} + \frac{7}{2}\sin^{-1}\left(\frac{2x+1}{7}\right) + c$ $\quad$ (solve same as above question)

Integral of the form: $- \quad \int f(e^x)\, dx$, $\quad$ Putting $e^x = z$

Example: $- \quad$ (a) $I = \int \frac{dx}{\sqrt{1-e^{2x}}}$

Let $1 - e^{2x} = z^2$ or $e^{2x} = 1 - z^2$ $\quad\therefore\quad -e^{2x}.2\, dx = 2z\, dz \quad\therefore\quad dx = -\frac{z\, dz}{e^{2x}} = -\frac{z\, dz}{1-z^2}$

or $I = -\int \frac{1}{z}.\frac{z\, dz}{(1-z^2)} = -\int \frac{dz}{1-z^2}$ $\quad$ use formula, $\quad \int \frac{dx}{a^2-x^2} = \frac{1}{2a}\log\left|\frac{a+x}{a-x}\right|$, when $x < a$

or $I = -\frac{1}{2.1}\log\left[\frac{1+z}{1-z}\right] + c = -\frac{1}{2}\log\left[\frac{1+\sqrt{1-e^{2x}}}{1-\sqrt{1-e^{2x}}}\right] + c = -\frac{1}{2}\log\left[\frac{\left(1+\sqrt{1-e^{2x}}\right)\left(1+\sqrt{1-e^{2x}}\right)}{\left(1-\sqrt{1-e^{2x}}\right)\left(1+\sqrt{1-e^{2x}}\right)}\right] + c = -\frac{1}{2}\log\left[\frac{\left(1+\sqrt{1-e^{2x}}\right)^2}{1-1+e^{2x}}\right] + c$

$$= -\frac{1}{2}\log\left[\frac{1+\sqrt{1-e^{2x}}}{e^x}\right]^2 + c = -\log\left(\frac{1+\sqrt{1-e^{2x}}}{e^x}\right) + c$$

or $I = \log\left(\frac{1+\sqrt{1-e^{2x}}}{e^x}\right)^{-1} + c = \log\left(\frac{e^x}{1+\sqrt{1-e^{2x}}}\right) + c$ $\quad$ Ans. $\quad$ or $I = -\log\left(\frac{1}{e^x} + \sqrt{\frac{1}{e^{2x}} - 1}\right) + c$ $\quad$ Ans.

IInd Method: $- \quad I = \int \frac{dx}{\sqrt{1-e^{2x}}}$, $\quad$ Putting $e^x = z$ $\quad\therefore\quad e^x\, dx = dz \quad$ or $\quad dx = \frac{dz}{e^x} = \frac{dz}{z}$

$I = \int \frac{1}{\sqrt{1-z^2}}.\frac{dz}{z} = \int \frac{dz}{z.\sqrt{1-z^2}}$ $\quad$ use formula, $\quad \int \frac{dx}{P\sqrt{Q}}$ where P is linear and Q is quadratic, putting $P = \frac{1}{z}$

Putting, $z = \frac{1}{t}$ $\quad$ or $\quad t = \frac{1}{z}$ $\quad\therefore\quad dz = -\frac{1}{t^2}\, dt$

$\therefore\ I = \int \frac{-\frac{1}{t^2}\, dt}{\frac{1}{t}.\sqrt{1-\frac{1}{t^2}}} = -\int \frac{dt}{\sqrt{t^2-1}}$ $\quad$ use formula, $\quad \int \frac{dx}{\sqrt{x^2-a^2}} = \log\left|x + \sqrt{x^2-a^2}\right|$

$\therefore\ I = -\log\left(t + \sqrt{t^2-1}\right) + c = -\log\left(\frac{1}{z} + \sqrt{\frac{1}{z^2} - 1}\right) + c = -\log\left[\frac{1}{e^x} + \frac{\sqrt{1-e^{2x}}}{e^x}\right] + c$ $\quad$ Ans.

(b) $I = \int \dfrac{e^x}{e^{2x}+1}\, dx$, Putting $e^x = z$ $\therefore e^x dx = dz$ $\therefore I = \int \dfrac{dz}{z^2+1} = \tan^{-1} z + c = \tan^{-1} e^x + c$ Ans.

(c) $I = \int \dfrac{e^x.\log(e^x+1)}{e^x+1}\, dx$, Putting $1 + e^x = z$ $\therefore e^x\, dx = dz$

$\therefore I = \int \dfrac{\log z}{z}\, dz$, Let $\log z = t$ $\therefore \dfrac{1}{z} dz = dt$ $\therefore I = \int t\, dt = \dfrac{t^2}{2} + c = \dfrac{(\log z)^2}{2} + c = \dfrac{[\log(1+e^x)]^2}{2} + c$ Ans.

Integral of the form: $-$ $\int e^x[f(x) + f'(x)]\, dx = e^x f(x) + c$

Proof: $-$ Let $I = \int e^x[f(x) + f'(x)]\, dx = \int e^x f(x)\, dx + \int e^x f'(x)\, dx$

$I = f(x)\int e^x\, dx - \int f'(x).dx \int e^x\, dx + \int e^x f'(x)\, dx = f(x).e^x - \int e^x f'(x)\, dx + \int e^x f'(x)\, dx = e^x f(x) + c$ Proved.

Example: $-$ (a) $I = \int e^x(\sin x + \cos x)\, dx$, $f(x) = \sin x$ $\therefore f'(x) = \cos x$ $\therefore I = \int e^x[f(x) + f'(x)]\, dx = e^x f(x) + c = e^x \sin x + c$ Ans.

(b) $I = \int e^x\left(\log x + \dfrac{1}{x}\right) dx = e^x \log x + c$ Ans. $\left(\text{Let } f(x) = \log x\,, f'(x) = \dfrac{1}{x}\right)$

(c) $I = \int e^x(x^2 + 2x)\, dx = e^x.x^2 + c = x^2 e^x + c$ Ans. $\left(\text{Let } f(x) = x^2, f'(x) = 2x \text{ then } \int e^x[f(x) + f'(x)]\, dx = e^x f(x) + c\right)$

(d) $I = \int \dfrac{e^{\sin x}}{\cos^2 x}(x \cos^3 x - \sin x)\, dx$, Putting $\sin x = z$ or $x = \sin^{-1} z$ $\therefore dx = \dfrac{1}{\sqrt{1-z^2}} dz$

$I = \int \dfrac{e^z}{1-z^2}\left(\sin^{-1} z.(1-z^2)^{\frac{3}{2}} - z\right).\dfrac{1}{\sqrt{1-z^2}} dz = \int e^z\left(\sin^{-1} z - \dfrac{z}{(1-z^2)^{\frac{3}{2}}}\right) dz = \int e^z\left\{\left(\sin^{-1} z - \dfrac{1}{\sqrt{1-z^2}}\right) + \left(\dfrac{1}{\sqrt{1-z^2}} - \dfrac{z}{(1-z^2)^{\frac{3}{2}}}\right)\right\} dz$

$= \int e^z[f(z) + f'(z)]\, dz$

$I = e^z\left(\sin^{-1} z - \dfrac{1}{\sqrt{1-z^2}}\right) + c = e^{\sin x}\left(x - \dfrac{1}{\cos x}\right) + c = e^{\sin x}(x - \sec x) + c$ Ans.

(e) $I = \int \left[\dfrac{1}{\log x} - \dfrac{1}{(\log x)^2}\right] dx$, Putting $\log x = z$ or $x = e^z$ $\therefore dx = e^z dz$

$I = \int \left(\dfrac{1}{z} - \dfrac{1}{z^2}\right).e^z dz = \int e^z\left(\dfrac{1}{z} + \dfrac{-1}{z^2}\right) dz$,   Let $f(x) = \dfrac{1}{z}$ $\therefore f'(x) = -\dfrac{1}{z^2}$

use formula, $\int e^x[f(x) + f'(x)]\, dx = e^x f(x) + c$ or $I = e^z.\dfrac{1}{z} + c = \dfrac{e^{\log x}}{\log x} + c = \dfrac{x}{\log x} + c$ Ans.

Integral of the form: $-$

Expression	Substitution
$\sqrt{a^2 - x^2}$ or $a^2 - x^2$	$x = a \sin\theta$
$\sqrt{x^2 - a^2}$ or $x^2 - a^2$	$x = a \sec\theta$
$\sqrt{a^2 + x^2}$ or $a^2 + x^2$	$x = \tan\theta$
$\sqrt{\dfrac{a-x}{a+x}}$ or $\sqrt{\dfrac{a+x}{a-x}}$	$x = \cos 2\theta$
$\sqrt{\dfrac{x}{a-x}}$ or $\sqrt{\dfrac{a-x}{x}}$	$x = a \sin^2\theta$

$\sqrt{\dfrac{x}{a+x}}$ or $\sqrt{\dfrac{a+x}{x}}$ or $\dfrac{x}{\sqrt{x+a}}$	$x = a\tan\theta$ or $x = a\cot\theta$
$\sqrt{(a-x)(x-b)}$	$x = a\cos^2\theta + b\sin^2\theta$

Example: — (1) (a) $I = \displaystyle\int \dfrac{x}{\sqrt{1+x^4}}\,dx$

Putting $x = a\tan\theta$ ∴ $a = 1$ or $x = \tan\theta$ ∴ $dx = \sec^2\theta\,d\theta$

$I = \displaystyle\int \dfrac{\tan\theta}{\sqrt{1+\tan^4\theta}}\cdot\sec^2\theta\,d\theta = \int \dfrac{\tan\theta\cdot\sec^2\theta\,d\theta}{\sqrt{1+(\tan^2\theta)^2}}$, Let $\tan^2\theta = z$ ∴ $2\tan\theta\cdot\sec^2\theta\,d\theta = dz$

$I = \displaystyle\int \dfrac{dz}{2\sqrt{1+z^2}} = \dfrac{1}{2}\cdot\log\left[z+\sqrt{1+z^2}\right] + c$ use formula, $\displaystyle\int \dfrac{dx}{\sqrt{x^2+a^2}} = \log\left|x+\sqrt{x^2+a^2}\right|$

$I = \dfrac{1}{2}\log\left[\tan^2\theta + \sqrt{1+\tan^4\theta}\right] + c = \dfrac{1}{2}\log\left[x^2+\sqrt{1+x^4}\right] + c$ Ans.

(b) $I = \displaystyle\int \sqrt{1-x^2}\,dx$, Putting $x = a\sin\theta$ ∴ $a = 1$ or $x = \sin\theta$ ∴ $dx = \cos\theta\,d\theta$

$I = \displaystyle\int \sqrt{1-\sin^2\theta}\cdot\cos\theta\,d\theta = \int \cos^2\theta\,d\theta = \int \dfrac{1+\cos 2\theta}{2}\,d\theta = \dfrac{1}{2}\int d\theta + \dfrac{1}{2}\int \cos 2\theta\,d\theta = \dfrac{1}{2}\theta + \dfrac{1}{2}\dfrac{\sin 2\theta}{2} + c = \dfrac{1}{2}\sin^{-1}x + \dfrac{1}{4}\cdot 2\sin\theta\cos\theta + c$

$= \dfrac{1}{2}\sin^{-1}x + \dfrac{1}{2}x\sqrt{1-x^2} + c$ Ans.

(c) $I = \displaystyle\int \sqrt{\dfrac{1+\sqrt{x}}{1-\sqrt{x}}}\,dx$, Putting $\sqrt{x} = a\cos 2\theta$ ∴ $a = 1$ or $\sqrt{x} = \cos 2\theta$ or $\dfrac{1}{2\sqrt{x}}dx = -2\sin 2\theta\,d\theta$ or $dx = -4\cos 2\theta\sin 2\theta\,d\theta$

$I = \displaystyle\int \sqrt{\dfrac{1+\cos 2\theta}{1-\cos 2\theta}}\cdot -4\cos 2\theta\sin 2\theta\,d\theta = -4\int \sqrt{\dfrac{1+\cos^2\theta-\sin^2\theta}{1-\cos^2\theta+\sin^2\theta}}\cdot\cos 2\theta\sin 2\theta\,d\theta = -4\int \dfrac{\cos\theta}{\sin\theta}\cdot 2\sin\theta\cos\theta\cos 2\theta\,d\theta$

$= -8\displaystyle\int \cos 2\theta\cdot\cos^2\theta\,d\theta = -8\int \cos 2\theta\cdot\left(\dfrac{1+\cos 2\theta}{2}\right)d\theta$

$\left[\cos 2\theta = 2\cos^2\theta - 1 \text{ or } \cos^2\theta = \dfrac{1+\cos 2\theta}{2} \text{ or } \cos^2\theta - \sin^2\theta = \cos 2\theta \text{ or } \cos 2\theta = 1 - 2\sin^2\theta\right]$

$I = -4\displaystyle\int (\cos 2\theta + \cos^2 2\theta)\,d\theta = -4\int \cos 2\theta\,d\theta - 4\int \cos^2 2\theta\,d\theta = -4\cdot\dfrac{\sin 2\theta}{2} - 4\int \dfrac{1+\cos 4\theta}{2}\,d\theta = -2\sin 2\theta - 2\int d\theta - 2\int \cos 4\theta\,d\theta$

$= -2\sin 2\theta - 2\theta - 2\cdot\dfrac{\sin 4\theta}{4} + c$

∴ Put $\sqrt{x} = \cos 2\theta$ or $2\theta = \cos^{-1}\sqrt{x}$ or $\theta = \dfrac{\cos^{-1}\sqrt{x}}{2}$

$I = -2\sin 2\theta - 2\theta - \dfrac{\sin 4\theta}{2} + c = -2\sin(\cos^{-1}\sqrt{x}) - 2\cdot\dfrac{\cos^{-1}\sqrt{x}}{2} - \dfrac{\sin(2\cos^{-1}\sqrt{x})}{2} + c$

$= -2\sin(\cos^{-1}\sqrt{x}) - \dfrac{\sin(2\cos^{-1}\sqrt{x})}{2} - \cos^{-1}\sqrt{x} + c$ Ans.

(d) $I = \displaystyle\int \cos^{-1}\left(\dfrac{x+1}{\sqrt{x^2+2x+5}}\right)dx = \int \cos^{-1}\left(\dfrac{x+1}{\sqrt{(x+1)^2+2^2}}\right)$, Putting $x+1 = a\cot\theta$ ∴ $a = 2$

or $x+1 = 2\cot\theta$ ∴ $dx = -2\,\text{cosec}^2\theta\,d\theta$ or $\cot\theta = \dfrac{x+1}{2}$ or $\theta = \cot^{-1}\left(\dfrac{x+1}{2}\right)$

$I = \displaystyle\int \cos^{-1}\left(\dfrac{2\cot\theta}{\sqrt{(2\cot\theta)^2+4}}\right)\cdot -2\,\text{cosec}^2\theta\,d\theta = -2\int \cos^{-1}\left(\dfrac{2\cot\theta}{\sqrt{4(\cot^2\theta+1)}}\right)\cdot\text{cosec}^2\theta\,d\theta$

$$I = -2 \int \cos^{-1}\left(\frac{\cot\theta}{\sqrt{\csc^2\theta}}\right).\csc^2\theta \, d\theta = -2\int \cos^{-1}\left(\frac{\cot\theta}{\csc\theta}\right).\csc^2\theta \, d\theta \quad [\cot^2\theta + 1 = \csc^2\theta]$$

$$I = -2\int \cos^{-1}\left(\frac{\frac{\cos\theta}{\sin\theta}}{\frac{1}{\sin\theta}}\right).\csc^2\theta \, d\theta = -2\int \theta.\csc^2\theta \, d\theta$$

formula, $\int f(x).g(x)\,dx = f(x).\int g(x)dx - \int f'(x)dx \int g(x)dx$ and $\int \csc^2\theta \, d\theta = -\cot\theta + c$

and $\int \cot\theta \, d\theta = \log(\sin\theta) + c$

$$I = -2\left[\theta.\int\csc^2\theta \, d\theta - \int \frac{d(\theta)}{d\theta}.d\theta.\int\csc^2\theta \, d\theta\right] = -2\left[\theta.(-\cot\theta) + \int\cot\theta \, d\theta\right] = -2[\theta(-\cot\theta) + \log(\sin\theta)] + c$$
$$= 2\theta\cot\theta - 2\log(\sin\theta) + c$$

$$I = 2\cot^{-1}\left(\frac{x+1}{2}\right).\cot\left[\cot^{-1}\left(\frac{x+1}{2}\right)\right] - 2\log\left\{\sin\left[\cot^{-1}\left(\frac{x+1}{2}\right)\right]\right\} + c$$
$$= 2\cot^{-1}\left(\frac{x+1}{2}\right).\left(\frac{x+1}{2}\right) - 2\log\left\{\sin\left[\cot^{-1}\left(\frac{x+1}{2}\right)\right]\right\} + c \quad \text{Ans.}$$

(2) find the following integral:− (a) $I = \int \dfrac{x^3}{\sqrt{1+x^8}} dx$, Putting $x^4 = \tan\theta \quad \therefore 4x^3 dx = \sec^2\theta \, d\theta$

$$I = \frac{1}{4}\int \frac{\sec^2\theta \, d\theta}{\sqrt{1+\tan^2\theta}} \quad \text{Let } \tan\theta = z \quad \therefore \sec^2\theta \, d\theta = dz \quad \text{using formula, } \int \frac{dx}{\sqrt{a^2+x^2}} = \log\left|x + \sqrt{x^2+a^2}\right|$$

$$I = \frac{1}{4}\int \frac{dz}{\sqrt{1+z^2}} = \frac{1}{4}\log\left(z + \sqrt{1+z^2}\right) + c = \frac{1}{4}\log\left[\tan\theta + \sqrt{1+\tan^2\theta}\right] + c = \frac{1}{4}\log\left[x^4 + \sqrt{1+x^8}\right] + c \quad \text{Ans.}$$

(b) $I = \int \sin^{-1}\left(\dfrac{2x+2}{\sqrt{(2x)^2 + 2.2.2x + 4 + 9}}\right) dx$

Putting $2x+2 = a\tan\theta$ or $2x+2 = 3\tan\theta \quad \therefore 2dx = 3\sec^2\theta \, d\theta$ or $dx = \dfrac{3}{2}\sec^2\theta \, d\theta$ or $\tan\theta = \dfrac{2x+2}{3}$

$$I = \int \sin^{-1}\left(\frac{2x+2}{\sqrt{(2x+2)^2 + 9}}\right) dx = \int \sin^{-1}\left(\frac{3\tan\theta}{\sqrt{9\tan^2\theta + 9}}\right).\frac{3}{2}\sec^2\theta \, d\theta = \frac{3}{2}\int \sin^{-1}\left(\frac{3\tan\theta}{3\sqrt{1+\tan^2\theta}}\right).\sec^2\theta \, d\theta$$

$$= \frac{3}{2}\int \sin^{-1}\left(\frac{\sin\theta/\cos\theta}{1/\cos\theta}\right).\sec^2\theta \, d\theta = \frac{3}{2}\int \sin^{-1}(\sin\theta).\sec^2\theta \, d\theta = \frac{3}{2}\int \theta.\sec^2\theta \, d\theta$$

$$I = \frac{3}{2}\left[\theta.\int\sec^2\theta \, d\theta - \int\frac{d(\theta)}{d\theta}.d\theta\int\sec^2\theta \, d\theta\right] = \frac{3}{2}\left[\theta.\tan\theta - \int\tan\theta \, d\theta\right] = \frac{3}{2}[\theta.\tan\theta + \log(\cos\theta)] + c$$

formula, $\int f(x).g(x)\,dx = f(x).\int g(x)dx - \int f'(x)dx\int g(x)dx$ and $\int\sec^2\theta \, d\theta = \tan\theta + c$

and $\int\tan\theta \, d\theta = -\log(\cos\theta) + c$ Put $\tan\theta = \dfrac{2x+2}{3} \quad \therefore \theta = \tan^{-1}\left(\dfrac{2x+2}{3}\right)$

$$I = \frac{3}{2}[\theta.\tan\theta + \log(\cos\theta)] + c = \frac{3}{2}\left\{\tan^{-1}\left(\frac{2x+2}{3}\right).\tan\left[\tan^{-1}\left(\frac{2x+2}{3}\right)\right] + \log\left[\cos\tan^{-1}\left(\frac{2x+2}{3}\right)\right]\right\} + c$$
$$= \frac{3}{2}.\left(\frac{2x+2}{3}\right).\tan^{-1}\left(\frac{2x+2}{3}\right) + \frac{3}{2}.\log\left[\cos\tan^{-1}\left(\frac{2x+2}{3}\right)\right] + c$$

$$I = \frac{3}{2}.\frac{2}{3}(x+1).\tan^{-1}\left(\frac{2x+2}{3}\right) + \frac{3}{2}.\log\left[\cos\tan^{-1}\left(\frac{2x+2}{3}\right)\right] + c = (x+1)\tan^{-1}\left(\frac{2x+2}{3}\right) + \frac{3}{2}.\log\left[\cos\tan^{-1}\left(\frac{2x+2}{3}\right)\right] + c \quad \text{Ans.}$$

(c) $I = \int \sin^{-1}\left(\dfrac{x+1}{\sqrt{x^2+2x+2}}\right) dx = \int \sin^{-1}\left(\dfrac{x+1}{\sqrt{(x+1)^2 + 1}}\right)$

Putting $x+1 = a\tan\theta$ or $\theta = \tan^{-1}(x+1)$ or $x+1 = \tan\theta \quad \therefore dx = \sec^2\theta \, d\theta$

$$I = \int \sin^{-1}\left(\frac{\tan\theta}{\sqrt{\tan^2\theta + 1}}\right).\sec^2\theta\,d\theta = \int \sin^{-1}\left(\frac{\sin\theta/\cos\theta}{1/\cos\theta}\right).\sec^2\theta\,d\theta = \int \sin^{-1}(\sin\theta).\sec^2\theta\,d\theta$$

$$= \int \theta.\sec^2\theta\,d\theta \quad \text{(use integrating by part formula)}$$

$$I = \theta.\int \sec^2\theta\,d\theta - \int \frac{d(\theta)}{d\theta}.d\theta \int \sec^2\theta\,d\theta = \theta\tan\theta - \int \tan\theta\,d\theta = \theta\tan\theta + \log(\cos\theta) + c$$

$$I = \tan^{-1}(x+1).\tan[\tan^{-1}(x+1)] + \log\{\cos[\tan^{-1}(x+1)]\} + c = (x+1)\tan^{-1}(x+1) + \log\{\cos[\tan^{-1}(x+1)]\} + c \quad \text{Ans.}$$

Exercise – A8

(1) (a) $I = \int \dfrac{dx}{\cos x + \tan x}$ (b) $I = \int \dfrac{dx}{\sin x + \cot x}$ (c) $I = \int \dfrac{dx}{1 + \sin 3x}$ (d) $I = \int \dfrac{dx}{\sqrt{1 + \sin 5x}}$

(2) (a) $I = \int \sqrt{1 - \sin 2x}\,dx$ (b) $I = \int \dfrac{x^4}{x-1}\,dx$ (c) $I = \int \dfrac{4x^3 + 12x^2 + 9x - 1}{2x + 3}\,dx$ (d) $I = \int \dfrac{dx}{(\sin^{-1} x)^3.\sqrt{1 - x^2}}$

(3) (a) $I = \int \dfrac{x+1}{x^2 + x\log x}\,dx$ (b) $I = \int \dfrac{x^3}{\sqrt{1 - x^2}}\,dx$ (c) $I = \int \sin^5 x.\sqrt{\cos x}\,dx$ (d) $I = \int \dfrac{x^2}{(x\sin x + \cos x)^2}\,dx$

(4) (a) $I = \int \sin^{-1}\sqrt{x}\,dx$ (b) $I = \int \cos^{-1}\sqrt{x}\,dx$ (c) $I = \int \tan^{-1}\sqrt{x}\,dx$ (d) $I = \int \cot^{-1}\sqrt{x}\,dx$

(5) (a) $I = \int \sqrt{e^{2x} + 1}\,dx$ (b) $I = \int \dfrac{dx}{(x-4)\sqrt{x-2}}$ (c) $I = \int e^x \log x\,dx$ (d) $I = \int x^2 e^x\,dx$

(6) (a) $I = \int x^3 \log x\,dx$ (b) $I = \int e^x \sin x\,dx$ (c) $I = \int \dfrac{\cos^{-1}\sqrt{x} - \sin^{-1}\sqrt{x}}{\sin^{-1}\sqrt{x} + \cos^{-1}\sqrt{x}}\,dx$ (d) $I = \int \cos x.\log(\sin x)\,dx$

(7) (a) $I = \int \log\left(\sqrt{1 - x^2}\right)dx$ (b) $I = \int \sin x.\log(\cot x)\,dx$ (c) $I = \int \dfrac{x\tan x}{\sec^3 x}\,dx$ (d) $I = \int x(e^x + 1)\,dx$

(8) (a) $I = \int \dfrac{\log\{\log(1 + \sqrt{x})\}}{(x + \sqrt{x})}\,dx$ (b) $I = \int 2^{3\log_2 \sqrt[3]{x}}\,dx$ (c) $I = \int x^3 \cot^{-1} x\,dx$ (d) $I = \int \dfrac{(2^x - 3^x)^2}{2^x.3^x}\,dx$

(9) (a) $I = \int \dfrac{\sin 2x.\cos x + 2\cos^2 x}{(1 + 2\sin^{x}/_2 \cos^{x}/_2)}\,dx$ (b) $I = \int \dfrac{x^2 + 1}{x(x+3)(x-4)}\,dx$ (c) $I = \int \dfrac{(x^3 + 64)(x - 2)}{x^2 - 4x + 16}\,dx$

(d) $I = \int \dfrac{dx}{\sqrt{x - 1} + \sqrt{x + 2}}$ (e) $I = \int \cos^{-1}\left(\dfrac{2\cot x}{1 + \cot^2 x}\right)dx$ (f) $I = \int \dfrac{x}{x^4 + 4}\,dx$

(10) (a) $I = \int \dfrac{dx}{\sqrt{x} - \sqrt{x - 2}}$ (b) $I = \int \dfrac{x^8}{\sqrt{1 + x^3}}\,dx$ (c) $I = \int \dfrac{x + 3}{\sqrt{4x + 5}}\,dx$ (d) $I = \int \dfrac{x}{2x^2 - 3x + 5}\,dx$

(11) (a) $I = \int \dfrac{3x + 2}{x^2 + 2x + 3}\,dx$ (b) $I = \int \dfrac{4x + 1}{x^2 + 3x + 2}\,dx$ (c) $I = \int \dfrac{2x - 3}{x^2 + 4x - 5}\,dx$ (d) $I = \int \dfrac{x^5 + x^2}{x^6 + 16}\,dx$

(12) (a) $I = \int \dfrac{\sin 2\theta - 5\cos\theta}{7 - \cos^2\theta + 4\sin\theta}\,d\theta$ (b) $I = \int \sin(\log x)\,dx + \int \cos(\log x)\,dx$ (c) $I = \int \cos^{-1}\left(\dfrac{2x}{1 + x^2}\right)dx$

(13) (a) $I = \int \sin^{-1}\left(\dfrac{x}{\sqrt{1 + x^2}}\right)dx$ (b) $I = \int \tan^{-1}\left(\dfrac{x}{\sqrt{1 - x^2}}\right)dx$ (c) $I = \int \dfrac{\cos^{-1} x}{\sqrt{1 - x^2}}\,dx$ (d) $I = \int \dfrac{\cos^{-1} x}{(1 - x^2)^{\frac{3}{2}}}\,dx$

(14) (a) $I = \int \sqrt{\sec\theta - 1}\,d\theta$ (b) $I = \int \dfrac{x - \sqrt[3]{x^2}}{x(\sqrt[4]{x} + \sqrt[6]{x^2})}\,dx$ (c) $I = \int e^{3x}(3x^2 + 4x + 1)\,dx$ (d) $I = \int \dfrac{x^2 - 9}{x + 2}\,dx$

(15) (a) $I = \int e^{\sin 2x}(1 + \cos 2x)\,dx$ (b) $I = \int \sin^2 x \cot^2 x\,dx$ (c) $I = \int \dfrac{\cot x \csc x}{3 + \csc x}\,dx$ (d) $I = \int \tan^4 x\,dx$

(16) (a) $I = \int (\sin^4 x + \cos^4 x)\, dx$ (b) $I = \int \dfrac{dx}{1 - \sqrt{1 + x^2}}$ (c) $I = \int \dfrac{(\sin x - \sqrt[3]{\sin^2 x})\cdot \cos x}{\sqrt{\sin^3 x} + \sqrt[4]{\sin^6 x}}\, dx$ (d) $I = \int \dfrac{x - 1}{(x - 2)\sqrt{x + 3}}\, dx$

(17) (a) $I = \int \sin x \cdot \sin 3x \, dx$ (b) $I = \int \cos x \cdot \cos 3x \cdot \cos 5x \, dx$ (c) $I = \int \tan 2x \cdot \tan 3x \cdot \tan 5x \, dx$

(d) $I = \int \sin x \cdot \sin 3x \cdot \sin 5x \cdot \sin 7x \, dx$ (e) $I = \int \sin x \cdot \cos 3x \, dx$

(18) (a) $I = \int \cos 2x \cdot \sin 3x \, dx$ (b) $I = \int \sin x \cdot \sin 2x \cdot \cos x \cdot \cos 2x \, dx$ (c) $I = \int \dfrac{\sin 2x}{\cos(x + \theta)\cdot \cos(x - \theta)}\, dx$ (d) $I = \int \dfrac{dx}{\sin x \cdot \sin(x + \theta)}$

Answer

(1) (a) $I = \int \dfrac{dx}{\cos x + \tan x} = \int \dfrac{dx}{\cos x + \dfrac{\sin x}{\cos x}} = \int \dfrac{\cos x}{\cos^2 x + \sin x}\, dx = \int \dfrac{\cos x \, dx}{1 - \sin^2 x + \sin x}$

Let $\sin x = t$ $\therefore$ $\cos x \, dx = dt$ or $I = \int \dfrac{dt}{1 - t^2 + t} = \int \dfrac{dt}{1 + t - t^2} = \int \dfrac{dt}{\dfrac{5}{4} - \left(t - \dfrac{1}{2}\right)^2} = \int \dfrac{dt}{\left(\dfrac{\sqrt{5}}{2}\right)^2 - \left(t - \dfrac{1}{2}\right)^2}$

using formula, $\int \dfrac{dx}{a^2 - x^2} = \dfrac{1}{2a}\log\left|\dfrac{a + x}{a - x}\right|$ or $I = \dfrac{1}{2 \cdot \dfrac{\sqrt{5}}{2}}\log\left(\dfrac{\dfrac{\sqrt{5}}{2} + \left(t - \dfrac{1}{2}\right)}{\dfrac{\sqrt{5}}{2} - \left(t - \dfrac{1}{2}\right)}\right) + c = \dfrac{1}{\sqrt{5}}\log\left(\dfrac{\sqrt{5} + 2t - 1}{\sqrt{5} - 2t + 1}\right) + c$

Put $t = \sin x$ $\therefore$ $I = \dfrac{1}{\sqrt{5}}\log\left(\dfrac{\sqrt{5} + 2\sin x - 1}{\sqrt{5} - 2\sin x + 1}\right) + c$ Ans.

(b) $I = \int \dfrac{dx}{\sin x + \cot x} = \int \dfrac{\sin x}{\sin^2 x + \cos x}\, dx = \int \dfrac{\sin x}{1 - \cos^2 x + \cos x}\, dx$

Put $\cos x = t$ $\therefore$ $-\sin x \, dx = dt$ use formula, $\int \dfrac{dx}{x^2 - a^2} = \dfrac{1}{2a}\log\left|\dfrac{x - a}{x + a}\right|$, when $x > a$

$I = \int \dfrac{dt}{t^2 - t - 1} = \int \dfrac{dt}{\left(t - \dfrac{1}{2}\right)^2 - \left(\dfrac{\sqrt{5}}{2}\right)^2} = \dfrac{1}{\sqrt{5}}\log\left|\dfrac{2t - 1 - \sqrt{5}}{2t - 1 + \sqrt{5}}\right| + c = \dfrac{1}{\sqrt{5}}\log\left|\dfrac{2\cos x - 1 - \sqrt{5}}{2\cos x - 1 + \sqrt{5}}\right| + c$ Ans.

(c) $I = \int \dfrac{dx}{1 + \sin 3x} = \int \dfrac{dx}{1 + 2\sin \dfrac{3x}{2}\cos \dfrac{3x}{2}}$ divide above and below by $\cos^2 \dfrac{3x}{2}$

Do yourself Ans: $-\dfrac{2}{3}\cdot \dfrac{\cos \dfrac{3x}{2}}{\sin \dfrac{3x}{2} + \cos \dfrac{3x}{2}} + c$

IInd Method: $- I = \int \dfrac{dx}{1 + \sin 3x} = \int \dfrac{\cos 3x \, dx}{(1 + \sin 3x)\cos 3x} = \int \dfrac{\cos 3x \, dx}{(1 + \sin 3x)\cdot \sqrt{1 - \sin^2 3x}}$

Let $\sin 3x = t$ $\therefore$ $3\cos 3x \, dx = dt$ or $\cos 3x \, dx = \dfrac{dt}{3}$

$I = \int \dfrac{dt}{3(1 + t)\cdot \sqrt{1 - t^2}}$, Put $1 + t = z$ $\therefore$ $dt = dz$ or $t = z - 1$

$I = \dfrac{1}{3}\int \dfrac{dz}{z\cdot \sqrt{1 - (z - 1)^2}} = \dfrac{1}{3}\int \dfrac{dz}{z\cdot \sqrt{2z - z^2}} = \dfrac{1}{3}\int \dfrac{dz}{z^2 \cdot \sqrt{\dfrac{2}{z} - 1}}$, Let $\dfrac{2}{z} - 1 = u^2$ $\therefore$ $-\dfrac{2}{z^2}\, dz = 2u\, du$

$$I = -\frac{1}{3}\int \frac{u\,du}{u} = -\frac{1}{3}\int du = -\frac{1}{3}u + c = -\frac{1}{3}\sqrt{\frac{2}{z}-1} + c = -\frac{1}{3}\sqrt{\frac{2-z}{z}} + c = -\frac{1}{3}\sqrt{\frac{2-1-t}{1+t}} + c = -\frac{1}{3}\sqrt{\frac{1-t}{1+t}} + c = -\frac{1}{3}\sqrt{\frac{1-t}{1+t} \times \frac{1-t}{1-t}} + c$$

$$= -\frac{1}{3} \cdot \frac{1-t}{\sqrt{1-t^2}} + c = -\frac{1}{3} \cdot \frac{1-\sin 3x}{\sqrt{1-\sin^2 3x}} + c = -\frac{1}{3} \cdot \frac{1-\sin 3x}{\cos 3x} + c \quad \text{Ans.}$$

IIIrd Method: − $I = \int \frac{dx}{1+\sin 3x} = \int \frac{dx}{\sin^2\left(\frac{3x}{2}\right) + \cos^2\left(\frac{3x}{2}\right) + 2\sin\left(\frac{3x}{2}\right)\cos\left(\frac{3x}{2}\right)} = \int \frac{dx}{\left(\sin\left(\frac{3x}{2}\right) + \cos\left(\frac{3x}{2}\right)\right)^2}$

$$= \int \frac{dx}{\left[\sqrt{2}\left\{\sin\left(\frac{3x}{2}\right).\cos\left(\frac{\pi}{4}\right) + \cos\left(\frac{3x}{2}\right).\sin\left(\frac{\pi}{4}\right)\right\}\right]^2} = \int \frac{dx}{2\left[\sin\left(\frac{3x}{2}+\frac{\pi}{4}\right)\right]^2} = \frac{1}{2}\int \text{cosec}^2\left(\frac{3x}{2}+\frac{\pi}{4}\right) dx$$

$$I = -\frac{1}{2}\cot\left(\frac{3x}{2}+\frac{\pi}{4}\right).\frac{1}{\frac{3}{2}} + c = -\frac{1}{3}\cot\left(\frac{\pi}{4}+\frac{3x}{2}\right) + c \quad \text{Ans.} \quad \left[\int \text{cosec}^2 x\,dx = -\cot x\right]$$

(d) $I = \int \frac{dx}{\sqrt{1+\sin 5x}} = \int \frac{dx}{\sqrt{1 + 2\sin\left(\frac{5x}{2}\right)\cos\left(\frac{5x}{2}\right)}}$, Divide above and below by $\cos^4\left(\frac{5x}{2}\right)$

$$I = \int \frac{\sec^4\left(\frac{5x}{2}\right)}{\sqrt{\sec^2\left(\frac{5x}{2}\right) + 2\tan\left(\frac{5x}{2}\right)}}\,dx, \quad \text{Let } \tan\left(\frac{5x}{2}\right) = t \quad \therefore \quad \sec^2\left(\frac{5x}{2}\right).\frac{5}{2}\,dx = dt \text{ or } \sec^2\left(\frac{5x}{2}\right)dx = \frac{2}{5}\,dt$$

$$I = \int \frac{\sec^2\left(\frac{5x}{2}\right).\sec^2\left(\frac{5x}{2}\right)}{\sqrt{\sec^2\left(\frac{5x}{2}\right) + 2\tan\left(\frac{5x}{2}\right)}}\,dx = \int \frac{\left[1+\tan^2\left(\frac{5x}{2}\right)\right].\sec^2\left(\frac{5x}{2}\right)}{\sqrt{1+\tan^2\left(\frac{5x}{2}\right) + 2\tan\left(\frac{5x}{2}\right)}}\,dx = \frac{2}{5}\int \frac{(1+t^2)}{\sqrt{1+t^2+2t}}\,dt = \frac{2}{5}\int \frac{(1+t^2)}{\sqrt{(1+t)^2}}\,dt = \frac{2}{5}\int \frac{1+t^2}{1+t}\,dt$$

$$= \frac{2}{5}\int \frac{1}{1+t}\,dt + \frac{2}{5}\int \frac{t^2}{1+t}\,dt = \frac{2}{5}\int \frac{1}{1+t}\,dt + \frac{2}{5}\int \left(t-\frac{t}{1+t}\right)dt = \frac{2}{5}\int \frac{1}{1+t}\,dt + \frac{2}{5}\int t\,dt - \frac{2}{5}\int \frac{t}{1+t}\,dt$$

$$= \frac{2}{5}\int \frac{1}{1+t}\,dt + \frac{2}{5}\int t\,dt - \frac{2}{5}\int \left(1-\frac{1}{1+t}\right)dt$$

$$I = \frac{2}{5}\int \frac{1}{1+t}\,dt + \frac{2}{5}\int t\,dt - \frac{2}{5}\int dt + \frac{2}{5}\int \frac{1}{1+t}\,dt = \frac{2}{5}\log(1+t) + \frac{2}{5}.\frac{t^2}{2} - \frac{2}{5}.t + \frac{2}{5}\log(1+t) + c = \frac{4}{5}\log(1+t) + \frac{1}{5}t^2 - \frac{2}{5}t + c$$

$$= \frac{4}{5}\log\left[1+\tan\left(\frac{5x}{2}\right)\right] + \frac{\tan^2\left(\frac{5x}{2}\right)}{2} - \frac{2\tan\left(\frac{5x}{2}\right)}{5} + c \quad \text{Ans.}$$

IInd Method: − $I = \int \frac{dx}{\sqrt{1+\sin 5x}} = \int \frac{dx}{\sqrt{\sin^2\left(\frac{5x}{2}\right) + \cos^2\left(\frac{5x}{2}\right) + 2\sin\left(\frac{5x}{2}\right)\cos\left(\frac{5x}{2}\right)}} = \int \frac{dx}{\sqrt{\left[\sin\left(\frac{5x}{2}\right) + \cos\left(\frac{5x}{2}\right)\right]^2}}$

$$= \int \frac{dx}{\sin\left(\frac{5x}{2}\right) + \cos\left(\frac{5x}{2}\right)}$$

$$I = \int \frac{dx}{\sqrt{2}\left[\sin\left(\frac{5x}{2}\right).\cos\left(\frac{\pi}{4}\right) + \cos\left(\frac{5x}{2}\right).\sin\left(\frac{\pi}{4}\right)\right]} = \frac{1}{\sqrt{2}}\int \frac{dx}{\sin\left[\frac{5x}{2}+\frac{\pi}{4}\right]} = \frac{1}{\sqrt{2}}\int \text{cosec}\left[\frac{5x}{2}+\frac{\pi}{4}\right]dx$$

$$I = \frac{1}{\sqrt{2}}.\log\left\{\tan\left[\frac{\left(\frac{5x}{2}+\frac{\pi}{4}\right)}{2}\right]\right\}.\frac{1}{\frac{5}{2}} + c = \frac{2}{5\sqrt{2}}\log\left\{\tan\left[\frac{\left(\frac{5x}{2}+\frac{\pi}{4}\right)}{2}\right]\right\} + c = \frac{\sqrt{2}}{5}\log\left\{\tan\left[\frac{\left(\frac{5x}{2}+\frac{\pi}{4}\right)}{2}\right]\right\} + c \quad \text{Ans.}$$

$$\left[\text{use formula, } \int \text{cosec}\,x\,dx = \log\left[\tan\left(\frac{x}{2}\right)\right]\right]$$

(2) (a) $I = \int \sqrt{1-\sin 2x}\,dx = \int \sqrt{\sin^2 x + \cos^2 x - 2\sin x\cos x}\,dx = \int \sqrt{(\sin x - \cos x)^2}\,dx = \int (\sin x - \cos x)\,dx$

$$I = \int \sin x\,dx - \int \cos x\,dx = -\cos x - \sin x + c = -(\sin x + \cos x) + c \quad \text{Ans.}$$

(b) $I = \int \frac{x^4}{x-1}\,dx = \int \frac{(x^3+x^2+x+1)(x-1)+1}{(x-1)}\,dx$, factorised of x^4 is $(x^3+x^2+x+1)(x-1)+1$

$$I = \int \frac{(x^3 + x^2 + x + 1)(x - 1)}{(x - 1)} dx + \int \frac{1}{x - 1} dx = \int (x^3 + x^2 + x + 1) dx + \int \frac{1}{x - 1} dx = \int x^3 dx + \int x^2 dx + \int x dx + \int dx + \int \frac{1}{x - 1} dx$$

$$= \frac{x^4}{4} + \frac{x^3}{3} + \frac{x^2}{2} + x + \log(x - 1) + c \quad \text{Ans.}$$

(c) $I = \int \frac{4x^3 + 12x^2 + 9x - 1}{2x + 3} dx = \int \frac{(2x + 3)(2x^2 + 3x) - 1}{2x + 3} dx = \int \frac{(2x + 3)(2x^2 + 3x)}{(2x + 3)} dx - \int \frac{dx}{2x + 3}$

$$I = \int (2x^2 + 3x) dx - \int \frac{dx}{2x + 3} = \int 2x^2 dx + \int 3x dx - \int \frac{dx}{2x + 3} = 2.\frac{x^3}{3} + 3.\frac{x^2}{2} - \log(2x + 3).\frac{1}{\text{d. c of } 2x + 3} + c$$

$$\therefore \ I = \frac{2x^3}{3} + \frac{3x^2}{2} - \frac{\log(2x + 3)}{2} + c \qquad \text{Ans.}$$

(d) $I = \int \frac{dx}{(\sin^{-1} x)^3 . \sqrt{1 - x^2}}$, Put $\sin^{-1} x = z$ $\quad \therefore \ \frac{1}{\sqrt{1 - x^2}} dx = dz$

$$I = \int \frac{dz}{z^3} = \int z^{-3} dz = \frac{z^{-3+1}}{-3 + 1} + c = -\frac{1}{2} z^{-2} + c = -\frac{1}{2z^2} + c = -\frac{1}{2(\sin^{-1} x)^2} + c \quad \text{Ans.}$$

(3) (a) $I = \int \frac{x + 1}{x^2 + x \log x} dx = \int \frac{x + 1}{x(x + \log x)} dx$, Put $x + \log x = z$ $\quad \therefore \ \left(1 + \frac{1}{x}\right) dx = dz$ or $\left(\frac{x + 1}{x}\right) dx = dz$

$$I = \int \frac{1}{(x + \log x)}.\frac{x + 1}{x} dx = \int \frac{dz}{z} = \log z + c = \log[x + \log x] + c \quad \text{Ans.}$$

(b) $I = \int \frac{x^3}{\sqrt{1 - x^2}} dx$, Put $1 - x^2 = z^2$ or $x^2 = 1 - z^2$ $\quad \therefore \ -2x dx = 2z dz$ or $x dx = -z dz$

$$I = \int \frac{x^2}{\sqrt{1 - x^2}}.x dx = \int \frac{1 - z^2}{z}.(-z) dz = \int (z^2 - 1) dz = \int z^2 dz - \int dz = \frac{z^{2+1}}{2 + 1} - z + c = \frac{z^3}{3} - z + c$$

$$I = \frac{z^2.z}{3} - z + c = \frac{(1 - x^2)\sqrt{1 - x^2}}{3} - \sqrt{1 - x^2} + c = \frac{1}{3}(1 - x^2)^{\frac{3}{2}} - \sqrt{1 - x^2} + c \quad \text{Ans.}$$

(c) $I = \int \sin^5 x . \sqrt{\cos x} \, dx$, Put $\cos x = t$ $\quad \therefore \ -\sin x \, dx = dt$

$$I = \int \sin^4 x . \sqrt{\cos x} . \sin x \, dx = \int (1 - \cos^2 x)^2 . \sqrt{\cos x} . \sin x \, dx = -\int (1 - t^2)^2 . \sqrt{t} \, dt = -\int (1 - 2t^2 + t^4) . \sqrt{t} \, dt$$

$$I = -\int \sqrt{t} \, dt + 2 \int t^2 . \sqrt{t} \, dt - \int t^4 . \sqrt{t} \, dt = -\int t^{\frac{1}{2}} dt + 2 \int t^{\frac{5}{2}} dt - \int t^{\frac{9}{2}} dt = -\frac{t^{\frac{3}{2}}}{\frac{3}{2}} + 2.\frac{t^{\frac{7}{2}}}{\frac{7}{2}} - \frac{t^{\frac{11}{2}}}{\frac{11}{2}} + c$$

$$I = -\frac{2}{3} t^{\frac{3}{2}} + \frac{4}{7} t^{\frac{7}{2}} - \frac{2}{11} t^{\frac{11}{2}} + c = -\frac{2}{3}(\cos x)^{\frac{3}{2}} + \frac{4}{7}(\cos x)^{\frac{7}{2}} - \frac{2}{11}(\cos x)^{\frac{11}{2}} + c \quad \text{Ans.}$$

(d) $I = \int \frac{x^2}{(x \sin x + \cos x)^2} dx = \int \frac{x^2}{x \cos x}.\frac{x \cos x}{(x \sin x + \cos x)^2} dx = \int u.v \, dx$ (integrating by part)

where $u = \frac{x^2}{x \cos x}$ $\quad \therefore \ \frac{du}{dx} = \frac{x \cos x.2x - x^2(x. - \sin x + \cos x)}{(x \cos x)^2} = \frac{2x^2 \cos x + x^3 \sin x - x^2 \cos x}{x^2 \cos^2 x}$

or $\frac{du}{dx} = \frac{x^2 \cos x + x^3 \sin x}{x^2 \cos^2 x} = \frac{x^2(\cos x + x \sin x)}{x^2 \cos^2 x} = \frac{(\cos x + x \sin x)}{\cos^2 x}$

Now, $v = \int \frac{x \cos x}{(x \sin x + \cos x)^2} dx$, Put $x \sin x + \cos x = t$ $\therefore (x \cos x + \sin x - \sin x) dx = dt$ or $x \cos x \, dx = dt$

$$v = \int \frac{dt}{t^2} = \int t^{-2} dt = \frac{t^{-2+1}}{-2 + 1} = -\frac{1}{t} + c = -\frac{1}{(x \sin x + \cos x)} + c$$

$$I = \int u.v \; dx = u. \int v \, dx - \int \left[\frac{du}{dx}. \int v . dx \right] dx = \frac{x^2}{x \cos x} . \left[-\frac{1}{(x \sin x + \cos x)} \right] - \int \left[-\frac{1}{(x \sin x + \cos x)} . \frac{(\cos x + x \sin x)}{\cos^2 x} \right] dx$$

$$I = -\frac{x}{\cos x} . \left[\frac{1}{(x \sin x + \cos x)} \right] + \int \sec^2 x \; dx = -\frac{x}{\cos x} . \left[\frac{1}{(x \sin x + \cos x)} \right] + \tan x + c \quad \text{Ans.}$$

(4) (a) $I = \int \sin^{-1} \sqrt{x} \; dx$, Put $\sqrt{x} = t$ $\therefore \frac{1}{2\sqrt{x}} dx = dt$ or $dx = 2t \; dt$

$$I = \int \sin^{-1} t . 2t \; dt = 2 \int t. \sin^{-1} t \; dt \quad [\text{use integrating by part formula}]$$

$$I = 2 \left\{ \sin^{-1} t . \int t \; dt - \int \left[\frac{d(\sin^{-1} t)}{dt} . \int t \; dt \right] dt \right\} = 2 \sin^{-1} t . \frac{t^2}{2} - 2 \int \frac{1}{\sqrt{1-t^2}} . \frac{t^2}{2} . dt = t^2 \sin^{-1} t - \int \frac{t^2}{\sqrt{1-t^2}} \; dt$$

Let $1 - t^2 = z^2$ or $t = \sqrt{1-z^2}$ $\therefore -2t \; dt = 2z \; dz$ or $t \; dt = -z \; dz$

$$I = t^2 \sin^{-1} t + \int \frac{z. \sqrt{1-z^2} \; dz}{z} = t^2 \sin^{-1} t + \int \sqrt{1-z^2} \; dz \quad \text{using formula,} \quad \int \sqrt{a^2 - x^2} \; dx = \frac{x}{2} \sqrt{a^2 - x^2} + \frac{a^2}{2} \sin^{-1} \frac{x}{a}$$

$$I = t^2 \sin^{-1} t + \frac{z}{2} \sqrt{1-z^2} + \frac{1}{2} \sin^{-1} z + c \quad \text{put } t = \sqrt{x} \text{ and } z = \sqrt{1-t^2} = \sqrt{1-x}$$

$$I = x \sin^{-1} \sqrt{x} + \frac{\sqrt{1-x}}{2} \sqrt{1-1+x} + \frac{1}{2} \sin^{-1} \sqrt{1-x} + c = x \sin^{-1} \sqrt{x} + \frac{\sqrt{x}}{2} . \sqrt{1-x} + \frac{1}{2} \sin^{-1} \sqrt{1-x} + c \quad \text{Ans.}$$

(b) $I = \int \cos^{-1} \sqrt{x} \; dx = x \cos^{-1} \sqrt{x} - \frac{\sqrt{x}}{2} . \sqrt{1-x} - \frac{1}{2} \sin^{-1} \sqrt{1-x} + c \quad \text{Ans.}$ (see above question)

(c) $I = \int \tan^{-1} \sqrt{x} \; dx = (x+1) \tan^{-1} \sqrt{x} - \sqrt{x} + c \quad \text{Ans.}$ (Do yourself, see above question)

(d) $I = \int \cot^{-1} \sqrt{x} \; dx = x \cot^{-1} \sqrt{x} - \tan^{-1} \sqrt{x} + \sqrt{x} + c \quad \text{Ans.}$ (Do yourself, see above question)

(5) (a) $I = \int \sqrt{e^{2x} + 1} \; dx = \int \frac{\sqrt{e^{2x} + 1} . e^{2x}}{e^{2x}} dx$, Put $1 + e^{2x} = z^2$ $\therefore 2. e^{2x} \; dx = 2z \; dz$

or $e^{2x} \; dx = z \; dz$ $\therefore 1 + e^{2x} = z^2$ or $e^{2x} = z^2 - 1$

$$I = \int \frac{\sqrt{z^2} . z \; dz}{z^2 - 1} = \int \frac{z^2}{z^2 - 1} dz = \int \left(1 + \frac{1}{z^2 - 1} \right) dz = \int dz + \int \frac{dz}{z^2 - 1}$$

use formula, $\int \frac{dx}{x^2 - a^2} = \frac{1}{2a} \log \left| \frac{x-a}{x+a} \right|$ or $I = z + \frac{1}{2.1} \log \left| \frac{z-1}{z+1} \right| + c = \sqrt{e^{2x} + 1} + \frac{1}{2} \log \left| \frac{\sqrt{e^{2x} + 1} - 1}{\sqrt{e^{2x} + 1} + 1} \right| + c \quad \text{Ans.}$

(b) $I = \int \frac{dx}{(x-4)\sqrt{x-2}}$, Let $x - 2 = t^2$ or $t = \sqrt{x-2}$ or $x = t^2 + 2$ $\therefore dx = 2t \; dt$

$$I = \int \frac{2t}{(t^2 + 2 - 4).t} dt = 2 \int \frac{dt}{t^2 - 2} = 2 \int \frac{dt}{t^2 - (\sqrt{2})^2} \quad \text{use formula,} \quad \int \frac{dx}{x^2 - a^2} = \frac{1}{2a} \log \left| \frac{x-a}{x+a} \right|$$

$$I = 2. \frac{1}{2. \sqrt{2}} \log \left| \frac{t - \sqrt{2}}{t + \sqrt{2}} \right| + c = \frac{1}{\sqrt{2}} \log \left| \frac{\sqrt{x-2} - \sqrt{2}}{\sqrt{x-2} + \sqrt{2}} \right| + c \quad \text{Ans.}$$

(c) $I = \int e^x \log x \; dx$ (use integrating by part formula)

(d) $I = \int x^2 e^x \; dx = \int u. v \; dx$ (use integrating by part formula)

$$I = x^2 . e^x - \int 2x. e^x \; dx \quad \text{Again use integrating by part formula}$$

$I = x^2 . e^x - 2\left[x . e^x - \int e^x \, dx\right] = x^2 . e^x - 2xe^x + 2e^x + c = e^x[x^2 - 2x + 2] + c = e^x[(x-1)^2 + 1] + c$ Ans.

(6) (a) $I = \int x^3 \log x \, dx = \dfrac{x^4 \log x}{4} - \dfrac{x^4}{16} + c$ Ans. [use integrating by part formula]

(b) $I = \int e^x \sin x \, dx = \dfrac{1}{2} e^x[\sin x - \cos x] + c$ Ans. [use integrating by part formula]

(c) $I = \int \dfrac{\cos^{-1} \sqrt{x} - \sin^{-1} \sqrt{x}}{\sin^{-1} \sqrt{x} + \cos^{-1} \sqrt{x}} \, dx = \dfrac{2}{\pi} \int \left(\cos^{-1} \sqrt{x} - \sin^{-1} \sqrt{x}\right) dx$ $\left[\sin^{-1}\sqrt{x} + \cos^{-1}\sqrt{x} = \dfrac{\pi}{2}\right]$

$I = \dfrac{2}{\pi} \int \left(\cos^{-1} \sqrt{x}\right) dx - \dfrac{2}{\pi} \int \left(\sin^{-1} \sqrt{x}\right) dx$ (see question no. $-(4)(a \, \& \, b)$)

$I = \dfrac{2}{\pi}\left[x \cos^{-1}\sqrt{x} - \dfrac{\sqrt{x}}{2} . \sqrt{1-x} - \dfrac{1}{2}\sin^{-1}\sqrt{1-x}\right] - \dfrac{2}{\pi}\left[x \sin^{-1}\sqrt{x} + \dfrac{\sqrt{x}}{2} . \sqrt{1-x} + \dfrac{1}{2}\sin^{-1}\sqrt{1-x}\right] + c$

(d) $I = \int \cos x . \log(\sin x) \, dx = \sin x \, [\log(\sin x) - 1] + c$ Ans. [use integrating by part formula]

(7) (a) $I = \int \log\left(\sqrt{1-x^2}\right) dx = x \log\left(\sqrt{1-x^2}\right) + \dfrac{1}{2}\log\left|\dfrac{1+x}{1-x}\right| - x + c$ Ans.

$\left[\text{use integrating by part formula and } \int \dfrac{dx}{a^2 - x^2} = \dfrac{1}{2a}\log\left|\dfrac{a+x}{a-x}\right|\right]$

(b) $I = \int \sin x . \log(\cot x) \, dx = -\cos x . \log(\cot x) + \log\left[\tan\left(\dfrac{x}{2}\right)\right] + c$ Ans.

$\left[\text{use integrating by part formula and } \int \text{cosec} \, x \, dx = \log\left[\tan\left(\dfrac{x}{2}\right)\right]\right]$

(c) $I = \int \dfrac{x \tan x}{\sec^3 x} \, dx = \int \dfrac{x . \frac{\sin x}{\cos x}}{\frac{1}{\cos^3 x}} \, dx = \int x \sin x . \cos^2 x \, dx$ Let $\cos x = t$ or $x = \cos^{-1} t$ $\therefore -\sin x \, dx = dt$

$I = -\int \cos^{-1} t . t^2 \, dt = -\left[\cos^{-1} t . \dfrac{t^3}{3} + \int \dfrac{1}{\sqrt{1-t^2}} . \dfrac{t^3}{3} . dt\right] = -\dfrac{t^3}{3} . \cos^{-1} t - \dfrac{1}{3}\int \dfrac{t^3}{\sqrt{1-t^2}} \, dt$

Put $1 - t^2 = z^2$ or $t = \sqrt{1-z^2}$ $\therefore -2t \, dt = 2z \, dz$ or $t \, dt = -z \, dz$

$I = -\dfrac{t^3}{3} . \cos^{-1} t + \dfrac{1}{3}\int \dfrac{z . (1-z^2)}{z} \, dz = -\dfrac{t^3}{3} . \cos^{-1} t + \dfrac{1}{3}\int (1-z^2) \, dz = -\dfrac{t^3}{3} . \cos^{-1} t + \dfrac{1}{3}\int dz - \dfrac{1}{3}\int z^2 \, dz$

$I = -\dfrac{t^3}{3} . \cos^{-1} t + \dfrac{1}{3}z - \dfrac{1}{3} . \dfrac{z^3}{3} + c$ Put $t = \cos x$ and $z = \sqrt{1-t^2} = \sqrt{1-\cos^2 x} = \sin x$

$\therefore \quad I = \dfrac{1}{3}\sin x - \dfrac{1}{9}\sin^3 x - \dfrac{\cos^3 x}{3} . \cos^{-1}(\cos x) + c = \dfrac{1}{3}\sin x - \dfrac{1}{9}\sin^3 x - \dfrac{x \cos^3 x}{3} + c$ Ans.

(d) $I = \int x(e^x + 1) \, dx = e^x(x-1) + \dfrac{x^2}{2} + c$ Ans.   [use integrating by part formula]

(8) (a) $I = \int \dfrac{\log\{\log(1+\sqrt{x})\}}{(x+\sqrt{x})} \, dx$ Let $\log(1+\sqrt{x}) = z$ $\therefore \dfrac{1}{1+\sqrt{x}} . \dfrac{1}{2\sqrt{x}} \, dx = dz$ or $\dfrac{1}{(x+\sqrt{x})} \, dx = 2 \, dz$

$I = \int \log z . 2 \, dz = 2\int \log z \, dz = 2\left[z \log z - \int \dfrac{1}{z} . z \, dz\right] = 2\left[z \log z - \int dz\right]$ [use integrating by part formula]

$I = 2z \log z - 2z + c = 2z(\log z - 1) = 2\log(1+\sqrt{x}) . \left[\log\log(1+\sqrt{x}) - 1\right] + c$ Ans.

(b) $I = \int 2^{3 \log_2 \sqrt[3]{x}} \, dx = \int 2^{\log_2 \left(\sqrt[3]{x}\right)^3} \, dx = \int 2^{\log_2 x} \, dx = \int x \, dx = \dfrac{x^2}{2} + c$ Ans. $\left[e^{\log_e x} = x\right]$

(c) $I = \int x^3 \cot^{-1} x \, dx = \dfrac{x^4}{4} \cot^{-1} x + \dfrac{x^3}{12} - \dfrac{x}{4} + \dfrac{1}{4} \tan^{-1} x + c$ Ans. [integrating by part]

(d) $I = \int \dfrac{(2^x - 3^x)^2}{2^x . 3^x} \, dx = \int \dfrac{2^{2x} + 3^{2x} - 2.2^x . 3^x}{2^x . 3^x} \, dx = \int \dfrac{2^{2x}}{2^x . 3^x} \, dx + \int \dfrac{3^{2x}}{2^x . 3^x} \, dx - 2 \int \dfrac{2^x . 3^x}{2^x . 3^x} \, dx = \int \dfrac{2^x}{3^x} \, dx + \int \dfrac{3^x}{2^x} \, dx - 2 \int dx$

$$= \int \left(\dfrac{2}{3}\right)^x dx + \int \left(\dfrac{3}{2}\right)^x dx - 2 \int dx$$

$I = \dfrac{\left(\dfrac{2}{3}\right)^x}{\log\left(\dfrac{2}{3}\right)} + \dfrac{\left(\dfrac{3}{2}\right)^x}{\log\left(\dfrac{3}{2}\right)} - 2x + c$ Ans. $\left[\text{use } \int a^x \, dx = \dfrac{a^x}{\log a} \right]$

(9) (a) $I = \int \dfrac{\sin 2x . \cos x + 2 \cos^2 x}{\left(1 + 2 \sin^x/_2 \cos^x/_2\right)} \, dx = x + \dfrac{\sin 2x}{2} + c$ Ans.   (Do yourself)

(b) $I = \int \dfrac{x^2 + 1}{x(x + 3)(x - 4)} \, dx$,   Let $\dfrac{x^2 + 1}{x(x + 3)(x - 4)} = \dfrac{A}{x} + \dfrac{B}{x + 3} + \dfrac{C}{x - 4}$ .... (i)

$\therefore A|_{x=0} = \dfrac{1}{3.(-4)} = -\dfrac{1}{12}$ and $B|_{x=-3} = \dfrac{10}{21}$ and $C|_{x=4} = \dfrac{17}{28}$

Put value of A, B and C in equation (i) and integrating, we get

$I = \int \dfrac{x^2 + 1}{x(x + 3)(x - 4)} \, dx = -\dfrac{1}{12} \int \dfrac{dx}{x} + \dfrac{10}{21} \int \dfrac{dx}{x + 3} + \dfrac{17}{28} \int \dfrac{dx}{x - 4} = -\dfrac{1}{12} \log x + \dfrac{10}{21} \log(x + 3) + \dfrac{17}{28} \log(x - 4) + c$ Ans.

(c) $I = \int \dfrac{(x^3 + 64)(x - 2)}{x^2 - 4x + 16} \, dx = \int \dfrac{(x^3 + 4^3)(x - 2)}{x^2 - 4x + 16} \, dx$ $[\, a^3 + b^3 = (a + b)(a^2 - ab + b^2) \,]$

$I = \int \dfrac{(x + 4)(x^2 - 4x + 16)(x - 2)}{(x^2 - 4x + 16)} \, dx = \int (x + 4)(x - 2) \, dx = \int (x^2 + 2x - 8) \, dx = \int x^2 \, dx + 2 \int x \, dx - 8 \int dx = \dfrac{x^3}{3} + 2\dfrac{x^2}{2} - 8x + c$

$$= \dfrac{x^3}{3} + x^2 - 8x + c \quad \text{Ans.}$$

(d) $I = \int \dfrac{dx}{\sqrt{x - 1} + \sqrt{x + 2}} = \int \dfrac{dx}{\sqrt{x - 1} + \sqrt{x + 2}} \times \dfrac{\sqrt{x - 1} - \sqrt{x + 2}}{\sqrt{x - 1} - \sqrt{x + 2}} = -\dfrac{1}{3} \int \left(\sqrt{x - 1} - \sqrt{x + 2}\right) dx = \dfrac{2}{9}(x + 2)^{\frac{3}{2}} - \dfrac{2}{9}(x - 1)^{\frac{3}{2}} + c$ Ans.

(e) $I = \int \cos^{-1}\left(\dfrac{2 \cot x}{1 + \cot^2 x}\right) dx = \int \cos^{-1}\left(\dfrac{2 \cot x}{\operatorname{cosec}^2 x}\right) dx = \int \cos^{-1}\left(\dfrac{2 \dfrac{\cos x}{\sin x}}{\dfrac{1}{\sin^2 x}}\right) dx = \int \cos^{-1}(2 \sin x \cos x) \, dx = \int \cos^{-1}(\sin 2x) \, dx$

$$= \int \cos^{-1}\left[\cos\left(\dfrac{\pi}{2} - 2x\right)\right] dx = \int \left(\dfrac{\pi}{2} - 2x\right) dx \quad [\cos^{-1}(\cos \theta) = \theta]$$

$I = \dfrac{\pi}{2} \int dx - 2 \int x \, dx = \dfrac{\pi}{2} x - 2 . \dfrac{x^2}{2} + c = -x^2 + \dfrac{\pi}{2} x + c$ Ans.

(f) $I = \int \dfrac{x}{x^4 + 4} \, dx = \int \dfrac{x}{(x^2)^2 + 2^2} \, dx$,   Let $x^2 = t$ $\therefore$ $2x \, dx = dt$ or $x \, dx = \dfrac{dt}{2}$

$I = \dfrac{1}{2} \int \dfrac{dt}{t^2 + 2^2} = \dfrac{1}{2} . \dfrac{1}{2} \tan^{-1}\left(\dfrac{t}{2}\right) + c = \dfrac{1}{4} \tan^{-1}\left(\dfrac{x^2}{2}\right) + c$ Ans. $\left[\text{formula, } \int \dfrac{dx}{x^2 + a^2} = \dfrac{1}{a} \tan^{-1} \dfrac{x}{a} \right]$

(10) (a) $I = \int \dfrac{dx}{\sqrt{x} - \sqrt{x - 2}} = \int \dfrac{dx}{\sqrt{x} - \sqrt{x - 2}} \times \dfrac{\sqrt{x} + \sqrt{x - 2}}{\sqrt{x} + \sqrt{x - 2}} = \dfrac{1}{2} \int \left(\sqrt{x} + \sqrt{x - 2}\right) dx = \dfrac{1}{2} \int \sqrt{x} \, dx + \dfrac{1}{2} \int \sqrt{x - 2} \, dx$

$$= \dfrac{1}{2} . \dfrac{x^{\frac{3}{2}}}{\dfrac{3}{2}} + \dfrac{1}{2} . \dfrac{(x - 2)^{\frac{3}{2}}}{\dfrac{3}{2}} + c = \dfrac{1}{3}(x)^{\frac{3}{2}} + \dfrac{1}{3}(x - 2)^{\frac{3}{2}} + c \quad \text{Ans.}$$

(b) $I = \int \dfrac{x^8}{\sqrt{1 + x^3}} \, dx$,   Let $1 + x^3 = t^2$ or $x^3 = t^2 - 1$ or $x^6 = (t^2 - 1)^2$ $\therefore$ $3x^2 \, dx = 2t \, dt$ or $x^2 \, dx = \dfrac{2t \, dt}{3}$

$I = \int \dfrac{x^6 . x^2 \, dx}{\sqrt{1 + x^3}} = \dfrac{2}{3}\int \dfrac{(t^2-1)^2 . t}{\sqrt{t^2}} dt = \dfrac{2}{3}\int (t^2-1)^2 \, dt = \dfrac{2}{3}\int (t^4 - 2t^2 + 1)\, dt = \dfrac{2}{3}\int t^4 \, dt - \dfrac{4}{3}\int t^2 \, dt + \dfrac{2}{3}\int dt = \dfrac{2}{3}.\dfrac{t^5}{5} - \dfrac{4}{3}.\dfrac{t^3}{3} + \dfrac{2}{3}t + c$

$\qquad = \dfrac{2}{15}\left(\sqrt{1+x^3}\right)^5 - \dfrac{4}{9}\left(\sqrt{1+x^3}\right)^3 + \dfrac{2}{3}\sqrt{1+x^3} + c \quad$ Ans.

(c) $I = \int \dfrac{x+3}{\sqrt{4x+5}} dx,$ Let $4x + 5 = t^2$ or $t = \sqrt{4x+5}$ or $x = \dfrac{t^2-5}{4}$ $\quad \therefore \ 4\, dx = 2t\, dt$ or $dx = \dfrac{t\, dt}{2}$

$I = \dfrac{1}{2}\int \dfrac{\frac{t^2-5}{4} + 3}{\sqrt{t^2}} . t\, dt = \dfrac{1}{2}\int \dfrac{t^2 - 5 + 12}{4}\, dt = \dfrac{1}{8}\int (t^2 + 7)\, dt = \dfrac{1}{8}\int t^2 \, dt + \dfrac{7}{8}\int dt = \dfrac{1}{8}.\dfrac{t^3}{3} + \dfrac{7}{8}t + c = \dfrac{1}{24}t^3 + \dfrac{7}{8}t + c$

$\qquad = \dfrac{1}{24}\left(\sqrt{4x+5}\right)^3 + \dfrac{7}{8}.\sqrt{4x+5} + c = \dfrac{1}{24}(4x+5)^{\frac{3}{2}} + \dfrac{7}{8}(4x+5)^{\frac{1}{2}} + c \quad$ Ans.

(d) $I = \int \dfrac{x}{2x^2 - 3x + 5} dx = \int \left(\dfrac{\frac{1}{4}.(4x-3) + \frac{3}{4}}{2x^2 - 3x + 5}\right) dx = \dfrac{1}{4}\int \dfrac{4x-3}{2x^2 - 3x + 5}\, dx + \dfrac{3}{4}\int \dfrac{dx}{2x^2 - 3x + 5}$

$I = \dfrac{1}{4}\int \dfrac{f'(x)}{f(x)} dx + \dfrac{3}{4}\int \dfrac{dx}{\left(\sqrt{2}x - \frac{3}{2\sqrt{2}}\right)^2 + \left(\frac{\sqrt{31}}{2\sqrt{2}}\right)^2}$ using formula $\left[\int \dfrac{f'(x)}{f(x)} dx = \log f(x) \text{ and } \int \dfrac{dx}{x^2 + a^2} = \dfrac{1}{a}\tan^{-1}\left(\dfrac{x}{a}\right) . \dfrac{1}{\text{d. c of } x}\right]$

$I = \dfrac{1}{4}\log(2x^2 - 3x + 5) + \dfrac{3}{4}.\dfrac{1}{\frac{\sqrt{31}}{2\sqrt{2}}}\tan^{-1}\left(\dfrac{\sqrt{2}x - \frac{3}{2\sqrt{2}}}{\frac{\sqrt{31}}{2\sqrt{2}}}\right).\dfrac{1}{\sqrt{2}} + c = \dfrac{1}{4}\log(2x^2 - 3x + 5) + \dfrac{3}{2\sqrt{31}}\tan^{-1}\left(\dfrac{4x-3}{\sqrt{31}}\right) + c \quad$ Ans.

(11) (a) $I = \int \dfrac{3x+2}{x^2 + 2x + 3} dx,$ Let $3x + 2 = l(\text{d. c of } x^2 + 2x + 3) + m = l(2x+2) + m \ \ldots\ldots\ldots .\text{(A)}$

or $2l = 3 \quad \therefore \ l = \dfrac{3}{2}$ and $2l + m = 2 \quad \therefore \ m = 2 - 2.\dfrac{3}{2} = 2 - 3 = -1$

Put value of l and m in equation (A), we get $\quad \therefore \ 3x + 2 = \dfrac{3}{2}(2x+2) - 1 = 3(x+1) - 1$

Divide by $x^2 + 2x + 3$ both of sides, we get $\quad \therefore \ \dfrac{3x+2}{x^2 + 2x + 3} = \dfrac{3}{2}.\dfrac{(2x+2)}{x^2 + 2x + 3} - \dfrac{1}{x^2 + 2x + 3}$

Integrating both of sides, $\int \dfrac{3x+2}{x^2 + 2x + 3} dx = \dfrac{3}{2}\int \dfrac{(2x+2)}{x^2 + 2x + 3} dx - \int \dfrac{1}{x^2 + 2x + 3} dx$

$I = \dfrac{3}{2}\int \dfrac{f'(x)}{f(x)} dx - \int \dfrac{dx}{(x+1)^2 + \left(\sqrt{2}\right)^2}$ using $\int \dfrac{f'(x)}{f(x)} dx = \log f(x)$ and $\int \dfrac{dx}{x^2 + a^2} = \dfrac{1}{a}\tan^{-1}\left(\dfrac{x}{a}\right)$

where $f(x) = x^2 + 2x + 3$, $f'(x) = 2x + 2 \quad \therefore \ I = \dfrac{3}{2}\log(x^2 + 2x + 3) - \dfrac{1}{\sqrt{2}}\tan^{-1}\left(\dfrac{x+1}{\sqrt{2}}\right) + c \quad$ Ans.

(b) Do yourself. $\quad I = \int \dfrac{4x+1}{x^2 + 3x + 2} dx = 2\log|x^2 + 3x + 2| - 5\log\left|\dfrac{x+1}{x+2}\right| + c \quad$ Ans.

(c) $I = \int \dfrac{2x-3}{x^2 + 4x - 5} dx = \log|x^2 + 4x - 5| - \dfrac{7}{6}\log\left|\dfrac{x-1}{x+5}\right| + c \quad$ Ans. [Do yourself, see question no. $-$(11) (a)]

(d) $I = \int \dfrac{x^5 + x^2}{x^6 + 16} dx = \int \dfrac{x^5}{x^6 + 16} dx + \int \dfrac{x^2}{x^6 + 16} dx = I_1 + I_2 \ \text{ (say)}$

Now, $I_1 = \int \dfrac{x^5}{x^6 + 16} dx,$ Let $x^6 + 16 = t \quad \therefore \ 6x^5 \, dx = dt$ or $x^5 \, dx = \dfrac{dt}{6} \quad \therefore \ I_1 = \dfrac{1}{6}\int \dfrac{dt}{t} = \dfrac{1}{6}\log|t| + c = \dfrac{1}{6}\log|x^6 + 16| + c$

Now, $I_2 = \int \dfrac{x^2}{x^6 + 16} dx = \int \dfrac{x^2}{(x^3)^2 + 4^2} dx,$ Let $x^3 = z \quad \therefore \ 3x^2 \, dx = dz$ or $x^2 \, dx = \dfrac{dz}{3}$

or $I_2 = \dfrac{1}{3}\displaystyle\int \dfrac{dz}{z^2 + 4^2} = \dfrac{1}{3}\cdot\dfrac{1}{4}\tan^{-1}\left(\dfrac{z}{4}\right) + c = \dfrac{1}{12}\tan^{-1}\left(\dfrac{x^3}{4}\right) + c \qquad \left[\displaystyle\int \dfrac{dx}{x^2 + a^2} = \dfrac{1}{a}\tan^{-1}\left(\dfrac{x}{a}\right)\right]$

Then, $I = I_1 + I_2 = \dfrac{1}{6}\log|x^6 + 16| + \dfrac{1}{12}\tan^{-1}\left(\dfrac{x^3}{4}\right) + k$ \qquad Ans.

(12) (a) $I = \displaystyle\int \dfrac{\sin 2\theta - 5\cos\theta}{7 - \cos^2\theta + 4\sin\theta}\,d\theta$, Let $\sin 2\theta - 5\cos\theta = l(\text{d.c of } 7 - \cos^2\theta + 4\sin\theta) + m$

or $\sin 2\theta - 5\cos\theta = l(2\sin\theta\cos\theta + 4\cos\theta) + m = l(\sin 2\theta + 4\cos\theta) + m\ \dots\dots\dots\dots\dots\dots\dots.. (A)$

$$\therefore\ l = 1 \quad \text{and}\quad 4l + m = -5 \quad \therefore\ m = -5 - 4 = -9$$

Put value of l and m in equation (A), we get \quad or \quad $\sin 2\theta - 5\cos\theta = (\sin 2\theta + 4\cos\theta) - 9$

Divide by $7 - \cos^2\theta + 4\sin\theta$ both of sides and integrating, we get

$$\int \dfrac{\sin 2\theta - 5\cos\theta}{7 - \cos^2\theta + 4\sin\theta}\,d\theta = \int \dfrac{\sin 2\theta + 4\cos\theta}{7 - \cos^2\theta + 4\sin\theta}\,d\theta - 9\int \dfrac{d\theta}{7 - \cos^2\theta + 4\sin\theta} = I_1 - I_2 \quad \text{(say)}$$

Now, $I_1 = \displaystyle\int \dfrac{\sin 2\theta + 4\cos\theta}{7 - \cos^2\theta + 4\sin\theta}\,d\theta$, Let $f(\theta) = 7 - \cos^2\theta + 4\sin\theta$, $f'(\theta) = \sin 2\theta + 4\cos\theta$

$$\therefore\ I_1 = \int \dfrac{f'(\theta)}{f(\theta)}\,d\theta = \log[f(\theta)] + c = \log[7 - \cos^2\theta + 4\sin\theta] + c$$

Now, $I_2 = 9\displaystyle\int \dfrac{d\theta}{7 - \cos^2\theta + 4\sin\theta} = 9\int \dfrac{d\theta}{7 - (1 - \sin^2\theta) + 4\sin\theta} = 9\int \dfrac{d\theta}{\sin^2\theta + 4\sin\theta + 6}$

$I_2 = 9\displaystyle\int \dfrac{d\theta}{(\sin\theta + 2)^2 + \left(\sqrt{2}\right)^2}$ \qquad $\left[\text{formula,}\ \displaystyle\int \dfrac{dx}{(bx)^2 + a^2} = \dfrac{1}{a}\tan^{-1}\left(\dfrac{bx}{a}\right).\dfrac{1}{\text{d.c of } bx}\right]$

$$\therefore\ I_2 = 9.\dfrac{1}{\sqrt{2}}\tan^{-1}\left(\dfrac{\sin\theta + 2}{\sqrt{2}}\right).\dfrac{1}{\text{d.c of }(\sin\theta + 2)} + c = \dfrac{9}{\sqrt{2}}\sec\theta\tan^{-1}\left(\dfrac{\sin\theta + 2}{\sqrt{2}}\right) + c$$

Then, $I = I_1 - I_2 = \log[7 - \cos^2\theta + 4\sin\theta] - \dfrac{9}{\sqrt{2}}\sec\theta\tan^{-1}\left(\dfrac{\sin\theta + 2}{\sqrt{2}}\right) + k$ \quad Ans.

(b) $I = \displaystyle\int \sin(\log x)\,dx + \int \cos(\log x)\,dx = x\sin(\log x) + c$ Ans. [use integrating by part formula]

(c) $I = \displaystyle\int \cos^{-1}\left(\dfrac{2x}{1 + x^2}\right)dx$, Put $x = \tan\theta$ \quad $\therefore\ dx = \sec^2\theta\,d\theta = (1 + \tan^2\theta)\,d\theta$

$I = \displaystyle\int \cos^{-1}\left(\dfrac{2\tan\theta}{1 + \tan^2\theta}\right).\sec^2\theta\,d\theta = \int \cos^{-1}(\sin 2\theta).\sec^2\theta\,d\theta = \int \cos^{-1}\left[\cos\left(\dfrac{\pi}{2} - 2\theta\right)\right].\sec^2\theta\,d\theta = \int \left(\dfrac{\pi}{2} - 2\theta\right).\sec^2\theta\,d\theta$

$$= \int \dfrac{\pi}{2}.\sec^2\theta\,d\theta - 2\int \theta.\sec^2\theta\,d\theta$$

$I = \dfrac{\pi}{2}.\tan\theta - 2\left[\theta.\displaystyle\int \sec^2\theta\,d\theta - \int \dfrac{d(\theta)}{d\theta}.d\theta.\int \sec^2\theta\,d\theta\right] = \dfrac{\pi}{2}.\tan\theta - 2\theta\tan\theta + 2\int \tan\theta\,d\theta = \dfrac{\pi}{2}.\tan\theta - 2\theta\tan\theta - 2\log|\cos\theta| + c$

Put $x = \tan\theta$ \quad $\left[\displaystyle\int \tan\theta\,d\theta = -\log|\cos\theta|\right]$ or $I = \dfrac{\pi}{2}x - 2x\tan^{-1}x - 2\log\left|\dfrac{1}{\sqrt{1 + x^2}}\right| + c$ \quad Ans. \quad $\left[\cos\theta = \dfrac{1}{\sqrt{1 + \tan^2\theta}} = \dfrac{1}{\sqrt{1 + x^2}}\right]$

(13) (a) $I = \displaystyle\int \sin^{-1}\left(\dfrac{x}{\sqrt{1 + x^2}}\right)dx$, Put $x = \tan\theta$ \quad or \quad $\theta = \tan^{-1}x$ \quad $\therefore\ dx = \sec^2\theta\,d\theta$

$I = \displaystyle\int \sin^{-1}\left(\dfrac{\tan\theta}{\sqrt{1 + \tan^2\theta}}\right).\sec^2\theta\,d\theta = \int \sin^{-1}\left(\dfrac{\tan\theta}{\sec\theta}\right).\sec^2\theta\,d\theta = \int \sin^{-1}(\sin\theta).\sec^2\theta\,d\theta = \int \theta.\sec^2\theta\,d\theta$

$I = \theta.\displaystyle\int \sec^2\theta\,d\theta - \int \dfrac{d(\theta)}{d\theta}.d\theta.\int \sec^2\theta\,d\theta = \theta.\tan\theta - \int \tan\theta\,d\theta = \theta\tan\theta + \log|\cos\theta| + c$

Put $\theta = \tan^{-1} x$ then, $I = \tan^{-1} x \cdot \tan(\tan^{-1} x) + \log\left|\dfrac{1}{\sqrt{1+x^2}}\right| + c = x\tan^{-1} x + \log\left|\dfrac{1}{\sqrt{1+x^2}}\right| + c$ Ans.

Put $\cos\theta = \dfrac{1}{\sqrt{1+\tan^2\theta}} = \dfrac{1}{\sqrt{1+x^2}}$ $\left[\displaystyle\int \sec^2\theta \ d\theta = \tan\theta \ \text{and} \ \int \tan\theta \ d\theta = -\log|\cos\theta|\right]$

(b) $I = \displaystyle\int \tan^{-1}\left(\dfrac{x}{\sqrt{1-x^2}}\right) dx = x\sin^{-1} x + \sqrt{1-x^2} + c$ Ans. [Put $x = \sin\theta$ $\therefore$ $dx = \cos\theta \ d\theta$]

(c) $I = \displaystyle\int \dfrac{\cos^{-1} x}{\sqrt{1-x^2}} dx$, Put $\cos^{-1} x = z$ $\therefore$ $-\dfrac{1}{\sqrt{1-x^2}} dx = dz$ $\therefore$ $I = -\displaystyle\int z \ dz = -\dfrac{z^2}{2} + c = -\dfrac{(\cos^{-1} x)^2}{2} + c$ Ans.

(d) $I = \displaystyle\int \dfrac{\cos^{-1} x}{(1-x^2)^{\frac{3}{2}}} dx = \int \dfrac{\cos^{-1} x}{(1-x^2)\cdot\sqrt{1-x^2}} dx$, Put $\cos^{-1} x = z$ or $x = \cos z$ $\therefore$ $-\dfrac{1}{\sqrt{1-x^2}} dx = dz$

$I = -\displaystyle\int \dfrac{z}{1-\cos^2 z} \ dz = -\int \dfrac{z}{\sin^2 z} \ dz = -\int z\cdot\text{cosec}^2 z \ dz$ [use integrating by part formula]

$I = -\left[z\cdot\displaystyle\int \text{cosec}^2 z \ dz - \int \dfrac{d(z)}{dz}\cdot dz\cdot\int \text{cosec}^2 z \ dz\right] = -\left[z\cdot(-\cot z) + \int \cot z \ dz\right] = z\cdot\cot z - \log|\sin z| + c$

$I = z\cdot\dfrac{\cos z}{\sqrt{1-\cos^2 z}} - \log\left|\sqrt{1-\cos^2 z}\right| + c = \cos^{-1} x\cdot\dfrac{x}{\sqrt{1-x^2}} - \log\left|\sqrt{1-x^2}\right| + c$ Ans.

(14) (a) $I = \displaystyle\int \sqrt{\sec\theta - 1} \ d\theta = -\log\left|\dfrac{1}{2} + \cos\theta + \sqrt{\cos\theta + \cos^2\theta}\right| + c$ Ans.

$\left[\text{Do yourself, use formula } \displaystyle\int \dfrac{dx}{\sqrt{x^2 - a^2}} = \log\left|x + \sqrt{x^2 - a^2}\right|\right]$

(b) $I = \displaystyle\int \dfrac{x - \sqrt[3]{x^2}}{x\left(\sqrt[4]{x} + \sqrt[6]{x^2}\right)} dx$, Integral of the form $\displaystyle\int R\left(x, x^{\frac{2}{3}}, x^{\frac{1}{4}}, x^{\frac{1}{3}}\right) dx$

Put $x = t^\alpha$ where $\alpha = $ L. C. M of $(3,4,3) = 12$ or $x = t^{12}$ $\therefore$ $dx = 12t^{11} \ dt$

Ans: $-\dfrac{12}{5}(x)^{\frac{5}{12}} + \dfrac{12}{7}(x)^{\frac{7}{12}} - 2(x)^{\frac{1}{2}} - \dfrac{3}{2}(x)^{\frac{2}{3}} + c$

(c) $I = \displaystyle\int e^{3x}(3x^2 + 4x + 1) \ dx = e^{3x}\left[x^2 + \dfrac{2}{3}x + \dfrac{7}{9}\right] + c$ Ans. [use integrating by part formula]

(d) $I = \displaystyle\int \dfrac{x^2 - 9}{x + 2} dx = \dfrac{x^2}{2} - 2x - 5\log|x + 2| + c$ Ans.

(15) (a) $I = \displaystyle\int e^{\sin 2x}(1 + \cos 2x) \ dx = \dfrac{1}{2}e^{\sin 2x}(1 + \sec 2x) + c$ Ans.

(b) $I = \displaystyle\int \sin^2 x \cot^2 x \ dx = \dfrac{x}{2} + \dfrac{\sin 2x}{4} + c$ Ans.   (c) $I = \displaystyle\int \dfrac{\cot x \ \text{cosec} \ x}{3 + \text{cosec} \ x} dx = -\log|3 + \text{cosec} \ x| + c$ Ans.

(d) $I = \displaystyle\int \tan^4 x \ dx = \dfrac{\tan^3 x}{3} - \tan x + x + c$ Ans.

(16) (a) $I = \displaystyle\int (\sin^4 x + \cos^4 x) \ dx = \dfrac{3x}{4} + \dfrac{\sin 4x}{16} + c$ Ans.

(b) $I = \displaystyle\int \dfrac{dx}{1 - \sqrt{1+x^2}}$ Integral of the form $\displaystyle\int R\left(x, (1+x^2)^{\frac{1}{2}}\right)$, Put $\sqrt{1+x^2} = t - x\sqrt{a}$ $\therefore$ $a = 1$

or $\sqrt{1+x^2} = t - x$ $\therefore$ $\dfrac{1}{2\sqrt{1+x^2}}\cdot(2x) \ dx = dt - 1$ or $\left(\dfrac{x}{\sqrt{1-x^2}} + 1\right) dx = dt$

(c) $I = \displaystyle\int \dfrac{\left(\sin x - \sqrt[3]{\sin^2 x}\right)\cdot\cos x}{\sqrt{\sin^3 x} + \sqrt[4]{\sin^6 x}} dx$, Put $\sin x = z$ $\therefore$ $\cos x \ dx = dz$

$I = \int \dfrac{(z - \sqrt[3]{z^2})}{\sqrt{z^3} + \sqrt[4]{z^6}}\, dz$ Integral of the form $\int R\left(z, z^{\frac{2}{3}}, z^{\frac{3}{2}}, z^{\frac{6}{4}}\right) dz$,   Put $z = t^\alpha$ where $\alpha = $ L.C.M of $(3,2,4) = 12$

or $z = t^{12}$ $\therefore$ $dz = 12t^{11}\, dt$ then, $I = \int \dfrac{(t^{12} - t^8)}{(t^{18} + t^{18})} \cdot 12t^{11}\, dt = 6 \int \dfrac{(t^{12} - t^8)}{t^7} \cdot dt$

$I = 6 \int t^5\, dt - 6 \int t\, dt = 6 \cdot \dfrac{t^6}{6} - 6 \cdot \dfrac{t^2}{2} + c = t^6 - 3t^2 + c = \sqrt{z} - 3\sqrt[6]{z} + c = \sqrt{\sin x} - 3 \cdot \sqrt[6]{\sin x} + c$ Ans.

(d) $I = \int \dfrac{x - 1}{(x - 2)\sqrt{x + 3}}\, dx$,   Put $x + 3 = z^2$ or $x = z^2 - 3$ $\therefore$ $dx = 2z\, dz$

$I = \int \dfrac{z^2 - 3 - 1}{(z^2 - 3 - 2) \cdot \sqrt{z^2}} \cdot 2z\, dz = 2 \int \dfrac{z^2 - 4}{z^2 - 5}\, dz = 2 \int \left(1 + \dfrac{1}{z^2 - 5}\right) dz = 2 \int dz + 2 \int \dfrac{1}{z^2 - 5}\, dz = 2z + 2 \cdot \dfrac{1}{2 \cdot \sqrt{5}} \log\left|\dfrac{z - \sqrt{5}}{z + \sqrt{5}}\right| + c$

put $z = \sqrt{x + 3}$ use formula, $\int \dfrac{dx}{x^2 - a^2} = \dfrac{1}{2a} \log\left|\dfrac{x - a}{x + a}\right|$ or $I = 2\sqrt{x + 3} + \dfrac{1}{\sqrt{5}} \log\left|\dfrac{\sqrt{x + 3} - \sqrt{5}}{\sqrt{x + 3} + \sqrt{5}}\right| + c$ Ans.

(17) (a) $I = \int \sin x \cdot \sin 3x\, dx = \dfrac{1}{2} \int 2 \sin x \cdot \sin 3x\, dx$,   using formula, $2 \sin A \sin B = \cos(A - B) - \cos(A + B)$

$I = \dfrac{1}{2} \int [\cos(-2x) - \cos 4x]\, dx = \dfrac{1}{2} \int \cos 2x\, dx - \dfrac{1}{2} \int \cos 4x\, dx = \dfrac{1}{2} \cdot \dfrac{\sin 2x}{2} - \dfrac{1}{2} \cdot \dfrac{\sin 4x}{4} + c$

$$\therefore\ I = \dfrac{\sin 2x}{4} - \dfrac{\sin 4x}{8} + c \quad \text{Ans.} \quad [\cos(-\theta) = \cos\theta]$$

(b) $I = \int \cos x \cdot \cos 3x \cdot \cos 5x\, dx = \dfrac{1}{2} \int (2 \cos x \cdot \cos 3x) \cdot \cos 5x\, dx$ using formula $[\, 2 \cos A \cdot \cos B = \cos(A + B) + \cos(A - B)\,]$

$I = \dfrac{1}{2} \int [\cos 4x + \cos(-2x)] \cdot \cos 5x\, dx = \dfrac{1}{2} \int \cos 4x \cos 5x\, dx + \dfrac{1}{2} \int \cos 2x \cos 5x\, dx$

$\qquad = \dfrac{1}{4} \int 2 \cos 4x \cos 5x\, dx + \dfrac{1}{4} \int 2 \cos 2x \cos 5x\, dx$ Again use above formula

$I = \dfrac{1}{4} \int [\cos 9x + \cos x]\, dx + \dfrac{1}{4} \int [\cos 7x + \cos 3x]\, dx = \dfrac{1}{4} \int \cos 9x\, dx + \dfrac{1}{4} \int \cos x\, dx + \dfrac{1}{4} \int \cos 7x\, dx + \dfrac{1}{4} \int \cos 3x\, dx$

$\qquad = \dfrac{1}{4} \cdot \dfrac{\sin 9x}{9} + \dfrac{1}{4} \cdot \sin x + \dfrac{1}{4} \cdot \dfrac{\sin 7x}{7} + \dfrac{1}{4} \cdot \dfrac{\sin 3x}{3} + c = \dfrac{\sin 9x}{36} + \dfrac{\sin x}{4} + \dfrac{\sin 7x}{28} + \dfrac{\sin 3x}{12} + c$ Ans.

(c) $I = \int \tan 2x \cdot \tan 3x \cdot \tan 5x\, dx$,   Let $\tan 5x = \tan(3x + 2x) = \dfrac{\tan 3x + \tan 2x}{1 - \tan 3x \tan 2x}$

or $\tan 5x\,(1 - \tan 3x \tan 2x) = \tan 3x + \tan 2x$ or $\tan 5x - \tan 2x \cdot \tan 3x \cdot \tan 5x = \tan 3x + \tan 2x$

or $\tan 2x \cdot \tan 3x \cdot \tan 5x = \tan 5x - \tan 3x - \tan 2x$ Integrating both of sides

$\int \tan 2x \cdot \tan 3x \cdot \tan 5x\, dx = \int (\tan 5x - \tan 3x - \tan 2x)\, dx = \int \tan 5x\, dx - \int \tan 3x\, dx - \int \tan 2x\, dx$

$I = -\dfrac{\log(\cos 5x)}{5} + \dfrac{\log(\cos 3x)}{3} + \dfrac{\log(\cos 2x)}{2} + c$ $\left[\int \tan x\, dx = -\log \cos x = \log \sec x \right]$

$I = \dfrac{\log(\cos 3x)}{3} + \dfrac{\log(\cos 2x)}{2} - \dfrac{\log(\cos 5x)}{5} + c$ Ans. or $I = \dfrac{\log(\sec 5x)}{5} - \dfrac{\log(\sec 3x)}{3} - \dfrac{\log(\sec 2x)}{2} + c$ Ans.

(d) $I = \int \sin x \cdot \sin 3x \cdot \sin 5x \cdot \sin 7x\, dx$ use formula, $2 \sin A \sin B = \cos(A - B) - \cos(A + B)$

$I = \dfrac{1}{8}\left[x + \dfrac{\sin 4x}{4} - \dfrac{\sin 14x}{14} - \dfrac{\sin 10x}{10} - \dfrac{\sin 6x}{6} - \dfrac{\sin 2x}{2} + \dfrac{\sin 16x}{16} + \dfrac{\sin 8x}{8}\right] + c$ Ans.

(e) $I = \int \sin x \cdot \cos 3x\, dx$ use formula, $2 \sin A \cos B = \sin(A + B) + \sin(A - B)$

$I = \dfrac{1}{2} \displaystyle\int 2\sin x . \cos 3x \ dx = \dfrac{1}{2} \int [\sin 4x + \sin(-2x)] \ dx = \dfrac{1}{2} \int \sin 4x \ dx - \dfrac{1}{2} \int \sin 2x \ dx \quad [\sin(-\theta) = -\sin\theta]$

$I = -\dfrac{1}{2} . \dfrac{\cos 4x}{4} + \dfrac{1}{2} . \dfrac{\cos 2x}{2} + c = \dfrac{\cos 2x}{4} - \dfrac{\cos 4x}{8} + c \quad$ Ans.

(18) (a) $I = \displaystyle\int \cos 2x . \sin 3x \ dx \qquad$ use formula, $\ 2\cos A \sin B = \sin(A+B) - \sin(A-B)$

$I = \dfrac{1}{2} \displaystyle\int 2\cos 2x . \sin 3x \ dx = \dfrac{1}{2} \int [\sin 5x + \sin x] \ dx = \dfrac{1}{2} \int \sin 5x \ dx + \dfrac{1}{2} \int \sin x \ dx \quad [\sin(-\theta) = \sin\theta]$

$I = -\dfrac{1}{2} . \dfrac{\cos 5x}{5} - \dfrac{1}{2}\cos x + c = -\left(\cos x + \dfrac{\cos 5x}{5}\right) + c \quad$ Ans.

(b) $I = \displaystyle\int \sin x . \sin 2x . \cos x . \cos 2x \ dx = \dfrac{\sin^3 2x}{12} + c \quad$ Ans.

(c) $I = \displaystyle\int \dfrac{\sin 2x}{\cos(x+\theta) . \cos(x-\theta)} \ dx = \int \dfrac{\sin[(x+\theta)+(x-\theta)]}{\cos(x+\theta) . \cos(x-\theta)} \ dx = \int \dfrac{\sin(x+\theta) . \cos(x-\theta) + \cos(x+\theta) . \sin(x-\theta)}{\cos(x+\theta) . \cos(x-\theta)} \ dx$

$I = \displaystyle\int \dfrac{\sin(x+\theta) . \cos(x-\theta)}{\cos(x+\theta) . \cos(x-\theta)} \ dx + \int \dfrac{\cos(x+\theta) . \sin(x-\theta)}{\cos(x+\theta) . \cos(x-\theta)} \ dx = \int \dfrac{\sin(x+\theta)}{\cos(x+\theta)} \ dx + \int \dfrac{\sin(x-\theta)}{\cos(x-\theta)} \ dx = \int \tan(x+\theta) \ dx + \int \tan(x-\theta) \ dx$
$$= -\log[\cos(x+\theta)] - \log[\cos(x-\theta)] + c = -\{\log[\cos(x+\theta)] + \log[\cos(x-\theta)]\} + c$$
$$= -\{\log[\cos(x+\theta) . \cos(x-\theta)]\} + c \quad \text{Ans.}$$

(d) $I = \displaystyle\int \dfrac{dx}{\sin x . \sin(x+\theta)} = \dfrac{1}{\sin\theta} \int \dfrac{\sin(x+\theta-x)}{\sin x . \sin(x+\theta)} \ dx = \dfrac{1}{\sin\theta} \int \dfrac{\sin(x+\theta) . \cos x - \cos(x+\theta) . \sin x}{\sin x . \sin(x+\theta)} \ dx$

$I = \dfrac{1}{\sin\theta} \displaystyle\int \dfrac{\sin(x+\theta) . \cos x}{\sin x . \sin(x+\theta)} \ dx - \dfrac{1}{\sin\theta} \int \dfrac{\cos(x+\theta) . \sin x}{\sin x . \sin(x+\theta)} \ dx = \dfrac{1}{\sin\theta} \int \dfrac{\cos x}{\sin x} \ dx - \dfrac{1}{\sin\theta} \int \dfrac{\cos(x+\theta)}{\sin(x+\theta)} \ dx$
$$= \dfrac{1}{\sin\theta} \int \cot x \ dx - \dfrac{1}{\sin\theta} \int \cot(x+\theta) \ dx = \dfrac{1}{\sin\theta} \{\log(\sin x) - \log[\sin(x+\theta)]\} + c = \dfrac{1}{\sin\theta} \left\{\log\left[\dfrac{\sin x}{\sin(x+\theta)}\right]\right\} + c$$
$$= \dfrac{1}{\sin\theta} . \log\left[\dfrac{\sin x}{\sin(x+\theta)}\right] + c \quad \text{Ans.}$$

Definite Integral

Definition: $-$ If $f(x)$ be a single valued continuous function in the interval (a, b), where $b > a$ and if the interval (a, b) be divided

into n equal part of length h by the points $\quad a + h, a + 2h, a + 3h, \dots\dots\dots\dots\dots\dots, a + (\overline{n-1})h$

so that, $\ a + nh = b \ $ or $\ nh = b - a \ $ then sum of limit: $-$

$$\lim_{h\to 0} h[f(a) + f(a+h) + f(a+2h) + f(a+3h) + \cdots \dots\dots. + f(a + (\overline{n-1})h)]$$

written, $\ \displaystyle\lim_{h\to 0} h \sum_{r=0}^{n-1} f(a+rh) \ $ or $\ \lim_{n\to\infty} h \sum_{r=0}^{n-1} f(a+rh) \ $ where $h = \dfrac{b-a}{n} \ $ and when $h \to 0, n \to \infty$

$f(x)$ is defined is a functions whose differentiation is $f(x)$ or $F(x)$ is the integral of $f(x)$ with respect to x.

Summation of series: $-$ (i) $\displaystyle\sum_{r=1}^{n} r = 1 + 2 + 3 + \cdots \dots\dots\dots\dots\dots.. + n = \dfrac{n(n+1)}{2}$

(ii) $\displaystyle\sum_{r=1}^{n} r^2 = 1 + 4 + 9 + \cdots \dots\dots\dots\dots\dots\dots. + n^2 = \dfrac{n(n+1)(2n+1)}{6}$

(iii) $\displaystyle\sum_{r=1}^{n} r^3 = 1 + 2^3 + 3^3 + \cdots \dots\dots\dots\dots\dots.. + n^3 = \left[\dfrac{n(n+1)}{2}\right]^2 = \dfrac{n^2(n+1)^2}{4}$

(iv) $\sum 1 = n$

(v) $\displaystyle\sum_{r=1}^{n} x^r = x + x^2 + x^3 + \cdots \ldots \ldots \ldots \ldots \ldots \ldots + x^n = \text{sum of a G. P} = \dfrac{x(x^n - 1)}{x - 1}$

(vi) $\displaystyle\sum_{r=1}^{n} ax^{r-1} = a + ax + ax^2 + \cdots \ldots \ldots \ldots \ldots \ldots + ax^{n-1} = \dfrac{a(x^n - 1)}{x - 1}$

(vii) we know that $\displaystyle\int_a^b f(x)\, dx = \lim_{n\to\infty} h \sum_{r=1}^{n} f(a + rh)$, where $nh = b - a$ or $h = \dfrac{b-a}{n}$

Now, Put $a = 0, b = 1, nh = 1 - 0 = 1$ $\therefore$ $h = \dfrac{1}{n}$

$\displaystyle\int_0^1 f(x)\, dx = \lim_{n\to\infty} \frac{1}{n} \sum f\left(\frac{r}{n}\right)$ Replace $\dfrac{r}{n}$ by x and $\dfrac{1}{n}$ by dx and the limit of the sum is $\displaystyle\int_0^1 f(x)\, dx$

Example: $-$ $I = \displaystyle\int_a^b x^2\, dx = \frac{1}{3}(b^3 - a^3)$

Solution: $-$ $\displaystyle\int_a^b x^2\, dx$, Here $f(x) = x^2, f(a) = a^2, f(a + h) = (a + h)^2, f(a + 2h) = (a + 2h)^2$ etc.

Now, $\displaystyle\int_a^b f(x)\, dx = \lim_{n\to 0} \sum_{r=0}^{n-1} h.f(a + rh) = \lim_{n\to\infty} \sum_{r=0}^{n-1} h.f(a + rh)$ where $nh = b - a$

$I = \displaystyle\int_a^b x^2\, dx = \lim_{h\to\infty} h\big[f(a) + f(a + h) + f(a + 2h) + f(a + 3h) + \cdots \ldots \ldots + f(a + (\overline{n-1})h)\big]$

$\qquad\qquad = \lim_{h\to\infty} h\left[a^2 + (a + h)^2 + (a + 2h)^2 + \cdots \ldots \ldots + \left[a + (\overline{n-1})h\right]^2\right]$

Grouping the terms of $a^2, 2a^2h$ and h^2 to the above, we get

$I = \lim_{n\to\infty} h\left[a^2 \sum 1 + 2ah.\{1 + 2 + 3 + \cdots \ldots \ldots + (n - 1)\}\right] + h^2[1^2 + 2^2 + \cdots \ldots \ldots + (n - 1)^2]$

Now, $\displaystyle\sum 1 = n$, $\displaystyle\sum_{r=1}^{n} r = \frac{n(n + 1)}{2}$ and $\displaystyle\sum_{r=1}^{n} r^2 = \frac{n(n + 1)(2n + 1)}{6}$

Replacing n by $n - 1$ in the above $\displaystyle\sum_{r=1}^{n-1} r = \frac{n(n - 1)}{2}$ and $\displaystyle\sum_{r=1}^{n-1} r^2 = \frac{n(n - 1)(2n - 1)}{6}$

$I = \lim_{n\to\infty} h\left\{a^2.n + 2ah.\frac{(n - 1).n}{2} + h^2.\frac{(n - 1)n(2n - 1)}{6}\right\}$

we have to take the limit $n \to \infty$ or $h \to 0$ and $nh = b - a$

$I = \lim_{n\to\infty} \left\{a^2(nh) + a(nh)^2\left(1 - \frac{1}{n}\right) + \frac{1}{6}(nh)^3\left(1 - \frac{1}{n}\right)\left(2 - \frac{1}{n}\right)\right\}$ when $n \to \infty$, $\frac{1}{n} \to 0$ and $nh = b - a$

$I = \left\{a^2(b - a) + a(b - a)^2.1 + \frac{1}{6}(b - a)^3.1.2\right\} = (b - a)\left[a^2 + a(b - a) + \frac{1}{3}(b - a)^2\right] = \frac{(b - a)}{3}[3a^2 + 3(ab - a^2) + (b^2 - 2ab + a^2)]$

$\qquad\qquad = \frac{(b - a)}{3}[b^2 + ab + a^2] = \frac{1}{3}(b^3 - a^3)$ Proved.

Note: $-$ $\displaystyle\int_a^b f(x)\, dx = \lim_{n\to\infty} \sum_{r=1}^{n} f(a + rh)$, where $nh = b - a$ or $h = \dfrac{b-a}{n}$

Exercise – A9

(1) (a) $\lim_{n\to\infty}\left[\dfrac{1}{n+1}+\dfrac{1}{n+2}+\dfrac{1}{n+3}+\cdots\ldots\ldots\ldots+\dfrac{1}{2n}\right]$ (b) $\lim_{n\to\infty}\left[\dfrac{1}{n+1}+\dfrac{1}{n+2}+\dfrac{1}{n+3}+\cdots\ldots\ldots\ldots+\dfrac{1}{8n}\right]$

(c) $\lim_{n\to\infty}\left[\dfrac{1}{\sqrt{1+n^2}}+\dfrac{1}{\sqrt{4+n^2}}+\dfrac{1}{\sqrt{9+n^2}}+\cdots\ldots\ldots\ldots+\dfrac{1}{\sqrt{49+n^2}}\right]$

(2) (a) $\lim_{n\to\infty}\dfrac{1}{n}\sum_{r=1}^{n}\cos\left(\dfrac{r\pi}{n}\right)$ (b) $\lim_{n\to\infty}\dfrac{1}{n}\sum_{r=1}^{2n}\cos^2\left(\dfrac{r\pi}{2n}\right)$ (c) $\lim_{n\to\infty}\dfrac{1}{n}\left[\sec^2\left(\dfrac{\pi}{3n}\right)+\sec^2\left(\dfrac{2\pi}{3n}\right)+\cdots\ldots\ldots+\sec^2\left(\dfrac{n\pi}{3n}\right)\right]$

(3) (a) If $S_n=1+\dfrac{n-1}{n+1}+\dfrac{n-2}{n+2}+\cdots\ldots\ldots\ldots+0$ then find $\lim_{n\to\infty}\dfrac{S_n}{n}$.

(b) If $S_n=\left[\left(1+\dfrac{1}{n}\right)\left(1+\dfrac{2}{n}\right)\left(1+\dfrac{3}{n}\right)\ldots\ldots\ldots\ldots\left(1+\dfrac{n}{n}\right)\right]^{\frac{1}{n}}$ then find $\lim_{n\to\infty}S_n$.

(4) (a) $\lim_{n\to\infty}\dfrac{1}{n}\sum_{r=1}^{n}\dfrac{r}{\sqrt{n^2+r^2}}$ (b) $\lim_{n\to\infty}\dfrac{1}{n^{m+1}}[1^m+2^m+3^m+\cdots\ldots\ldots\ldots+n^m]\,,m>-1$ (c) $\lim_{n\to\infty}\left[\dfrac{n!}{n^n}\right]^{\frac{1}{n}}$ (d) $\lim_{n\to\infty}\left[\dfrac{(2n)!}{n^n\cdot n!}\right]^{\frac{1}{n}}$

(5) (a) $\lim_{n\to\infty}\dfrac{1}{n}\left[e^{\left(1+\frac{1}{n}\right)+\left(1+\frac{2}{n}\right)+\left(1+\frac{3}{n}\right)+\cdots\ldots\ldots\ldots+\left(1+\frac{n}{n}\right)}\right]$ (b) $\lim_{n\to\infty}\dfrac{1}{n}\left[\sin\left(\dfrac{n+1}{n}\right)+\sin\left(\dfrac{n+2}{n}\right)+\cdots\ldots\ldots+\sin\left(\dfrac{n+n}{n}\right)\right]$

(6) use $\displaystyle\int_a^b f(x)\,dx=\lim_{n\to\infty}\sum_{r=1}^{n}f(a+rh)$, where $nh=b-a$ and find the following integrals: –

(a) $\displaystyle\int_1^2 x^3\,dx=\dfrac{15}{4}$ (b) $\displaystyle\int_a^b\sin x\,dx=\cos a-\cos b$ (c) $\displaystyle\int_a^b e^{2x}\,dx=\dfrac{e^{2b}-e^{2a}}{2}$ (d) $\displaystyle\int_a^b\dfrac{1}{x}\,dx=\log\left(\dfrac{b}{a}\right)$

(7) (a) $\displaystyle\int_0^3(x-3)\,dx=-4$ (b) $\displaystyle\int_1^3(2x^2-3x)\,dx=\dfrac{16}{3}$ (c) $\displaystyle\int_0^2(x^3+3)\,dx=10$ (d) $\displaystyle\int_0^1 x\,dx$ (e) $\displaystyle\int_0^{\frac{\pi}{2}}\cos x\,dx=1$

(8) (a) $\displaystyle\int_a^b\cos^2 x\,dx=\dfrac{1}{2}\left[(b-a)+\dfrac{\sin 2b-\sin 2a}{2}\right]$ (b) $\displaystyle\int_0^1\sqrt{x}\,dx$ (c) $\displaystyle\int_2^3 e^{x+1}\,dx=e^4-e^3$ (d) $\displaystyle\int_1^2\dfrac{2}{\sqrt{x}}\,dx=4(\sqrt{2}-1)$

Answer

(1) (a) $\lim_{n\to\infty}\left[\dfrac{1}{n+1}+\dfrac{1}{n+2}+\dfrac{1}{n+3}+\cdots\ldots\ldots\ldots+\dfrac{1}{2n}\right]=\lim_{n\to\infty}\dfrac{1}{n}\left[\dfrac{1}{1+\frac{1}{n}}+\dfrac{1}{1+\frac{2}{n}}+\cdots\ldots\ldots+\dfrac{1}{1+\frac{n}{n}}\right]=\lim_{n\to\infty}\dfrac{1}{n}\sum_{r=1}^{n}\dfrac{1}{\left(1+\frac{r}{n}\right)}$ standard form

Put $\dfrac{r}{n}=x$ and $\dfrac{1}{n}=dx$ when $r=1,x=\dfrac{1}{n}\to 0$ when $r=n,x=\dfrac{n}{n}=1$ as $n\to\infty$

$I=\displaystyle\int_0^1\dfrac{1}{1+x}\,dx=[\log(1+x)]_0^1=\log(1+1)-\log 1=\log 2-0=\log 2$ Ans.

(b) Do yourself. $\lim_{n\to\infty}\left[\dfrac{1}{n+1}+\dfrac{1}{n+2}+\dfrac{1}{n+3}+\cdots\ldots\ldots\ldots+\dfrac{1}{8n}\right]=\log 8=3\log 2$ Ans.   (solve same as above question)

(c) Do yourself. $\lim_{n\to\infty}\left[\dfrac{1}{\sqrt{1+n^2}}+\dfrac{1}{\sqrt{4+n^2}}+\dfrac{1}{\sqrt{9+n^2}}+\cdots\ldots\ldots\ldots+\dfrac{1}{\sqrt{49+n^2}}\right]=\log\left(1+\sqrt{2}\right)$ or $\sinh^{-1}(1)$ Ans.

(2) (a) $\lim_{n\to\infty}\dfrac{1}{n}\sum_{r=1}^{n}\cos\left(\dfrac{r\pi}{n}\right)=0$ Ans.   (Do yourself)

(b) $\lim_{n\to\infty}\dfrac{1}{n}\sum_{r=1}^{2n}\cos^2\left(\dfrac{r\pi}{2n}\right)=\dfrac{\pi^2}{4}$ Ans. (c) $\lim_{n\to\infty}\dfrac{1}{n}\left[\sec^2\left(\dfrac{\pi}{3n}\right)+\sec^2\left(\dfrac{2\pi}{3n}\right)+\cdots\ldots\ldots+\sec^2\left(\dfrac{n\pi}{3n}\right)\right]=\dfrac{3\sqrt{3}}{\pi}$ Ans.

(3) (a) $2\log 2-1$ or $\log 4-1$ Ans.   (b) $A=\dfrac{4}{e}$ Let $A=S_n$

---

(4) (a) $\sqrt{2} - 1$ Ans. (b) $\dfrac{1}{m+1}$ Ans. (c) $\dfrac{1}{e}$ Ans. (d) $\dfrac{4}{e}$ Ans.

(5) (a) $e^2 - e$ or $e(e-1)$ Ans. (b) $\cos 1 - \cos 2$ Ans.

Properties of Definite Integrals: − Property I: − $\displaystyle\int_a^b f(x)\,dx = \int_a^b f(t)\,dt$ change of variable does not make any difference

Proof: − L.H.S $= \displaystyle\int_a^b f(x)\,dx$, Put $x = t$ ∴ $dx = dt$ or $x = a, x = b$ then $t = a, t = b$

L.H.S $= \displaystyle\int_a^b f(x)\,dx = \int_a^b f(t)\,dt = $ R.H.S Proved

Property II: − $\displaystyle\int_a^b f(x)\,dx = -\int_a^b f(x)\,dx$ Interchanging the limit to change of sign.

Property III: − $\displaystyle\int_a^b f(x)\,dx = \int_a^c f(x)\,dx + \int_c^b f(x)\,dx$, If $\displaystyle\int f(x)\,dx = F(x) + c$

Proof: − L.H.S $= \displaystyle\int_a^b f(x)\,dx = [F(x) + c]_a^b = F(b) - F(a) + c - c = F(b) - F(a)$

R.H.S $= \displaystyle\int_a^c f(x)\,dx + \int_c^b f(x)\,dx = [F(x) + c]_a^c + [F(x) + c]_c^b = F(c) + c - F(a) - c + F(b) + c - F(c) - c = F(b) - F(a)$

∴ L.H.S $= F(b) - F(a) = $ R.H.S Proved.

Property IV: − $\displaystyle\int_0^a f(x)\,dx = \int_0^a f(a - x)\,dx$ $\left\{\displaystyle\int_0^a f(x)\,dx,\ \text{Put } t = a + 0 - x \text{ or } t = a - x \text{ or } x = a - t\right\}$

Proof: − L.H.S $= \displaystyle\int_0^a f(x)\,dx$, Put $x = a - t$ ∴ $dx = -dt$

If $x = 0$ then $x = a - t$ or $t = a - x = a$ and If $x = a$ then $t = a - x = a - a = 0$

∴ $\displaystyle\int_0^a f(x)\,dx = \int_a^0 f(a - t)(-dt) = -\int_a^0 f(a - t)\,dt = \int_0^a f(a - t)\,dt$ [form property I]

$\displaystyle\int_0^a f(a - t)\,dt = \int_0^a f(a - x)\,dx$ Proved. [from property I]

Example: − (a) If $f(x) = \displaystyle\int_0^x \sin^2 t\,dt$ then find $f(x + \pi)$.

Solution: − $f(x) = \displaystyle\int_0^x \sin^2 t\,dt$

$f(x + \pi) = \displaystyle\int_0^{x+\pi} \sin^2 t\,dt = \int_0^{\pi} \sin^2 t\,dt + \int_{\pi}^{x+\pi} \sin^2 t\,dt = f(\pi) + I_2$ by prop. I

or $I_2 = \displaystyle\int_{\pi}^{x+\pi} \sin^2 t\,dt$, Put $t = \theta + \pi$ or $t = \pi,\ \theta = 0$ and $t = x + \pi,\ \theta = x$

∴ $I_2 = \displaystyle\int_{\pi}^{x+\pi} \sin^2 t\,dt = \int_0^x \sin^2 \theta\,d\theta = f(x)$ by property I.

∴ $f(x + \pi) = f(\pi) + I_2 = f(\pi) + f(x)$ Ans.

(b) $I = \displaystyle\int_0^{\frac{\pi}{2}} \dfrac{\cos x}{\sin x + \cos x}\,dx = \int_0^{\frac{\pi}{2}} \dfrac{\cos\left(\frac{\pi}{2} - x\right)}{\sin\left(\frac{\pi}{2} - x\right) + \cos\left(\frac{\pi}{2} - x\right)}\,dx = \int_0^{\frac{\pi}{2}} \dfrac{\sin x}{\sin x + \cos x}\,dx = I$ (say) by property IV.

Adding, $2I = \int_0^{\frac{\pi}{2}} \left(\dfrac{\cos x}{\sin x + \cos x} + \dfrac{\sin x}{\sin x + \cos x} \right) dx = \int_0^{\frac{\pi}{2}} \left(\dfrac{\sin x + \cos x}{\sin x + \cos x} \right) dx = \int_0^{\frac{\pi}{2}} dx = [x]_0^{\frac{\pi}{2}} = \dfrac{\pi}{2}$

$$\therefore \ 2I = \frac{\pi}{2} \quad \text{or} \quad I = \frac{\pi}{4} \qquad \text{Ans.}$$

(c) $I = \int_0^{\frac{\pi}{2}} \dfrac{e^{\sin x}}{e^{\sin x} + e^{\cos x}} \, dx = \int_0^{\frac{\pi}{2}} \dfrac{e^{\sin\left(\frac{\pi}{2}-x\right)}}{e^{\sin\left(\frac{\pi}{2}-x\right)} + e^{\cos\left(\frac{\pi}{2}-x\right)}} \, dx = \int_0^{\frac{\pi}{2}} \dfrac{e^{\cos x}}{e^{\cos x} + e^{\sin x}} \, dx = I$ (say) $\left\{ \text{by property IV. } \int_0^a f(x) \, dx = \int_0^a f(a-x) \, dx \right\}$

Adding, $2I = \int_0^{\frac{\pi}{2}} \left(\dfrac{e^{\sin x}}{e^{\sin x} + e^{\cos x}} + \dfrac{e^{\cos x}}{e^{\cos x} + e^{\sin x}} \right) dx = \int_0^{\frac{\pi}{2}} \left(\dfrac{e^{\sin x} + e^{\cos x}}{e^{\sin x} + e^{\cos x}} \right) dx = \int_0^{\frac{\pi}{2}} dx = [x]_0^{\frac{\pi}{2}} = \dfrac{\pi}{2}$ $\therefore \ 2I = \dfrac{\pi}{2}$ or $I = \dfrac{\pi}{4}$ Ans.

Property V: — $\displaystyle\int_{-a}^{a} f(x) \, dx = \begin{cases} 2 \displaystyle\int_0^{a} f(x) \, dx \, , & \text{if } f(-x) = f(x) \ \ \text{[Even function]} \\ 0 \, , & \text{if } f(-x) = -f(x) \ \ \text{[Odd function]} \end{cases}$

Proof: — $\displaystyle\int_{-a}^{a} f(x) \, dx = \int_{-a}^{0} f(x) \, dx + \int_{0}^{a} f(x) \, dx$, Put $x = -t$ $\therefore$ $dx = -dt$

$\displaystyle\int_{-a}^{a} f(x) \, dx = -\int_{a}^{0} f(-t) \, dt + \int_{0}^{a} f(x) \, dx = \int_{0}^{a} f(-x) \, dx + \int_{0}^{a} f(x) \, dx$ [by property I and II]

$\therefore \ I = \displaystyle\int_{-a}^{a} f(x) \, dx = \begin{cases} 2 \displaystyle\int_0^{a} f(x) \, dx \, , & \text{if } f(-x) = f(x) \ \ \text{[Even function]} \\ 0 \, , & \text{if } f(-x) = -f(x) \ \ \text{[Odd function]} \end{cases}$ \qquad Proved.

Example: — (a) Evaluate $\displaystyle\int_{-\frac{\pi}{3}}^{\frac{\pi}{3}} \sin^3 x \ dx$

Let $f(x) = \sin^3 x$, $f(-x) = [\sin(-x)]^3 = [-\sin x]^3 = -\sin^3 x = -f(x)$

$\therefore$ $f(x)$ is an odd function then $\displaystyle\int_{-\frac{\pi}{3}}^{\frac{\pi}{3}} \sin^3 x \ dx = 0$ Ans. [by property V.]

(b) $I = \displaystyle\int_{-\pi}^{\pi} \cos^2 x \ dx$, Let $f(x) = \cos^2 x$, $f(-x) = \cos^2(-x) = \cos^2 x = f(x)$ $\therefore$ $f(x)$ is an even function

$I = \displaystyle\int_{-\pi}^{\pi} \cos^2 x \ dx = 2 \int_0^{\pi} \cos^2 x \ dx = 2 \int_0^{\pi} \dfrac{1 - \cos 2x}{2} \ dx = 2 \int_0^{\pi} \dfrac{1}{2} \ dx - 2 \int_0^{\pi} \dfrac{\cos 2x}{2} \ dx$ [by property V.]

$I = 2 \cdot \dfrac{1}{2} [x]_0^{\pi} - 2 \cdot \dfrac{1}{2} \left[\dfrac{\sin 2x}{2} \right]_0^{\pi} = \pi - \dfrac{1}{2}[0 - 0] = \pi - 0 = \pi$ Ans.

(c) $I = \displaystyle\int_{-\frac{\pi}{4}}^{\frac{\pi}{4}} \cos^3 x \ dx$, Let $f(x) = \cos^3 x$, $f(-x) = \cos^3(-x) = \cos^3 x = f(x)$ [$\cos(-\theta) = \cos\theta$] $\therefore$ $f(x)$ is an even function

$I = \displaystyle\int_{-\frac{\pi}{4}}^{\frac{\pi}{4}} \cos^3 x \ dx = 2 \int_0^{\frac{\pi}{4}} \dfrac{\cos 3x + 3\cos x}{4} \ dx = \dfrac{1}{2} \int_0^{\frac{\pi}{4}} (\cos 3x + 3\cos x) \ dx = \dfrac{1}{2} \int_0^{\frac{\pi}{4}} \cos 3x \ dx + \dfrac{3}{2} \int_0^{\frac{\pi}{4}} \cos x \ dx$ [by prop. V]

$I = \dfrac{1}{2} \left[\dfrac{\sin 3x}{3} \right]_0^{\frac{\pi}{4}} + \dfrac{3}{2} [\sin x]_0^{\frac{\pi}{4}} = \dfrac{1}{6} \left[\sin \left(\dfrac{3\pi}{4} \right) - \sin 0 \right] + \dfrac{3}{2} \left[\sin \left(\dfrac{\pi}{4} \right) - \sin 0 \right] = \dfrac{1}{6} \cdot \dfrac{1}{\sqrt{2}} + \dfrac{3}{2} \cdot \dfrac{1}{\sqrt{2}} = \dfrac{5}{3\sqrt{2}}$ Ans.

Property VI: — $\displaystyle\int_a^b f(x) \, dx = \int_a^b f(a + b - x) \, dx$

Example: $-$ $I = \int_{-\frac{\pi}{4}}^{\frac{\pi}{2}} \cos x \, dx = \int_{-\frac{\pi}{4}}^{\frac{\pi}{2}} \cos\left(-\frac{\pi}{4} + \frac{\pi}{2} - x\right) dx = \int_{-\frac{\pi}{4}}^{\frac{\pi}{2}} \cos\left(\frac{\pi}{4} - x\right) dx$ $\quad [\cos(A - B) = \cos A \cos B + \sin A \sin B]$

$I = \int_{-\frac{\pi}{4}}^{\frac{\pi}{2}} \left[\cos\left(\frac{\pi}{4}\right) \cos x + \sin\left(\frac{\pi}{4}\right) \sin x\right] dx = \frac{1}{\sqrt{2}} \int_{-\frac{\pi}{4}}^{\frac{\pi}{2}} [\cos x + \sin x] \, dx = \frac{1}{\sqrt{2}} [\sin x - \cos x]_{-\frac{\pi}{4}}^{\frac{\pi}{2}} = \frac{1}{\sqrt{2}}\left[(1 - 0) - \frac{1}{\sqrt{2}} - \frac{1}{\sqrt{2}}\right) = \frac{1}{\sqrt{2}}\left[1 + \frac{1}{\sqrt{2}} + \frac{1}{\sqrt{2}}\right]$

$= \frac{1}{\sqrt{2}}\left[\frac{\sqrt{2} + 1 + 1}{\sqrt{2}}\right] = \frac{1}{\sqrt{2}} \cdot \frac{\sqrt{2} + 2}{\sqrt{2}}$

$I = \frac{\sqrt{2} + 2}{2} = \frac{\sqrt{2}}{2} + \frac{2}{2} = \frac{\sqrt{2}}{\sqrt{2}\sqrt{2}} + 1 = \frac{1}{\sqrt{2}} + 1 = \frac{1 + \sqrt{2}}{\sqrt{2}}$ $\qquad$ Ans.

Property VII: $-$ $\int_{0}^{2a} f(x) \, dx = \begin{cases} 2 \int_{0}^{a} f(x) \, dx, & \text{if } f(2a - x) = f(x) \\ 0, & \text{if } f(2a - x) = -f(x) \end{cases}$

Proof: $-$ $\int_{0}^{2a} f(x) \, dx = \int_{0}^{a} f(x) \, dx + \int_{a}^{2a} f(x) \, dx \ldots \ldots \ldots \ldots \ldots \ldots \ldots . (A)$

Putting, $x = 2a - t$ in last integral $\therefore$ $dx = -dt$ If $x = a$ then $t = 2a - x = 2a - a = a$ and if $x = 2a$ then $t = 2a - x = 2a - 2a = 0$

From equation (A), $\int_{0}^{2a} f(x) \, dx = \int_{0}^{a} f(x) \, dx + \int_{a}^{0} f(2a - t) \, (-dt) = \int_{0}^{a} f(x) \, dx + \int_{0}^{a} f(2a - t) \, dt$

$\Rightarrow \int_{0}^{2a} f(x) \, dx = \int_{0}^{a} f(x) \, dx + \int_{0}^{a} f(2a - x) \, dx$ $\quad$ by property I. $\quad \left[\int_{a}^{0} f(x) \, dx = -\int_{0}^{a} f(x) \, dx \quad \text{and} \quad \int_{a}^{b} f(x) \, dx = \int_{a}^{b} f(t) \, dt\right]$

$\Rightarrow \int_{0}^{2a} f(x) \, dx = \begin{cases} \int_{0}^{a} f(x) \, dx + \int_{0}^{a} f(x) \, dx, & \text{if } f(2a - x) = f(x) \\ \int_{0}^{a} f(x) \, dx + \int_{0}^{a} -f(x) \, dx, & \text{if } f(2a - x) = -f(x) \end{cases} = \begin{cases} 2 \int_{0}^{a} f(x) \, dx, & \text{if } f(2a - x) = f(x) \\ 0, & \text{if } f(2a - x) = -f(x) \end{cases}$ $\quad$ Proved.

Example: $-$ (i) Evaluate: $-$ $I = \int_{-\frac{\pi}{2}}^{\frac{\pi}{2}} \cos^3 x \, dx$

Solution: $-$ $I = \int_{-\frac{\pi}{2}}^{\frac{\pi}{2}} \cos^3 x \, dx$, Let $f(x) = \cos^3 x$, $f(-x) = \cos^3(-x) = \cos^3 x = f(x)$ $\therefore$ $f(x)$ is an even function

$I = \int_{-\frac{\pi}{2}}^{\frac{\pi}{2}} \cos^3 x \, dx = 2 \int_{0}^{\frac{\pi}{2}} \cos^3 x \, dx = 2 \int_{0}^{\frac{\pi}{2}} \left(\frac{\cos 3x + 3 \cos x}{4}\right) dx = \frac{1}{2} \int_{0}^{\frac{\pi}{2}} (\cos 3x + 3 \cos x) \, dx = \frac{1}{2} \left[\frac{\sin 3x}{3} + 3 \sin x\right]_{0}^{\frac{\pi}{2}}$

$= \frac{1}{2}\left[\frac{\sin\left(\frac{3\pi}{2}\right)}{3} + 3 \sin\left(\frac{\pi}{2}\right) - \frac{\sin 0}{3} - 3 \sin 0\right] = \frac{1}{2}\left[-\frac{1}{3} + 3\right] = \frac{1}{2} \cdot \left[\frac{-1 + 9}{3}\right] = \frac{1}{2} \cdot \frac{8}{3} = \frac{4}{3}$ $\quad$ Ans.

(ii) Evaluate: $-$ $I = \int_{-\frac{\pi}{4}}^{\frac{\pi}{4}} \sin^3 x \, dx$

Solution: $-$ Let $f(x) = \sin^3 x$, $f(-x) = \sin^3(-x) = -\sin^3 x = -f(x)$

$\therefore$ $f(x)$ is an odd function then $I = \int_{-\frac{\pi}{4}}^{\frac{\pi}{4}} \sin^3 x \, dx = 0$ $\quad$ Ans.

(iii) Evaluate: $-$ $I = \int_{-\frac{\pi}{3}}^{\frac{\pi}{3}} \cos^2 x \, dx$,

Solution: $-$ Let $f(x) = \cos^2 x$, $f(-x) = \cos^2(-x) = \cos^2 x = f(x)$ $\quad \therefore$ $f(x)$ is an even function.

$\left[\cos 2x = \cos^2 x - \sin^2 x, \ \cos 2x = \cos^2 x - 1 + \cos^2 x = 2 \cos^2 x - 1, \ 2 \cos^2 x = 1 + \cos 2x \ \text{or} \ \cos^2 x = \frac{1 + \cos 2x}{2}\right]$

$I = \int_{-\frac{\pi}{3}}^{\frac{\pi}{3}} \cos^2 x \, dx = 2\int_0^{\frac{\pi}{3}} \cos^2 x \, dx = 2\int_0^{\frac{\pi}{3}} \left(\frac{1+\cos 2x}{2}\right) dx = 2.\frac{1}{2}\int_0^{\frac{\pi}{3}} dx + 2.\frac{1}{2}\int_0^{\frac{\pi}{3}} \cos 2x \, dx = (x)_0^{\frac{\pi}{3}} + \left(\frac{\sin 2x}{2}\right)_0^{\frac{\pi}{3}} = \frac{\pi}{3} - 0 + \frac{\sqrt{3}}{4} - 0 = \frac{\pi}{3} + \frac{\sqrt{3}}{4}$

$$= \frac{4\pi + 3\sqrt{3}}{12} \quad \text{Ans.}$$

(iv) Evaluate: $- \quad I = \int_{-\frac{\pi}{6}}^{\frac{\pi}{6}} \sin^2 x \, dx$

Solution: $-$ Let $f(x) = \sin^2 x$, $f(-x) = \sin^2(-x) = (-\sin x)^2 = \sin^2 x = f(x)$ $\therefore f(x)$ is an even function.

$\left[\cos 2x = \cos^2 x - \sin^2 x, \ \cos 2x = 1 - \sin^2 x - \sin^2 x = 1 - 2\sin^2 x, \ 2\sin^2 x = 1 - \cos 2x \ \text{or} \ \sin^2 x = \frac{1-\cos 2x}{2}\right]$

$I = \int_{-\frac{\pi}{6}}^{\frac{\pi}{6}} \sin^2 x \, dx = 2\int_0^{\frac{\pi}{6}} \sin^2 x \, dx = 2\int_0^{\frac{\pi}{6}} \left(\frac{1-\cos 2x}{2}\right) dx = \int_0^{\frac{\pi}{6}} dx - \int_0^{\frac{\pi}{6}} \cos 2x \, dx = (x)_0^{\frac{\pi}{6}} - \left(\frac{\sin 2x}{2}\right)_0^{\frac{\pi}{6}} = \frac{\pi}{6} - \frac{\sqrt{3}}{4} = \frac{2\pi - 3\sqrt{3}}{12}$ Ans.

(v) Evaluate: $- \quad I = \int_0^{\pi} \cos^2 x \, dx$

Solution: $-$ Let $f(x) = \cos^2 x$, $f(\pi - x) = \cos^2(\pi - x) = [\cos(\pi - x)]^2 = \cos^2 x = f(x)$ [use property VII.]

$\int_0^{\pi} \cos^2 x \, dx = 2\int_0^{\frac{\pi}{2}} \cos^2 x \, dx = 2\int_0^{\frac{\pi}{2}} \left(\frac{1+\cos 2x}{2}\right) dx = 2.\frac{1}{2}\int_0^{\frac{\pi}{2}} dx + 2.\frac{1}{2}\int_0^{\frac{\pi}{2}} \cos 2x \, dx \quad \left[\therefore 2a = \pi \quad \therefore a = \frac{\pi}{2}\right]$

$I = (x)_0^{\frac{\pi}{2}} + \left(\frac{\sin 2x}{2}\right)_0^{\frac{\pi}{2}} = \frac{\pi}{2} - 0 + 0 - 0 = \frac{\pi}{2}$ Ans.

(vi) Evaluate: $- \quad I = \int_{-\pi}^{\pi} \sin^4 x \, dx$

Solution: $-$ Let $f(x) = \sin^4 x$, $f(-x) = [\sin(-x)]^4 = [-\sin x]^4 = \sin^4 x = f(x)$ $\therefore$ $f(x)$ is an even function [use property V.]

$I = \int_{-\pi}^{\pi} \sin^4 x \, dx = 2\int_0^{\pi} \sin^4 x \, dx = 2\int_0^{\pi} [\sin^2 x]^2 \, dx = 2\int_0^{\pi} \left[\frac{1-\cos 2x}{2}\right]^2 dx = \frac{2}{4}\int_0^{\pi} [1 - 2\cos 2x + \cos^2 2x] \, dx$

$\qquad = \frac{1}{2}\int_0^{\pi} dx - \int_0^{\pi} \cos 2x \, dx + \frac{1}{2}\int_0^{\pi} \cos^2 2x \, dx = \frac{1}{2}\int_0^{\pi} dx - \int_0^{\pi} \cos 2x \, dx + \frac{1}{2}\int_0^{\pi} \left(\frac{1+\cos 4x}{2}\right) dx$

$I = \frac{1}{2}\int_0^{\pi} dx - \int_0^{\pi} \cos 2x \, dx + \frac{1}{4}\int_0^{\pi} dx + \frac{1}{4}\int_0^{\pi} \cos 4x \, dx = \frac{1}{2}(x)_0^{\pi} - \left(\frac{\sin 2x}{2}\right)_0^{\pi} + \frac{1}{4}(x)_0^{\pi} + \frac{1}{4}\left(\frac{\sin 4x}{4}\right)_0^{\pi}$

$\qquad = \frac{1}{2}(\pi - 0) - \frac{1}{2}(0 - 0) + \frac{1}{4}(\pi - 0) + \frac{1}{16}(0 - 0) = \frac{\pi}{2} + \frac{\pi}{4} = \frac{2\pi + \pi}{4} = \frac{3\pi}{4}$ Ans.

(vii) Evaluate: $- \quad I = \int_0^{\pi} \sin^5 x \, dx$

Solution: $-$ Let $f(x) = \sin^5 x$, $f(\pi - x) = [\sin(\pi - x)]^5 = \sin^5 x = f(x)$ [use property VII.]

$I = \int_0^{\pi} \sin^5 x \, dx = 2\int_0^{\frac{\pi}{2}} \sin^5 x \, dx = 2\int_0^{\frac{\pi}{2}} \sin^4 x \cdot \sin x \, dx = 2\int_0^{\frac{\pi}{2}} (\sin^2 x)^2 \cdot \sin x \, dx = 2\int_0^{\frac{\pi}{2}} (1 - \cos^2 x)^2 \cdot \sin x \, dx$

Put $\cos x = t$ $\therefore -\sin x \, dx = dt$ when $x = \frac{\pi}{2}$, $t = 0$ and $x = 0$, $t = 1$ upper and lower limit is $t = (0,1)$

$I = -2\int_1^0 [1 - t^2]^2 \, dt = 2\int_0^1 (1 - 2t^2 + t^4) \, dt = 2\int_0^1 dt - 4\int_0^1 t^2 \, dt + 2\int_0^1 t^4 \, dt = 2(t)_0^1 - 4\left(\frac{t^3}{3}\right)_0^1 + 2\left(\frac{t^5}{5}\right)_0^1$

$\qquad = 2(1 - 0) - \frac{4}{3}(1 - 0) + \frac{2}{5}(1 - 0) = 2 - \frac{4}{3} + \frac{2}{5} = \frac{30 - 20 + 6}{15} = \frac{16}{15}$ Ans.

(viii) Evaluate: $-\quad \displaystyle\int_{-\frac{\pi}{2}}^{\frac{\pi}{2}} \frac{\cos^2 x}{1-a^x}\, dx$

Solution: $-\quad$ Let $I = \displaystyle\int_{-\frac{\pi}{2}}^{\frac{\pi}{2}} \frac{\cos^2 x}{1-a^x}\, dx$ ……………… (A)

$I = \displaystyle\int_{-\frac{\pi}{2}}^{\frac{\pi}{2}} \frac{\cos^2\left(-\frac{\pi}{2}+\frac{\pi}{2}-x\right)}{1-a^{\left(-\frac{\pi}{2}+\frac{\pi}{2}-x\right)}}\, dx = \int_{-\frac{\pi}{2}}^{\frac{\pi}{2}} \frac{\cos^2(-x)}{1-a^{-x}}\, dx = \int_{-\frac{\pi}{2}}^{\frac{\pi}{2}} \frac{\cos^2(-x)}{\frac{a^x-1}{a^x}}\, dx = \int_{-\frac{\pi}{2}}^{\frac{\pi}{2}} \frac{a^x\cos^2 x}{a^x-1}\, dx$ …………… (B)

Adding equation (A) and (B), we get

$2I = \displaystyle\int_{-\frac{\pi}{2}}^{\frac{\pi}{2}} \left[\frac{\cos^2 x}{1-a^x}+\frac{a^x\cos^2 x}{a^x-1}\right] dx = \int_{-\frac{\pi}{2}}^{\frac{\pi}{2}} \left[\frac{\cos^2 x}{1-a^x}-\frac{a^x\cos^2 x}{1-a^x}\right] dx = \int_{-\frac{\pi}{2}}^{\frac{\pi}{2}} \left[\frac{\cos^2 x-a^x\cos^2 x}{1-a^x}\right] dx = \int_{-\frac{\pi}{2}}^{\frac{\pi}{2}} \cos^2 x.\left[\frac{1-a^x}{1-a^x}\right] dx = \int_{-\frac{\pi}{2}}^{\frac{\pi}{2}} \cos^2 x\, dx$

$= \displaystyle\int_{-\frac{\pi}{2}}^{\frac{\pi}{2}} \left(\frac{1+\cos 2x}{2}\right) dx = \frac{1}{2}\int_{-\frac{\pi}{2}}^{\frac{\pi}{2}} dx + \frac{1}{2}\int_{-\frac{\pi}{2}}^{\frac{\pi}{2}} \cos 2x\, dx = \frac{1}{2}(x)_{-\frac{\pi}{2}}^{\frac{\pi}{2}} + \frac{1}{2}\left(\frac{\sin 2x}{2}\right)_{-\frac{\pi}{2}}^{\frac{\pi}{2}}$

$2I = \dfrac{1}{2}\left(\dfrac{\pi}{2}-\dfrac{-\pi}{2}\right) + \dfrac{1}{4}(\sin\pi - \sin(-\pi)) = \dfrac{1}{2}\left(\dfrac{\pi}{2}+\dfrac{\pi}{2}\right) + \dfrac{1}{4}(0+0) = \dfrac{1}{2}.\dfrac{2\pi}{2} = \dfrac{\pi}{2}\quad$ or $\quad \boxed{I = \dfrac{\pi}{4}\quad \text{Ans.}}$

(ix) Evaluate: $-\quad \displaystyle\int_{-\pi}^{\pi} \frac{\sin^2 x}{1+a^x}\, dx$

Solution: $-\quad$ Let $I = \displaystyle\int_{-\pi}^{\pi} \frac{\sin^2 x}{1+a^x}\, dx$ …………….. (A)

$I = \displaystyle\int_{-\pi}^{\pi} \frac{\sin^2(-\pi+\pi-x)}{1+a^{(-\pi+\pi-x)}}\, dx = \int_{-\pi}^{\pi} \frac{\sin^2(-x)}{1+a^{(-x)}}\, dx = \int_{-\pi}^{\pi} \frac{\sin^2 x}{1+\frac{1}{a^x}}\, dx = \int_{-\pi}^{\pi} \frac{\sin^2 x}{\frac{1+a^x}{a^x}}\, dx = \int_{-\pi}^{\pi} \frac{a^x\sin^2 x}{1+a^x}\, dx$ …………….. (B)

Adding equation (A) and (B), we get

$2I = \displaystyle\int_{-\pi}^{\pi} \left(\frac{\sin^2 x}{1+a^x}+\frac{a^x\sin^2 x}{1+a^x}\right) dx = \int_{-\pi}^{\pi} \left(\frac{\sin^2 x+a^x\sin^2 x}{1+a^x}\right) dx = \int_{-\pi}^{\pi} \sin^2 x.\left(\frac{1+a^x}{1+a^x}\right) dx = \int_{-\pi}^{\pi} \sin^2 x\, dx = \int_{-\pi}^{\pi} \left(\frac{1-\cos 2x}{2}\right) dx$

$= \dfrac{1}{2}\displaystyle\int_{-\pi}^{\pi} dx - \dfrac{1}{2}\int_{-\pi}^{\pi} \cos 2x\, dx = \dfrac{1}{2}(x)_{-\pi}^{\pi} - \dfrac{1}{2}\left(\dfrac{\sin 2x}{2}\right)_{-\pi}^{\pi} = \dfrac{1}{2}[\pi+\pi] - \dfrac{1}{4}[0-0]$

$2I = \dfrac{1}{2}.2\pi = \pi\quad \therefore\ I = \dfrac{\pi}{2}\quad$ Ans.

(x) Evaluate: $-\quad \displaystyle\int_{-\frac{\pi}{4}}^{\frac{\pi}{4}} (\cos^4 x + \sin^2 x)\, dx$

Solution: $-\quad$ Let $I = \displaystyle\int_{-\frac{\pi}{4}}^{\frac{\pi}{4}} (\cos^4 x + \sin^2 x)\, dx = \int_{-\frac{\pi}{4}}^{\frac{\pi}{4}} \cos^4 x\, dx + \int_{-\frac{\pi}{4}}^{\frac{\pi}{4}} \sin^2 x\, dx = I_1 + I_2\quad$ (say)

$I_1 = \displaystyle\int_{-\frac{\pi}{4}}^{\frac{\pi}{4}} \cos^4 x\, dx = \dfrac{3\pi+8}{16}\quad$ (solve do yourself) and $I_2 = \displaystyle\int_{-\frac{\pi}{4}}^{\frac{\pi}{4}} \sin^2 x\, dx = \dfrac{\pi-2}{4}\quad$ (solve do yourself)

$I = I_1 + I_2 = \dfrac{3\pi+8}{16}+\dfrac{\pi-2}{4} = \dfrac{7\pi}{16}\quad$ Ans.

Property VIII: $-$ If $f(x)$ is a periodic function with period T then $\displaystyle\int_{a}^{a+nT} f(x)\, dx = n\int_{0}^{T} f(x)\, dx$

If $f(x)$ is a periodic function with period T then, $f(x+nT) = f(x)$

Proof: $-\quad f(x+T) = f(x)$, Replace x by $x+T\quad \Rightarrow\ f(x+2T) = f(x+T) = f(x)$

$$f(x+3T) = f(x+2T) = f(x)$$

$$\ldots\ldots\ldots\ldots\ldots\ldots\ldots\ldots\ldots\ldots\ldots\ldots$$

$$\ldots\ldots\ldots\ldots\ldots\ldots\ldots\ldots\ldots\ldots\ldots\ldots$$

$$f(x + nT) = f(x) \qquad \text{Proved.}$$

Again, Proof: $-\displaystyle\int_a^{a+nT} f(x)\,dx = \int_a^{nT} f(x)\,dx + \int_{nT}^{a+nT} f(x)\,dx, \qquad$ Putting $x = nT + y$ in last integral.

$$\therefore \quad dx = dy \qquad \text{if } x = nT + y \text{ when } x = nT, \ y = 0 \text{ and } x = a + nT, \ y = a$$

$$\int_a^{a+nT} f(x)\,dx = \int_a^{nT} f(x)\,dx + \int_0^a f(nT + y)\,dy = \int_0^a f(nT + x)\,dx + \int_a^{nT} f(x)\,dx = \int_0^a f(x)\,dx + \int_a^{nT} f(x)\,dx \ \text{ by prop. I}$$

$$\int_a^{a+nT} f(x)\,dx = \int_0^{nT} f(x)\,dx = \int_0^T f(x)\,dx + \int_T^{2T} f(x)\,dx + \int_{2T}^{3T} f(x)\,dx + \cdots \ldots \ldots + \int_{(n-1)T}^{nT} f(x)\,dx$$

$$\therefore \quad \int_a^{a+nT} f(x)\,dx = I_1 + I_2 + I_3 + \cdots \ldots \ldots \ldots \ldots + I_n \quad \ldots \ldots \ldots \ldots \ldots \ldots (A)$$

where $I_1 = \displaystyle\int_0^T f(x)\,dx$ and $I_2 = \displaystyle\int_T^{2T} f(x)\,dx$, putting $x = T + y$ $\therefore$ $dx = dy$ when $x = T, \ y = 0$ and $x = 2T, \ y = T$

$$I_2 = \int_T^{2T} f(x)\,dx = \int_0^T f(T + y)\,dy = \int_0^T f(T + x)\,dx = \int_0^T f(x)\,dx = I_1 \qquad \text{by property I.}$$

$$I_3 = \int_{2T}^{3T} f(x)\,dx = I_1 \quad \text{and} \quad I_4 = \int_{3T}^{4T} f(x)\,dx = I_1 \ , \ldots \ldots \ldots \ldots \text{and } I_n = \int_{(n-1)T}^{nT} f(x)\,dx = I_1$$

From equation (A), $\Rightarrow \displaystyle\int_a^{a+nT} f(x)\,dx = nI_1 = n\int_0^T f(x)\,dx \qquad$ Proved.

 # Particular case: $-$ (i) $\displaystyle\int_0^{nT} f(x)\,dx = n\int_0^T f(x)\,dx$ (ii) $\displaystyle\int_a^{a+T} f(x)\,dx = \int_0^T f(x)\,dx$

Property IX: $-$ If $f(x)$ is a periodic function with period T then , $\displaystyle\int_{mT}^{nT} f(x)\,dx = (n - m)\int_0^T f(x)\,dx$

Proof: $-$ L.H.S $= \displaystyle\int_{mT}^{nT} f(x)\,dx = \int_{mT}^0 f(x)\,dx + \int_0^{nT} f(x)\,dx = -\int_0^{mT} f(x)\,dx + \int_0^{nT} f(x)\,dx = -m\int_0^T f(x)\,dx + n\int_0^T f(x)\,dx$

$$= (n - m)\int_0^T f(x)\,dx \qquad \text{Proved.}$$

Property X: $-$ If $f(x)$ is a periodic function with period T then, $\displaystyle\int_{a+nT}^{b+nT} f(x)\,dx = \int_a^b f(x)\,dx$

Proof: $-$ L.H.S $= \displaystyle\int_{a+nT}^{b+nT} f(x)\,dx = \int_{a+nT}^a f(x)\,dx + \int_a^b f(x)\,dx + \int_b^{b+nT} f(x)\,dx = -\int_a^{a+nT} f(x)\,dx + \int_a^b f(x)\,dx + \int_b^{b+nT} f(x)\,dx$

$$= -n\int_0^T f(x)\,dx + \int_a^b f(x)\,dx + n\int_0^T f(x)\,dx = \int_a^b f(x)\,dx \qquad \text{Proved.} \qquad (\text{ by property VIII. })$$

Example: $-$ (a) $\displaystyle\int_0^{200\pi} \sqrt{1 + \cos 2x}\ dx$

Solution: $-$ Let $I = \displaystyle\int_0^{200\pi} \sqrt{1 + \cos 2x}\ dx = \int_0^{200\pi} \sqrt{2\cos^2 x}\ dx = \int_0^{200\pi} \sqrt{2}|\cos x|\ dx = 200.\sqrt{2}\int_0^\pi \cos x\ dx = 200.\sqrt{2}(\sin x)_0^\pi$

$$= 200.\sqrt{2}(0 - 0) = 0 \qquad \text{Ans.} \quad \text{(b)} \ \int_0^{101} e^{x-[x]}\ dx$$

Solution: $-$ Let $I = \displaystyle\int_0^{101} e^{x-[x]}\ dx = 101\int_0^1 e^{x-[x]}\ dx = 101\int_0^1 e^x\ dx = 101(x)_0^1 = 101(e^1 - e^0) = 101(e - 1) \quad$ Ans.

(c) $\displaystyle\int_0^{50\pi} (\sin x + \cos x)\, dx$

Solution: − Let $I = \displaystyle\int_0^{50\pi} (\sin x + \cos x)\, dx = 50\int_0^{\pi}(\sin x + \cos x)\, dx = 50\int_0^{\pi}\sin x\, dx + 50\int_0^{\pi}\cos x\, dx = 50(-\cos x)_0^{\pi} + 50(\sin x)_0^{\pi}$

$\qquad = -50(-1-1) + 50.0 = 100 \quad$ Ans.

(d) show that $\displaystyle\int_0^{n\pi+v}|\cos x|\, dx = \sin v$ where n is a positive integer and $0 \le v < \pi$.

Proof: − L.H.S $= \displaystyle\int_0^{n\pi+v}|\cos x|\, dx = \int_0^{v}|\cos x|\, dx + \int_v^{n\pi+v}|\cos x|\, dx = \int_0^{v}|\cos x|\, dx + \int_0^{n\pi}|\cos x|\, dx = \int_0^{v}|\cos x|\, dx + n\int_0^{v}|\cos x|\, dx$

$\qquad = (\sin x)_0^{v} + n(\sin x)_0^{\pi} = \sin v \quad$ Proved.

(e) Let $T > 0$ be a fixed number suppose f is a continuous function for all $x \in R, f(x+T) = f(x)$.

if $I = \displaystyle\int_0^{T} f(x)\, dx$ then the value of $\displaystyle\int_4^{4+4T} f(3x)\, dx$.

Solution: − $\displaystyle\int_4^{4+4T} f(3x)\, dx = \int_4^{4} f(3x)\, dx + \int_4^{4+4T} f(3x)\, dx = \int_4^{4} f(3x)\, dx + \int_4^{4+12.\frac{T}{3}} f(3x)\, dx = 12\int_0^{\frac{T}{3}} f(3x)\, dx$

Putting, $3x = y \quad \therefore \; dx = \dfrac{dy}{3}$ when $x = 0, \; y = 0$ and $x = \dfrac{T}{3}, \; y = T$

$\Rightarrow \displaystyle\int_4^{4+4T} f(3x)\, dx = \dfrac{12}{3}\int_0^{T} f(y)\, dy = 4\int_0^{T} f(x)\, dx = 4I \quad$ Ans.

(f) If $f(x)$ is an odd function in $\left[-\dfrac{T}{2}, \dfrac{T}{2}\right]$ and has period equal to T. Prove that $\displaystyle\int_a^{x} f(y)\, dy$ is also periodic function with period T.

Solution: − $\displaystyle\int_{-\frac{T}{2}}^{\frac{T}{2}} f(x)\, dx = 0, \; f(x+T) = f(x)$ Let $g(x) = \displaystyle\int_a^{x} f(y)\, dy$ to prove $g(x+T) = g(x)$

or $\displaystyle\int_a^{x+T} f(y)\, dy = \int_a^{x} f(y)\, dy \quad$ or $\displaystyle\int_a^{x+T} f(y)\, dy - \int_a^{x} f(y)\, dy = 0$

L.H.S $= \displaystyle\int_a^{x+T} f(y)\, dy - \int_a^{x} f(y)\, dy = \int_a^{x} f(y)\, dy + \int_x^{x+T} f(y)\, dy - \int_a^{x} f(y)\, dy = \int_x^{x+T} f(y)\, dy = \int_x^{-\frac{T}{2}} f(y)\, dy + \int_{-\frac{T}{2}}^{\frac{T}{2}} f(y)\, dy + \int_{\frac{T}{2}}^{x+T} f(y)\, dy$

$\qquad = \displaystyle\int_x^{-\frac{T}{2}} f(y)\, dy + \int_{\frac{T}{2}}^{x+T} f(y)\, dy$

Putting, $y = z + T$ in last integral $\quad \therefore \; dy = dz$

y	$\dfrac{T}{2}$	$x + T$
z	$-\dfrac{T}{2}$	x

$\displaystyle\int_a^{x+T} f(y)\, dy = \int_x^{-\frac{T}{2}} f(y)\, dy + \int_{-\frac{T}{2}}^{x} f(z+T)\, dz = \int_x^{-\frac{T}{2}} f(y)\, dy + \int_{-\frac{T}{2}}^{x} f(z)\, dz = \int_x^{-\frac{T}{2}} f(z)\, dz + \int_{-\frac{T}{2}}^{x} f(z)\, dz = -\int_{-\frac{T}{2}}^{x} f(z)\, dz + \int_{-\frac{T}{2}}^{x} f(z)\, dz = 0$

$\qquad = $ R.H.S $\quad$ Proved.

IInd Method: − L.H.S $= \displaystyle\int_a^{x+T} f(y)\, dy - \int_a^{x} f(y)\, dy = \int_a^{x} f(y)\, dy + \int_x^{x+T} f(y)\, dy - \int_a^{x} f(y)\, dy = \int_x^{x+T} f(y)\, dy$

$\qquad \left\{ \therefore \; \displaystyle\int_x^{x+T} f(y)\, dy \text{ is independent of } x \right\}, \; \displaystyle\int_x^{x+T} f(y)\, dy = \int_{-\frac{T}{2}}^{\frac{T}{2}} f(y)\, dy = 0 = $ R.H.S $\quad$ Proved.

Property XI: — If $I(t) = \int_a^b f(x, t)\, dx$ then, $\dfrac{d[I(t)]}{dt} = \int_a^b \dfrac{\partial[f(x,t)]}{\partial t}\cdot dx$

Example: — (a) Evaluate: — $\displaystyle\int_0^1 \dfrac{x^b + 1}{\log x}\, dx \quad (b \geq 0)$

Solution: — Let $I(b) = \displaystyle\int_0^1 \dfrac{x^b + 1}{\log x}\, dx$, Differentiating both of sides with respect to b , we get

$\therefore \dfrac{d[I(b)]}{db} = \int_0^1 \dfrac{\partial}{\partial b}\left(\dfrac{x^b + 1}{\log x}\right)\cdot dx = \int_0^1 \left[\dfrac{1}{\log x}\cdot x^b\cdot \log x\right] dx = \int_0^1 x^b\, dx = \left(\dfrac{x^{b+1}}{b+1}\right)_0^1 = \dfrac{1}{b+1} - 0 = \dfrac{1}{b+1}$

$\therefore I'(b) = \dfrac{1}{b+1}$, Integrating $\displaystyle\int I'(b)\, db = \int \dfrac{db}{b+1}$ $\therefore I(b) = \log|1 + b| + c \ \dots\dots\dots\dots$ (i)

Putting, $b = 0$, $I(0) = c$ $\therefore c = 0$ from equation (i) $\Rightarrow I(b) = \log|1 + b| + 0 = \log|1 + b|$

$$\therefore \int_0^1 \dfrac{x^b + 1}{\log x}\, dx = \log|1 + b| \qquad \text{Ans.}$$

(b) $\displaystyle\int_{\frac{\pi}{6}}^{\frac{\pi}{3}} (\tan^2\theta + b^2\sec^2\theta)\, d\theta$

Solution: — Let $I(b) = \displaystyle\int_{\frac{\pi}{6}}^{\frac{\pi}{3}} (\tan^2\theta + b^2\sec^2\theta)\, d\theta$ Differentiating both sides w.r.t b, we get

$\dfrac{d[I(b)]}{db} = \displaystyle\int_{\frac{\pi}{6}}^{\frac{\pi}{3}} \dfrac{\partial}{\partial b}(\tan^2\theta + b^2\sec^2\theta)\, d\theta = \int_{\frac{\pi}{6}}^{\frac{\pi}{3}} 2b\sec^2\theta\, d\theta = 2b\int_{\frac{\pi}{6}}^{\frac{\pi}{3}} \sec^2\theta\, d\theta = 2b(\tan\theta)_{\frac{\pi}{6}}^{\frac{\pi}{3}} = 2b\left(\tan\dfrac{\pi}{3} - \tan\dfrac{\pi}{6}\right)$

$$I'(b) = 2b\left(\sqrt{3} - \dfrac{1}{\sqrt{3}}\right) = \dfrac{2b(3-1)}{\sqrt{3}} = \dfrac{4b}{\sqrt{3}}$$

Integrating, $\displaystyle\int I'(b)\, db = \int \dfrac{4b}{\sqrt{3}}\, db = \dfrac{4}{\sqrt{3}}\int b\, db = \dfrac{4}{\sqrt{3}}\cdot\dfrac{b^2}{2} + c = \dfrac{2b^2}{\sqrt{3}} + c$ $\therefore I(b) = \dfrac{2b^2}{\sqrt{3}} + c \ \dots\dots\dots\dots\dots$ (i)

Put $b = 0$, $c = \displaystyle\int_{\frac{\pi}{6}}^{\frac{\pi}{3}} \tan^2\theta\, d\theta = \int_{\frac{\pi}{6}}^{\frac{\pi}{3}} (\sec^2\theta - 1)\, d\theta = \int_{\frac{\pi}{6}}^{\frac{\pi}{3}} \sec^2\theta\, d\theta - \int_{\frac{\pi}{6}}^{\frac{\pi}{3}} d\theta = (\tan\theta)_{\frac{\pi}{6}}^{\frac{\pi}{3}} - (\theta)_{\frac{\pi}{6}}^{\frac{\pi}{3}} = \left(\tan\dfrac{\pi}{3} - \tan\dfrac{\pi}{6}\right) - \left(\dfrac{\pi}{3} - \dfrac{\pi}{6}\right) = \sqrt{3} - \dfrac{1}{\sqrt{3}} - \dfrac{\pi}{6}$

$$= \dfrac{2}{\sqrt{3}} - \dfrac{\pi}{6}$$

Put the value of c in equation (i), we get $\therefore I(b) = \dfrac{2b^2}{\sqrt{3}} + c = \dfrac{2b^2}{\sqrt{3}} + \dfrac{2}{\sqrt{3}} - \dfrac{\pi}{6}$ Ans.

Exercise – A10

(1) Evaluate the following definite integral: —   (a) $I = \displaystyle\int_1^{\sqrt{3}} \dfrac{x^3}{\sqrt{x^4 - 1}}\, dx$ (b) $I = \displaystyle\int_0^{\frac{\pi}{6}} \cos 3x\, dx$

(c) $I = \displaystyle\int_0^2 \dfrac{3}{x+2}\, dx$ (d) $I = \displaystyle\int_1^e \dfrac{3}{2\sqrt{1+x}}\, dx$ (e) $I = \displaystyle\int_0^{\frac{\pi}{2}} \tan^{-1}\left(\dfrac{x}{\sqrt{1-x^2}}\right) dx$ (f) $I = \displaystyle\int_0^1 \tan^{-1}\left(\dfrac{x-1}{x+1}\right) dx$

(2) Find the following integral: — (a) $I = \displaystyle\int_0^{\frac{\pi}{2}} \log(\sin x)\, dx$ (b) $I = \displaystyle\int_0^{\frac{\pi}{2}} \log(\cos x)\, dx$ (c) $I = \displaystyle\int_0^{\frac{\pi}{2}} \log(\tan x)\, dx$ (d) $I = \displaystyle\int_0^{\frac{\pi}{2}} \log(\cot x)\, dx$

(e) $I = \displaystyle\int_0^{\frac{\pi}{2}} \log(\sec x)\, dx$ (f) $I = \displaystyle\int_0^{\frac{\pi}{2}} \log(\csc x)\, dx$ (g) $I = \displaystyle\int_0^{\frac{\pi}{2}} \log x\, dx$ (h) $I = \displaystyle\int_0^{\frac{\pi}{2}} (\sin x + \cos x)\, dx$ (i) $I = \displaystyle\int_0^{\frac{\pi}{4}} (\tan x + \cot x)\, dx$

(3) Evaluate: – (a) $I = \int_0^1 \dfrac{x^2 + 3x + 2}{(x + 1)} \, dx$ (b) $I = \int_1^2 \dfrac{\log x}{x} \, dx$ (c) $I = \int_0^\pi \log(1 + \cos x) \cdot \sin x \, dx$ (d) $I = \int_0^1 \sin^{-1} x \, dx$

(e) $I = \int_0^1 \cos^{-1} x \, dx$ (f) $I = \int_0^1 \tan^{-1} x \, dx$ (g) $I = \int_0^1 \cot^{-1} x \, dx$ (h) $I = \int_0^1 \sec^{-1} x \, dx$ (i) $I = \int_0^1 \operatorname{cosec}^{-1} x \, dx$

(4) Evaluate: – (a) $I = \int_0^{2\pi} 2^{3\log_2 x} \, dx$ (b) $I = \int_0^1 \cos^{-1}\left(\dfrac{1 + \cos x}{2\cos^{x}/_2}\right) dx$ (c) $I = \int_1^2 x^2 \log x \, dx$ (d) $I = \int_0^1 \dfrac{1}{\sqrt{x + 1} - \sqrt{x + 2}} \, dx$

(e) $I = \int_0^5 |x - 3| \, dx$ (f) $I = \int_0^{\frac{\pi}{2}} \sin^5 x \, dx$ (g) $I = \int_1^2 \dfrac{1}{x\left(\log x + \dfrac{1}{\log x}\right)} \, dx$ (h) $I = \int_{-\frac{\pi}{2}}^{\frac{\pi}{2}} e^x \cdot \sin x \, dx$

(5) Evaluate: – (a) $I = \int_{-\frac{\pi}{2}}^{\frac{\pi}{2}} \dfrac{\cos x}{2 + \sin x} \, dx$ (b) $I = \int_{\frac{\pi}{4}}^{\frac{\pi}{2}} \sin(2x + 3) \, dx$ (c) $I = \int_1^3 (5x + 2)^5 \, dx$ (d) $I = \int_1^5 \dfrac{\sqrt{6 - x}}{\sqrt{x} + \sqrt{6 - x}} \, dx$

(e) $I = \int_0^{\frac{\pi}{2}} \dfrac{\cos x}{\sqrt{\sin^2 x + 2\sin x + 3}} \, dx$ (f) $I = \int_0^{\frac{\pi}{4}} \tan^2(x + 1) \, dx$ (g) $I = \int_0^{\frac{\pi}{2}} \dfrac{e^{\sqrt{\cos x}}}{e^{\sqrt{\sin x}} + e^{\sqrt{\cos x}}} \, dx$ (h) $I = \int_0^{\frac{\pi}{2}} \sin^2 x \cdot \cos^3 x \, dx$

(6) Evaluate: – (a) $I = \int_0^1 \dfrac{dx}{x^2 + x + 1}$ (b) $I = \int_0^2 |x^2 + x| \, dx$ (c) $I = \int_0^2 \dfrac{dx}{(x - 1) \cdot \sqrt{x + 2}}$ (d) $I = \int_1^e \dfrac{\log x}{x} \, dx$

(e) $I = \int_6^7 \dfrac{dx}{(x + 3) \cdot \sqrt{x - 5}}$ (f) $I = \int_2^3 \dfrac{dx}{(x + 2)(x + 3)}$ (g) $I = \int_{-1}^6 \dfrac{dx}{(x - 1)(x - 5)}$ (h) $I = \int_1^3 \dfrac{dx}{(x + 1)\sqrt{x^2 + 2x + 3}}$

(7) Evaluate: – (a) $I = \int_0^1 \dfrac{2x^2 - 5x + 1}{\sqrt{x}} \, dx$ (b) $I = \int_1^2 \dfrac{3x^3 + 4x^2 - x - 2}{(3x - 2)} \, dx$ (c) $I = \int_0^{\frac{\pi}{2}} \dfrac{\sin 2x}{2 - \cos^2 x} \, dx$ (d) $I = \int_0^{\log 2} e^{2x} \, dx$

(e) $I = \int_{\log 2}^{\log 3} (e^x - 1) \, dx$ (f) $I = \int_e^{e^2} |x - 4| \, dx$ (g) $I = \int_{-2}^2 \dfrac{e^x}{e^x + e^{-x}} \, dx$ (h) $I = \int_0^{\frac{\pi}{2}} \dfrac{\sin x}{\sin x + \cos x} \, dx$ (i) $I = \int_1^{\sin^2 \theta} \dfrac{\sin^{-1} \sqrt{x}}{\sqrt{1 - x}} \, dx$

(j) $I = \int_0^{\frac{\pi}{2}} \dfrac{dx}{1 + \cos x}$ (k) $I = \int_0^{\frac{\pi}{2}} \dfrac{dx}{1 + \tan x}$ (l) $I = \int_0^1 \dfrac{\sqrt{x} + \sqrt[3]{x}}{\sqrt[4]{x}} \, dx$ (m) $I = \int_0^{\frac{\pi}{2}} \dfrac{\sqrt{\cos x}}{\sqrt{\sin x} + \sqrt{\cos x}} \, dx$

(8) Evaluate: – (a) $I = \int_1^2 \dfrac{\sin(1 + \log x)}{x} \, dx$ (b) $I = \int_{-\frac{1}{3}}^{\frac{1}{3}} e^{\cos x} \cdot \sin x \, dx$ (c) $I = \int_{-\frac{\pi}{4}}^{\frac{\pi}{4}} e^{\sec x} \cdot \sec x \tan x \, dx$

(d) $I = \int_{-5}^5 \dfrac{3x^5 + 2x^4 + 5x^3 + 4x^2 + 2x + 1}{(x^2 + 2)} \, dx$ (e) $I = \int_{-3}^3 \dfrac{5x^3 + x}{x^2 + 1} \, dx$ (f) $I = \int_{-1}^1 \dfrac{x^3 \cos x}{x^2 + 1} \, dx$ (g) $I = \int_{\frac{\pi}{4}}^{\frac{3\pi}{4}} \dfrac{dx}{1 - \cos x}$

(9) Evaluate the following integral: – (a) $I = \int_0^{\frac{\pi}{2}} \dfrac{\sqrt{\sin x}}{\sqrt{\sin x} + \sqrt{\cos x}} \, dx$ (b) $I = \int_0^1 \dfrac{e^{1-x}}{e^x + e^{1-x}} \, dx$

(c) $I = \int_0^\pi \dfrac{e^{\cos x}}{e^{\cos x} + e^{-\cos x}} \, dx$ (d) $I = \int_0^{\frac{\pi}{2}} \dfrac{dx}{1 - \tan^2 x}$

(10) Evaluate: – (a) $I = \int_{\frac{\pi}{6}}^{\frac{\pi}{3}} \dfrac{dx}{1 + \sqrt{\tan x}}$ (b) $I = \int_{\sqrt{\log 2}}^{\sqrt{\log 3}} \dfrac{x \sin x^2}{\sin x^2 + \sin(\log 6 - x^2)} \, dx$ (c) $I = \int_0^1 \dfrac{x^4(1 - x)^4}{1 + x^2} \, dx$ (d) $I = \int_{e^{-1}}^{e^2} \left|\dfrac{\log_e x}{x}\right| \, dx$

(11) Evaluate: – (a) $\int_0^\pi [2\sin x] \, dx$ (b) $\int_0^{102} [\tan^{-1} x] \, dx$ (c) $\int_0^{2n\pi} [\sin x + \cos x] \, dx$ (d) $\int_0^{\frac{5\pi}{12}} [\tan x] \, dx$

(e) $\int_0^{n^2} [\sqrt{x}] \, dx, \ n \in N$ where [.] denotes the greatest integer functions (G. I. F).

(f) Prove that $\int_0^x [t] \, dt = \dfrac{[x]([x] - 1)}{2} + [x](x - [x])$ where [.] denotes the G. I. F.

Answer

(1) (a) $I = \int_1^{\sqrt{3}} \dfrac{x^3}{\sqrt{x^4 - 1}} dx = \dfrac{\sqrt{8}}{2} = \dfrac{2\sqrt{2}}{2} = \sqrt{2}$ or 1.414   Ans. [Do yourself,   Put $x^4 - 1 = t^2$]

(b) $I = \int_0^{\frac{\pi}{6}} \cos 3x\, dx = \left(\dfrac{\sin 3x}{3}\right)_0^{\frac{\pi}{6}} = \dfrac{\sin 3.\frac{\pi}{6}}{3} = \dfrac{\sin \frac{\pi}{2}}{3} = \dfrac{1}{3}$ or 0.33 or 0.334   Ans.

(c) $I = \int_0^2 \dfrac{3}{x + 2} dx = 3\int_0^2 \dfrac{dx}{x + 2} = 3[\log(x + 2)]_0^2 = 3[\log 4 - \log 2] = 3[\log 2^2 - \log 2] = 3[2\log 2 - \log 2] = 3\log 2$ Ans.

(d) $I = \int_1^e \dfrac{3}{2\sqrt{1 + x}} dx = 3(\sqrt{e + 1} - \sqrt{2})$ Ans.   (e) $I = \int_0^{\frac{\pi}{2}} \tan^{-1}\left(\dfrac{x}{\sqrt{1 - x^2}}\right) dx = \dfrac{\pi}{2} - 1$ Ans.

$\left(\text{Do yourself, put } x = \sin\theta \text{ and use integration by part formula Q. No. } -(e)\right)$

(f) $I = \int_0^1 \tan^{-1}\left(\dfrac{x - 1}{x + 1}\right) dx = \int_0^1 \tan^{-1}\left(\dfrac{x - 1}{1 + x.1}\right) dx = \int_0^1 (\tan^{-1} x - \tan^{-1} 1)\, dx = \int_0^1 \tan^{-1} x\, dx - \int_0^1 \tan^{-1} 1\, dx$

Let $I_1 = \int_0^1 \tan^{-1} x\, dx = \left[\tan^{-1} x . x - \int \dfrac{x}{1 + x^2} dx\right]_0^1$,   Put $1 + x^2 = t$ $\therefore$ $2x\, dx = dt$ or $x\, dx = \dfrac{dt}{2}$

$I_1 = \left[\tan^{-1} x . x - \dfrac{1}{2}\int \dfrac{dt}{t}\right]_0^1 = \left[\tan^{-1} x . x - \dfrac{1}{2}\log t\right]_0^1 = \left[\tan^{-1} x . x - \dfrac{1}{2}\log(1 + x^2)\right]_0^1 = \tan^{-1} 1 - \dfrac{1}{2}\log 2 = \dfrac{\pi}{4} - \dfrac{1}{2}\log 2$

Let $I_2 = \int_0^1 \tan^{-1} 1\, dx = \tan^{-1} 1 . (x)_0^1 = \tan^{-1} 1 = \dfrac{\pi}{4}$

$I = I_1 - I_2 = \dfrac{\pi}{4} - \dfrac{1}{2}\log 2 - \dfrac{\pi}{4} = -\dfrac{1}{2}\log 2$ or $-\log\sqrt{2}$ or $\log\left(\dfrac{1}{\sqrt{2}}\right)$ Ans.

(2) (a) $I = \int_0^{\frac{\pi}{2}} \log(\sin x)\, dx$ $\qquad \left[\text{use by property,}\qquad \int_0^a f(x)\, dx = \int_0^a f(a - x)\, dx\right]$

or $I = \int_0^{\frac{\pi}{2}} \log\left[\sin\left(\dfrac{\pi}{2} - x\right)\right] dx = \int_0^{\frac{\pi}{2}} \log(\cos x)\, dx = I$

Adding, $2I = \int_0^{\frac{\pi}{2}} \log(\sin x)\, dx + \int_0^{\frac{\pi}{2}} \log(\cos x)\, dx = \int_0^{\frac{\pi}{2}} [\log(\sin x) + \log(\cos x)]\, dx = \int_0^{\frac{\pi}{2}} \log(\sin x . \cos x)\, dx$

or $2I = \int_0^{\frac{\pi}{2}} \log\left(\dfrac{2\sin x . \cos x}{2}\right) dx = \int_0^{\frac{\pi}{2}} \log\left(\dfrac{\sin 2x}{2}\right) dx = \int_0^{\frac{\pi}{2}} \log(\sin 2x)\, dx - \int_0^{\frac{\pi}{2}} \log 2\, dx = I_1 - I_2$ (say) .... (i)

where $I_1 = \int_0^{\frac{\pi}{2}} \log(\sin 2x)\, dx$ and $I_2 = \int_0^{\frac{\pi}{2}} \log 2\, dx$

Solve, $I_1 = \int_0^{\frac{\pi}{2}} \log(\sin 2x)\, dx$ Put $2x = t$ $\therefore$ $2dx = dt$ or $dx = \dfrac{dt}{2}$ and limit become 0 to π.

$I_1 = \int_0^{\pi} \log(\sin t) . \dfrac{dt}{2} = \dfrac{1}{2}\int_0^{\pi} \log(\sin t)\, dt$ use property, $\int_0^{2a} f(x)\, dx = \begin{cases} 2\displaystyle\int_0^a f(x)\, dx, & \text{if } f(2a - x) = f(x) \\ 0, & \text{if } f(2a - x) = -f(x) \end{cases}$

$\therefore$ $a = \pi$, $f(t) = \log(\sin t)$, $f(2a - t) = f(2\pi - t) = \log[\sin(2\pi - t)] = \log(\sin t) = f(t)$

$\therefore$ $I_1 = \dfrac{1}{2}\int_0^{\pi} \log(\sin t)\, dt = \dfrac{1}{2} . 2\int_0^{\frac{\pi}{2}} \log(\sin t)\, dt = \int_0^{\frac{\pi}{2}} \log(\sin t)\, dt = I$

Now, solve $I_2 = \int_0^{\frac{\pi}{2}} \log 2 \ dx = \log 2 \int_0^{\frac{\pi}{2}} dx = \log 2 \ (x)_0^{\frac{\pi}{2}} = \log 2 \left(\frac{\pi}{2} - 0\right) = \frac{\pi}{2}\log 2$

Put value I_1 and I_2 in equation (i), we get $\quad \therefore \ 2I = I_1 - I_2 = I - \frac{\pi}{2}\log 2 \quad$ or $\ 2I - I = -\frac{\pi}{2}\log 2 \quad \therefore \ I = \frac{\pi}{2}\log\frac{1}{2} \quad$ Ans.

(b) $I = \int_0^{\frac{\pi}{2}} \log(\cos x) \ dx = \frac{\pi}{2}\log\frac{1}{2} \quad$ Ans. (Do yourself, to be solve same as above question)

(c) $I = \int_0^{\frac{\pi}{2}} \log(\tan x) \ dx = \int_0^{\frac{\pi}{2}} \log\left(\frac{\sin x}{\cos x}\right) dx = \int_0^{\frac{\pi}{2}} \log(\sin x) \ dx - \int_0^{\frac{\pi}{2}} \log(\cos x) \ dx = \frac{\pi}{2}\log\frac{1}{2} - \frac{\pi}{2}\log\frac{1}{2} = 0 \quad$ Ans.

(d) $I = \int_0^{\frac{\pi}{2}} \log(\cot x) \ dx = 0 \quad$ Ans. $\qquad \boxed{\therefore \ \int_0^{\frac{\pi}{2}} \log(\tan x) \ dx = \int_0^{\frac{\pi}{2}} \log(\cot x) \ dx = 0 \quad \text{Ans.}}$

(e) $I = \int_0^{\frac{\pi}{2}} \log(\sec x) \ dx = \int_0^{\frac{\pi}{2}} \log\left(\frac{1}{\cos x}\right) dx = \int_0^{\frac{\pi}{2}} \log 1 \ dx - \int_0^{\frac{\pi}{2}} \log(\cos x) \ dx = 0 - \int_0^{\frac{\pi}{2}} \log(\cos x) \ dx = -\int_0^{\frac{\pi}{2}} \log(\cos x) \ dx = -\frac{\pi}{2}\log\frac{1}{2}$

$\qquad\qquad = \frac{\pi}{2}\log 2 \quad$ Ans. [see question no. $-$(1)(b)]

(f) $I = \int_0^{\frac{\pi}{2}} \log(\csc x) \ dx = \int_0^{\frac{\pi}{2}} \log\left(\frac{1}{\sin x}\right) dx = \int_0^{\frac{\pi}{2}} \log 1 \ dx - \int_0^{\frac{\pi}{2}} \log(\sin x) \ dx = 0 - \int_0^{\frac{\pi}{2}} \log(\sin x) \ dx = -\int_0^{\frac{\pi}{2}} \log(\sin x) \ dx = -\frac{\pi}{2}\log\frac{1}{2}$

$\qquad\qquad = \frac{\pi}{2}\log 2 \quad$ Ans. [see question no. $-$(1)(a)]

$$\boxed{\therefore \ \int_0^{\frac{\pi}{2}} \log(\sec x) \ dx = \int_0^{\frac{\pi}{2}} \log(\csc x) \ dx = \frac{\pi}{2}\log 2 \quad \text{Ans.}}$$

(g) $I = \int_0^{\frac{\pi}{2}} \log x \ dx = \int_0^{\frac{\pi}{2}} \log x \cdot 1 \ dx \quad$ use integration by part formula, $\int u.v \ dx = u.\int v \ dx - \int \left[\frac{du}{dx}.\int v \ dx\right] dx$

$I = \log x \int 1 \ dx - \int \left\{\frac{d(\log x)}{dx}.\int 1 \ dx\right\} dx = \log x . x - \int \frac{1}{x}.x \ dx = x\log x - \int dx = x\log x - x = x(\log x - 1)$

Put limit 0 to $\frac{\pi}{2}$ then $\ I = [x(\log x - 1)]_0^{\frac{\pi}{2}} = \left[\frac{\pi}{2}\left(\log\frac{\pi}{2} - 1\right) - 0\right] = \frac{\pi}{2}\left(\log\frac{\pi}{2} - 1\right) \quad$ Ans.

(h) $I = \int_0^{\frac{\pi}{2}} (\sin x + \cos x) \ dx = \int_0^{\frac{\pi}{2}} \sin x \ dx + \int_0^{\frac{\pi}{2}} \cos x \ dx = (-\cos x)_0^{\frac{\pi}{2}} + (\sin x)_0^{\frac{\pi}{2}} = -\cos\left(\frac{\pi}{2}\right) + \cos 0 + \sin\left(\frac{\pi}{2}\right) - \sin 0 = -0 + 1 + 1 - 0$

$\qquad\qquad = 1 + 1 = 2 \quad$ Ans.

(i) $I = \int_0^{\frac{\pi}{4}} (\tan x + \cot x) \ dx = \int_0^{\frac{\pi}{4}} \left(\tan x + \frac{1}{\tan x}\right) dx = \int_0^{\frac{\pi}{4}} \left(\frac{\tan^2 x + 1}{\tan x}\right) dx = \int_0^{\frac{\pi}{4}} \frac{\sec^2 x}{\tan x} \ dx$

Put $\tan x = t \quad \therefore \ \sec^2 x \ dx = dt$

$I = \int_0^{\frac{\pi}{4}} \frac{dt}{t} = (\log t)_0^{\frac{\pi}{4}} = [\log(\tan x)]_0^{\frac{\pi}{4}} = \left[\log\left(\tan\frac{\pi}{4}\right) - \log(\tan 0)\right] = \log 1 - \log 0 = 0 \quad$ Ans.

(3) (a) $I = \int_0^1 \frac{x^2 + 3x + 2}{(x + 1)} \ dx = \int_0^1 \frac{x^2 + 2x + x + 2}{(x + 1)} \ dx = \int_0^1 \frac{x(x + 2) + 1(x + 2)}{(x + 1)} \ dx = \int_0^1 \frac{(x + 1)(x + 2)}{(x + 1)} \ dx = \int_0^1 (x + 2) \ dx = \left[\frac{x^2}{2} + 2x\right]_0^1$

$\qquad\qquad = \left[\frac{1}{2} + 2 - 0\right] = \frac{1}{2} + 2 = \frac{5}{2} \quad$ Ans.

x	1	2

(b) $I = \int_1^2 \dfrac{\log x}{x}\ dx$, Put $\log x = t$ $\quad \therefore \dfrac{1}{x} dx = dt$ and limit become 0 to $\log 2$

t	0	log 2

$I = \int_0^{\log 2} t\ dt = \left(\dfrac{t^2}{2}\right)_0^{\log 2} = \dfrac{(\log 2)^2}{2} - 0 = \dfrac{[\log 2]^2}{2}$ $\qquad$ Ans.

(c) $I = \int_0^{\pi} \log(1 + \cos x).\sin x\ dx$

Put $1 + \cos x = t$ $\quad \therefore -\sin x\ dx = dt$ $\quad \therefore \sin x\ dx = -dt$ $\qquad$ Limit $1 + \cos x = t$

x	0	π
t	2	0

$I = -\int_2^0 \log t\ dt = \int_0^2 \log t\ dt$ $\quad$ use integration by part formula,

$I = [t(\log t - 1)]_0^2 = 2(\log 2 - 1) - 0 = 2(\log 2 - 1)$ $\quad$ Ans. $\quad$ [see question (1)(g).]

(d) $I = \int_0^1 \sin^{-1} x\ dx$ $\quad \left\{\text{use integration by part formula, } \int u.v\ dx = u.\int v\ dx - \int \left[\dfrac{du}{dx}.\int v\ dx\right] dx\right\}$

$I = \sin^{-1} x. \int dx - \int \left[\dfrac{d(\sin^{-1} x)}{dx}.\int dx\right] dx = \sin^{-1} x.x - \int \dfrac{x}{\sqrt{1 - x^2}}\ dx$

Put $1 - x^2 = t^2$ $\quad \therefore -2x\ dx = 2t\ dt$ $\quad \therefore x\ dx = -t\ dt$

$I = x \sin^{-1} x + \int \dfrac{t\ dt}{t} = x \sin^{-1} x + \int dt = x \sin^{-1} x + t = x \sin^{-1} x + \sqrt{1 - x^2}$

Put limit 0 to 1 then, $I = \left(x \sin^{-1} x + \sqrt{1 - x^2}\right)_0^1 = (1.\sin^{-1} 1 + 0 - 0 - 1) = \dfrac{\pi}{2} - 1$ $\qquad$ Ans.

(e) $I = \int_0^1 \cos^{-1} x\ dx = 1$ $\quad$ Ans. (see above question)

(f) $I = \int_0^1 \tan^{-1} x\ dx$ $\quad \left\{\text{use integration by part formula, } \int u.v\ dx = u.\int v\ dx - \int \left[\dfrac{du}{dx}.\int v\ dx\right] dx\right\}$

$I = \left\{\tan^{-1} x. \int dx - \int \left[\dfrac{d(\tan^{-1} x)}{dx}.\int dx\right] dx\right\}_0^1 = \left[x \tan^{-1} x - \int \dfrac{x}{1 + x^2}\ dx\right]_0^1$

Put $1 + x^2 = t$ $\quad$ or $\quad 2x\ dx = dt$

$I = \left[x \tan^{-1} x - \dfrac{1}{2}\int \dfrac{dt}{t}\right]_0^1 = \left[x \tan^{-1} x - \dfrac{1}{2}\log t\right]_0^1 = \left[x \tan^{-1} x - \dfrac{1}{2}\log(1 + x^2)\right]_0^1 = \left[1.\tan^{-1} 1 - \dfrac{1}{2}\log(1 + 1)\right] - [0 - 0] = \tan^{-1} 1 - \dfrac{1}{2}\log 2$

$\qquad\qquad = \dfrac{\pi}{4} - \dfrac{1}{2}\log 2$ $\quad$ or $\quad \dfrac{\pi}{4} - \log \sqrt{2}$ $\qquad$ Ans.

(g) $I = \int_0^1 \cot^{-1} x\ dx = \dfrac{\pi}{4} + \dfrac{1}{2}\log 2$ $\quad$ or $\quad \dfrac{\pi}{4} + \log \sqrt{2}$ $\quad$ Ans. (Do yourself, To be solve same as above question)

(h) $I = \int_0^1 \sec^{-1} x\ dx$, doyourself $\left[x \sec^{-1} x - \log\left(x + \sqrt{x^2 - 1}\right)\right]_0^1$

(4) (a) $I = \int_0^{2\pi} 2^{3\log_2 x}\ dx = \int_0^{2\pi} 2^{\log_2 x^3}\ dx = \int_0^{2\pi} x^3\ dx = \left(\dfrac{x^4}{4}\right)_0^{2\pi} = \dfrac{(2\pi)^4}{4} - 0 = \dfrac{16\pi^4}{4} = 4\pi^4$ $\quad$ Ans.

(b) $I = \int_0^1 \cos^{-1}\left(\dfrac{1 + \cos x}{2\cos^x/_2}\right) dx = \dfrac{1}{4}$ $\quad$ Ans. (Do yourself)

$\left[\text{Hint:} - 1 + \cos x = 2\cos^2 x/_2\ ,\quad I = \int_0^1 \cos^{-1}\left(\dfrac{2\cos^2 x/_2}{2\cos^x/_2}\right) dx = \int_0^1 \cos^{-1}\left(\cos x/_2\right) dx = \int_0^1 \dfrac{x}{2} dx\right]$

(c) $I = \int_1^2 x^2 \log x \; dx = \dfrac{8}{3}\log 2 - \dfrac{7}{9}$ Ans.   (Do yourself, use integration by part formula)

(d) $I = \int_0^1 \dfrac{1}{\sqrt{x+1} - \sqrt{x+2}} \; dx = \int_0^1 \left(\dfrac{1}{\sqrt{x+1} - \sqrt{x+2}} \times \dfrac{\sqrt{x+1} + \sqrt{x+2}}{\sqrt{x+1} + \sqrt{x+2}} \right) dx = \int_0^1 \dfrac{\sqrt{x+1} + \sqrt{x+2}}{x + 1 - x - 2} \; dx$

$I = -\int_0^1 \left(\sqrt{x+1} + \sqrt{x+2} \right) dx = -\int_0^1 \sqrt{x+1} \; dx - \int_0^1 \sqrt{x+2} \; dx = -I_1 - I_2 \; \dots\dots\dots\dots\dots. \text{(A)}$

solve, $I_1 = \int_0^1 \sqrt{x+1} \; dx$, Put $x + 1 = t^2$ or $t = \sqrt{x+1}$ $\therefore$ $dx = 2t \, dt$

$\therefore \; I_1 = \int t. \, 2t \; dt = 2 \int t^2 \; dt = 2.\dfrac{t^3}{3} = \dfrac{2}{3} . (x+1) . \sqrt{x+1}$,   Put limit 0 to 1

$\therefore \; I_1 = \dfrac{2}{3} . \left((x+1) . \sqrt{x+1} \right)_0^1 = \dfrac{2}{3}\left(2\sqrt{2} - 1 \right)$

x	0	1
z	2	3

Now, solve $I_2 = \int_0^1 \sqrt{x+2} \; dx$, Put $x + 2 = z$ $\therefore$ $dx = dz$ and limit become 2 to 3

$\therefore \; I_2 = \int_2^3 \sqrt{z} \; dz = \int_2^3 z^{\frac{1}{2}} \; dz = \left(\dfrac{z^{\frac{3}{2}}}{\frac{3}{2}} \right)_2^3 = \dfrac{2}{3}\left[z\sqrt{z} \right]_2^3 = \dfrac{2}{3}\left(3\sqrt{3} - 2\sqrt{2} \right)$

Put value of I_1 and I_2 in equation (A), we have $\therefore \; I = -I_1 - I_2 = -\dfrac{2}{3}\left(2\sqrt{2} - 1 \right) - \dfrac{2}{3}\left(3\sqrt{3} - 2\sqrt{2} \right)$

$\therefore \; I = -\dfrac{2}{3}\left(2\sqrt{2} - 1 + 3\sqrt{3} - 2\sqrt{2} \right) = -\dfrac{2}{3}\left(3\sqrt{3} - 1 \right) = \dfrac{2}{3}\left(1 - 3\sqrt{3} \right)$ or $\left(\dfrac{2}{3} - 2\sqrt{3} \right)$ Ans.

(e) $I = \int_0^5 |x - 3| \; dx$, Put $f(x) = x - 3$, $f(x) = 0$ or $x - 3 = 0$ or $x = 3$

case I: $-$ If $x > 3$ $then$ $f(x)$ is positive in the interval $3 < x < 5$

case II: $-$ If $x < 3$ $then$ $f(x)$ is negative in the interval $0 < x < 3$

$$\therefore \; f(x) = \begin{cases} -(x - 3), & \text{if } x < 3 \\ (x - 3), & \text{if } x > 3 \end{cases}$$

$I = \int_0^3 |x - 3| \; dx + \int_3^5 |x - 3| \; dx = -\int_0^3 (x - 3) \; dx + \int_3^5 (x - 3) \; dx = -\int_0^3 x \, dx + 3\int_0^3 dx + \int_3^5 x \, dx - 3 \int_3^5 dx$

$= -\left(\dfrac{x^2}{2} \right)_0^3 + 3(x)_0^3 + \left(\dfrac{x^2}{2} \right)_3^5 - 3(x)_3^5 = -\dfrac{9}{2} + 9 + \dfrac{25}{2} - \dfrac{9}{2} - 3(5 - 3) = -\dfrac{9}{2} + 9 + \dfrac{25}{2} - \dfrac{9}{2} - 6 = \dfrac{-9 + 18 + 25 - 9 - 12}{2}$

$= \dfrac{13}{2}$ Ans.

(f) $I = \int_0^{\frac{\pi}{2}} \sin^5 x \; dx = \int_0^{\frac{\pi}{2}} \sin^4 x . \sin x \; dx = \int_0^{\frac{\pi}{2}} (\sin^2 x)^2 . \sin x \; dx = \int_0^{\frac{\pi}{2}} (1 - \cos^2 x)^2 . \sin x \; dx$

Put $\cos x = t$ $\therefore$ $-\sin x \; dx = dt$ or $\sin x \; dx = -dt$ and limit becomes

x	0	$\dfrac{\pi}{2}$
t	1	0

$\therefore \; I = -\int_1^0 (1 - t^2)^2 \; dt = \int_0^1 (1 - t^2)^2 \; dt = \int_0^1 (1 - 2t^2 + t^4) \; dt$

$\therefore \; I = \int_0^1 dt - 2\int_0^1 t^2 \; dt + \int_0^1 t^4 \; dt = (t)_0^1 - 2\left(\dfrac{t^3}{3} \right)_0^1 + \left(\dfrac{t^5}{5} \right)_0^1 = (1 - 0) - 2\left(\dfrac{1}{3} - 0 \right) + \left(\dfrac{1}{5} - 0 \right) = 1 - \dfrac{2}{3} + \dfrac{1}{5} = \dfrac{15 - 10 + 3}{15} = \dfrac{18 - 10}{15}$

$= \dfrac{8}{15}$ Ans.

(g) $I = \int_1^2 \dfrac{1}{x\left(\log x + \frac{1}{\log x}\right)}\, dx$

Put $\log x = t$ $\quad \therefore \dfrac{1}{x}\, dx = dt$ $\quad$ and limit becomes

x	1	2
t	0	log 2

$\therefore I = \int_1^2 \dfrac{1}{\left(\log x + \frac{1}{\log x}\right)} \cdot \dfrac{1}{x}\, dx = \int_0^{\log 2} \dfrac{dt}{\left(t + \frac{1}{t}\right)} = \int_0^{\log 2} \dfrac{t}{t^2 + 1}\, dt$

Put $1 + t^2 = z$ $\quad \therefore 2t\, dt = dz$ $\quad$ or $t\, dt = \dfrac{dz}{2}$

$1 + t^2 = z$		
t	0	log 2
z	1	$1 + (\log 2)^2$

$\therefore I = \dfrac{1}{2}\int_1^{1+(\log 2)^2} \dfrac{dz}{z} = \dfrac{1}{2}(\log z)_1^{1+(\log 2)^2} = \dfrac{1}{2}\{\log[1 + (\log 2)^2] - 0\}$

$\therefore I = \dfrac{1}{2}\log[1 + (\log 2)^2]$ or $\log[1 + (\log 2)^2]^{\frac{1}{2}}$ or $\log\left(\sqrt{1 + (\log 2)^2}\right)$ $\quad$ Ans.

(h) $I = \int_{-\frac{\pi}{2}}^{\frac{\pi}{2}} e^x \cdot \sin x\, dx$

(5) (a) $I = \int_{-\frac{\pi}{2}}^{\frac{\pi}{2}} \dfrac{\cos x}{2 + \sin x}\, dx$ Put $2 + \sin x = t$ $\quad \therefore \cos x\, dx = dt$ $\quad$ and limit becomes 1 to 3

x	$-\frac{\pi}{2}$	$\frac{\pi}{2}$
t	1	3

$\therefore I = \int_1^3 \dfrac{dt}{t} = (\log t)_1^3 = \log 3 - \log 1 = \log 3 - 0 = \log 3$ $\quad$ Ans.

(b) $I = \int_{\frac{\pi}{4}}^{\frac{\pi}{2}} \sin(2x + 3)\, dx$, Put $2x + 3 = t$ $\quad \therefore 2\, dx = dt$ $\quad$ or $dx = \dfrac{dt}{2}$

$\therefore I = \dfrac{1}{2}\int_{\frac{\pi}{4}}^{\frac{\pi}{2}} \sin t\, dt = -\dfrac{1}{2} \cdot (\cos t)_{\frac{\pi}{4}}^{\frac{\pi}{2}} = -\dfrac{1}{2}[\cos(2x + 3)]_{\frac{\pi}{4}}^{\frac{\pi}{2}} = -\dfrac{1}{2}\left[\cos(\pi + 3) - \cos\left(\frac{\pi}{2} + 3\right)\right] = -\dfrac{1}{2}[-\cos 3 + \sin 3]$

$\qquad\qquad = \dfrac{1}{2}[\cos 3 - \sin 3]$ $\quad$ Ans. $\quad$ or $\quad \dfrac{1}{\sqrt{2}}\left[\sin\left(\dfrac{3\pi}{4} + 3\right)\right]$ $\quad$ Ans.

(c) $I = \int_1^3 (5x + 2)^5\, dx$, Put $5x + 2 = t$ $\quad \therefore 5\, dx = dt$ $\quad$ or $dx = \dfrac{dt}{5}$ and limit become 7 to 17.

$\therefore I = \int_7^{17} t^5 \cdot \dfrac{dt}{5} = \dfrac{1}{5}\int_7^{17} t^5\, dt = \dfrac{1}{5}\left(\dfrac{t^6}{6}\right)_7^{17} = \dfrac{1}{30}[17^6 - 7^6]$ or $\dfrac{1}{30}[24137569 - 117649] = \dfrac{24019920}{30} = 800664$ Ans.

(d) $I = \int_1^5 \dfrac{\sqrt{6 - x}}{\sqrt{x} + \sqrt{6 - x}}\, dx$ $\qquad \left[\text{use property VI,}\ \int_b^a f(x)\, dx = \int_b^a f(a + b - x)\, dx\right]$

$\therefore I = \int_1^5 \dfrac{\sqrt{6 - x}}{\sqrt{x} + \sqrt{6 - x}}\, dx = \int_1^5 \dfrac{\sqrt{6 - (6 - x)}}{\sqrt{6 - x} + \sqrt{6 - (6 - x)}}\, dx = \int_1^5 \dfrac{\sqrt{x}}{\sqrt{6 - x} + \sqrt{x}}\, dx = I$ (say)

Adding, $2I = \int_1^5 \left(\dfrac{\sqrt{6 - x}}{\sqrt{x} + \sqrt{6 - x}} + \dfrac{\sqrt{x}}{\sqrt{6 - x} + \sqrt{x}}\right) dx = \int_1^5 \left(\dfrac{\sqrt{x} + \sqrt{6 - x}}{\sqrt{x} + \sqrt{6 - x}}\right) dx = \int_1^5 dx = (x)_1^5 = 5 - 1 = 4$

$\qquad\qquad\qquad\qquad \therefore 2I = 4 \quad \therefore I = 2$ $\quad$ Ans.

(e) $I = \int_0^{\frac{\pi}{2}} \dfrac{\cos x}{\sqrt{\sin^2 x + 2\sin x + 3}}\, dx$, Put $\sin x = t$ $\quad \therefore \cos x\, dx = dt$ and limit become 0 to 1

x	0	$\frac{\pi}{2}$

or $I = \int_0^1 \dfrac{dt}{\sqrt{t^2 + 2t + 3}} = \int_0^1 \dfrac{dt}{\sqrt{(t+1)^2 + 2}} = \int_0^1 \dfrac{dt}{\sqrt{(t+1)^2 + \left(\sqrt{2}\right)^2}}$

t	0	1

use formula, $\displaystyle\int \dfrac{dx}{\sqrt{x^2 + a^2}} = \log\left|x + \sqrt{x^2 + a^2}\right|$

or $I = \left[\log\left|(t+1) + \sqrt{(t+1)^2 + 2}\right|\right]_0^1 = \log|2 + \sqrt{6}| - \log|1 + \sqrt{3}| = \log\left|\dfrac{2 + \sqrt{6}}{1 + \sqrt{3}}\right|$ Ans.

(f) $I = \displaystyle\int_0^{\frac{\pi}{4}} \tan^2(x + 1)\ dx$, Put $x + 1 = t$ $\therefore\ dx = dt$

$\therefore\ I = \displaystyle\int_0^{\frac{\pi}{4}} \tan^2 t\ dt = \int_0^{\frac{\pi}{4}} (\sec^2 t - 1)\ dt = \int_0^{\frac{\pi}{4}} \sec^2 t\ dt - \int_0^{\frac{\pi}{4}} dt = [\tan t - t]_0^{\frac{\pi}{4}}$ Put $t = x + 1$

$\therefore\ I = [\tan(x + 1) - (x + 1)]_0^{\frac{\pi}{4}} = \tan\left(\dfrac{\pi}{4} + 1\right) - \left(\dfrac{\pi}{4} + 1\right) - \tan 1 + 1 = \tan\left(\dfrac{\pi}{4} + 1\right) - \tan 1 - \dfrac{\pi}{4}$ Ans.

(g) $I = \displaystyle\int_0^{\frac{\pi}{2}} \dfrac{e^{\sqrt{\cos x}}}{e^{\sqrt{\sin x}} + e^{\sqrt{\cos x}}}\ dx = \dfrac{\pi}{4}$ Ans.   [Do yourself, see question no. $-(4)(d)$ and use property IV.]

(h) $I = \displaystyle\int_0^{\frac{\pi}{2}} \sin^2 x \cdot \cos^3 x\ dx = \int_0^{\frac{\pi}{2}} \sin^2 x \cdot \cos^2 x \cdot \cos x\ dx = \int_0^{\frac{\pi}{2}} \sin^2 x\,(1 - \sin^2 x) \cdot \cos x\ dx$

Put $\sin x = t$ $\therefore\ \cos x\ dx = dt$ and limit become 0 to 1

x	0	$\dfrac{\pi}{2}$
t	0	1

$\therefore I = \displaystyle\int_0^1 t^2(1 - t^2)\ dt = \int_0^1 (t^2 - t^4)\ dt = \left[\dfrac{t^3}{3} - \dfrac{t^5}{5}\right]_0^1 = \left(\dfrac{1}{3} - \dfrac{1}{5} - 0\right) = \dfrac{5 - 3}{15} = \dfrac{2}{15}$ Ans.

(6) (a) $I = \displaystyle\int_0^1 \dfrac{dx}{x^2 + x + 1} = \int_0^1 \dfrac{dx}{\left(x + \dfrac{1}{2}\right)^2 + \dfrac{3}{4}}$

$= \displaystyle\int_0^1 \dfrac{dx}{\left(x + \dfrac{1}{2}\right)^2 + \left(\dfrac{\sqrt{3}}{2}\right)^2}$ $\left[\text{use formula,}\ \displaystyle\int \dfrac{dx}{x^2 + a^2} = \dfrac{1}{a}\tan^{-1}\left(\dfrac{x}{a}\right)\right]$

$\therefore I = \dfrac{1}{\frac{\sqrt{3}}{2}}\tan^{-1}\left(\dfrac{x + \frac{1}{2}}{\frac{\sqrt{3}}{2}}\right) = \dfrac{2}{\sqrt{3}}\tan^{-1}\left(\dfrac{2x + 1}{\sqrt{3}}\right)$,   Put limit 0 to 1   and $\tan^{-1}\sqrt{3} = \dfrac{\pi}{3}$, $\tan^{-1}\dfrac{1}{\sqrt{3}} = \dfrac{\pi}{6}$

$\therefore I = \left[\dfrac{2}{\sqrt{3}}\tan^{-1}\left(\dfrac{2x + 1}{\sqrt{3}}\right)\right]_0^1 = \dfrac{2}{\sqrt{3}}\tan^{-1}\left(\dfrac{3}{\sqrt{3}}\right) - \dfrac{2}{\sqrt{3}}\tan^{-1}\left(\dfrac{1}{\sqrt{3}}\right) = \dfrac{2}{\sqrt{3}}\left[\tan^{-1}(\sqrt{3}) - \tan^{-1}\left(\dfrac{1}{\sqrt{3}}\right)\right] = \dfrac{2}{\sqrt{3}}\left[\dfrac{\pi}{3} - \dfrac{\pi}{6}\right] = \dfrac{2}{\sqrt{3}}\left[\dfrac{2\pi - \pi}{6}\right] = \dfrac{2}{\sqrt{3}} \cdot \dfrac{\pi}{6}$

$= \dfrac{\pi}{3\sqrt{3}}$ Ans.

(b) $I = \displaystyle\int_0^2 |x^2 + x|\ dx$ $\therefore\ x^2 + x = 0$ or $x(x + 1) = 0$ $\therefore\ x = 0, -1$

If limit between $0 < x < 2$ then $|x^2 + x|$ is positive.

If limit between $-1 < x < 0$ then $|x^2 + x|$ is negative.

$I = \displaystyle\int_0^2 |x^2 + x|\ dx = \int_{-1}^0 -(x^2 + x)\ dx + \int_0^2 (x^2 + x)\ dx = -\int_{-1}^0 x^2\ dx - \int_{-1}^0 x\ dx + \int_0^2 x^2\ dx + \int_0^2 x\ dx = -\left(\dfrac{x^3}{3}\right)_{-1}^0 - \left(\dfrac{x^2}{2}\right)_{-1}^0 + \left(\dfrac{x^3}{3}\right)_0^2 + \left(\dfrac{x^2}{2}\right)_0^2$

$= -\left(0 + \dfrac{1}{3}\right) - \left(0 - \dfrac{1}{2}\right) + \left(\dfrac{8}{3} - 0\right) + \left(\dfrac{4}{2} - 0\right) = -\dfrac{1}{3} + \dfrac{1}{2} + \dfrac{8}{3} + 2 = \dfrac{-2 + 3 + 16 + 12}{6} = \dfrac{29}{6}$ Ans.

(c) $I = \displaystyle\int_0^2 \dfrac{dx}{(x - 1)\cdot\sqrt{x + 2}}$,   Put $x + 2 = t^2$ or $x = t^2 - 2$ $\therefore\ dx = 2t\ dt$

$I = \int_0^2 \dfrac{2t}{(t^2 - 2 - 1)\sqrt{t^2}}\, dt = \int_0^2 \dfrac{2t}{(t^2 - 3)\cdot t}\, dt = 2\int_0^2 \dfrac{dt}{(t^2 - 3)} = 2\int_0^2 \dfrac{dt}{\left[t^2 - (\sqrt{3})^2\right]}$ formula $\int \dfrac{dx}{x^2 - a^2} = \dfrac{1}{2a}\log\left|\dfrac{x-a}{x+a}\right|$

$I = \left[2\cdot\dfrac{1}{2\sqrt{3}}\log\left|\dfrac{t-\sqrt{3}}{t+\sqrt{3}}\right|\right]_0^2 = \left[\dfrac{1}{\sqrt{3}}\log\left|\dfrac{\sqrt{x+2}-\sqrt{3}}{\sqrt{x+2}+\sqrt{3}}\right|\right]_0^2 = \dfrac{1}{\sqrt{3}}\left[\log\left|\dfrac{\sqrt{2+2}-\sqrt{3}}{\sqrt{2+2}+\sqrt{3}}\right| - \log\left|\dfrac{\sqrt{0+2}-\sqrt{3}}{\sqrt{0+2}+\sqrt{3}}\right|\right] = \dfrac{1}{\sqrt{3}}\left[\log\left|\dfrac{2-\sqrt{3}}{2+\sqrt{3}}\right| - \log\left|\dfrac{\sqrt{2}-\sqrt{3}}{\sqrt{2}+\sqrt{3}}\right|\right]$

$= \dfrac{1}{\sqrt{3}}\log\left|\dfrac{2-\sqrt{3}}{2+\sqrt{3}} \times \dfrac{2-\sqrt{3}}{2-\sqrt{3}}\right| - \dfrac{1}{\sqrt{3}}\log\left|\dfrac{\sqrt{2}-\sqrt{3}}{\sqrt{2}+\sqrt{3}} \times \dfrac{\sqrt{2}-\sqrt{3}}{\sqrt{2}-\sqrt{3}}\right|$

$\therefore\ I = \dfrac{1}{\sqrt{3}}\log\left|\dfrac{(2-\sqrt{3})^2}{4-3}\right| - \dfrac{1}{\sqrt{3}}\log\left|\dfrac{(\sqrt{2}-\sqrt{3})^2}{2-3}\right| = \dfrac{1}{\sqrt{3}}\log\left|(2-\sqrt{3})^2\right| - \dfrac{1}{\sqrt{3}}\log\left|(\sqrt{2}-\sqrt{3})^2\right|$ Ans.

(d) $I = \int_1^e \dfrac{\log x}{x}\, dx$, Put $\log x = t$ $\therefore$ $\dfrac{1}{x}\, dx = dt$ and limit become 0 to 1. or $I = \int_0^1 t\, dt = \left(\dfrac{t^2}{2}\right)_0^1 = \dfrac{1}{2} - 0 = \dfrac{1}{2}$ Ans.

(e) $I = \int_6^7 \dfrac{dx}{(x+3)\cdot\sqrt{x-5}} = \dfrac{1}{\sqrt{2}}\left[\tan^{-1}\left(\dfrac{1}{2}\right) - \tan^{-1}\left(\dfrac{1}{2\sqrt{2}}\right)\right]$ Ans. [Do yourself, see question no. $-(5)(c)$]

(f) $I = \int_2^3 \dfrac{dx}{(x+2)(x+3)} = \int_2^3 \dfrac{dx}{x^2+5x+6} = \int_2^3 \dfrac{dx}{\left(x+\frac{5}{2}\right)^2 - \frac{1}{4}} = \int_2^3 \dfrac{dx}{\left(x+\frac{5}{2}\right)^2 - \left(\frac{1}{2}\right)^2}$ use formula $\int \dfrac{dx}{x^2 - a^2} = \dfrac{1}{2a}\log\left|\dfrac{x-a}{x+a}\right|$

$I = \left[\dfrac{1}{2\cdot\frac{1}{2}}\cdot\log\left|\dfrac{x+\frac{5}{2}-\frac{1}{2}}{x+\frac{5}{2}+\frac{1}{2}}\right|\right]_2^3 = \left[\log\left|\dfrac{2x+5-1}{2x+5+1}\right|\right]_2^3 = \log\left(\dfrac{2\cdot3+5-1}{2\cdot3+5+1}\right) - \log\left(\dfrac{2\cdot2+5-1}{2\cdot2+5+1}\right) = \log\left(\dfrac{10}{12}\right) - \log\left(\dfrac{8}{10}\right) = \log\left(\dfrac{5}{6}\right) - \log\left(\dfrac{4}{5}\right)$

$= \log\left[\dfrac{\frac{5}{6}}{\frac{4}{5}}\right] = \log\left[\dfrac{5}{6} \times \dfrac{5}{4}\right] = \log\left[\dfrac{25}{24}\right]$ Ans.

(g) $I = \int_{-1}^6 \dfrac{dx}{(x-1)(x-5)} = \dfrac{1}{4}\log\left(\dfrac{1}{15}\right)$ Ans. (Do yourself, see above question)

(h) $I = \int_1^3 \dfrac{dx}{(x+1)\sqrt{x^2+2x+3}}$, Put $x+1 = t$ $\therefore$ $dx = dt$ and limit become tend to 2 to 4.

$I = \int_1^3 \dfrac{dx}{(x+1)\sqrt{(x+1)^2+2}} = \int_2^4 \dfrac{dt}{t\sqrt{t^2+2}}$

Again put $t^2 + 2 = z^2$ $\therefore$ $2t\, dt = 2z\, dz$ or $t\, dt = z\, dz$ or $dt = \dfrac{z\, dz}{\sqrt{z^2-2}}$

$I = \int_2^4 \dfrac{z\, dz}{z\cdot\sqrt{z^2-2}\cdot\sqrt{z^2-2}} = \int_2^4 \dfrac{dz}{z^2-2} = \int_2^4 \dfrac{dz}{z^2 - (\sqrt{2})^2}$ use formula, $\int \dfrac{dx}{x^2-a^2} = \dfrac{1}{2a}\log\left|\dfrac{x-a}{x+a}\right|$

$I = \left[\dfrac{1}{2\cdot\sqrt{2}}\log\left|\dfrac{z-\sqrt{2}}{z+\sqrt{2}}\right|\right]_2^4 = \dfrac{1}{2\sqrt{2}}\left[\log\left|\dfrac{\sqrt{t^2+2}-\sqrt{2}}{\sqrt{t^2+2}+\sqrt{2}}\right|\right]_2^4 = \dfrac{1}{2\sqrt{2}}\left[\log\left|\dfrac{\sqrt{16+2}-\sqrt{2}}{\sqrt{16+2}+\sqrt{2}}\right| - \log\left|\dfrac{\sqrt{4+2}-\sqrt{2}}{\sqrt{4+2}+\sqrt{2}}\right|\right]$

$= \dfrac{1}{2\sqrt{2}}\left[\log\left|\dfrac{\sqrt{18}-\sqrt{2}}{\sqrt{18}+\sqrt{2}}\right| - \log\left|\dfrac{\sqrt{6}-\sqrt{2}}{\sqrt{6}+\sqrt{2}}\right|\right] = \dfrac{1}{2\sqrt{2}}\log\left|\dfrac{(\sqrt{18}-\sqrt{2})(\sqrt{6}+\sqrt{2})}{(\sqrt{18}+\sqrt{2})(\sqrt{6}-\sqrt{2})}\right|$ Ans.

(7) (a) $I = \int_0^1 \dfrac{2x^2 - 5x + 1}{\sqrt{x}}\, dx = \int_0^1 \left(\dfrac{2x^2}{\sqrt{x}} - \dfrac{5x}{\sqrt{x}} + \dfrac{1}{\sqrt{x}}\right) dx = \int_0^1 \dfrac{2x^2}{\sqrt{x}}\, dx - \int_0^1 \dfrac{5x}{\sqrt{x}}\, dx + \int_0^1 \dfrac{1}{\sqrt{x}}\, dx$

$I = 2\int_0^1 x^{3/2}\, dx - 5\int_0^1 x^{1/2}\, dx + \int_0^1 x^{-1/2}\, dx = 2\left(\dfrac{x^{5/2}}{5/2}\right)_0^1 - 5\left(\dfrac{x^{3/2}}{3/2}\right)_0^1 + \left(\dfrac{x^{1/2}}{1/2}\right)_0^1 = \dfrac{4}{5}(1-0) - \dfrac{10}{3}(1-0) + 2(1-0) = \dfrac{4}{5} - \dfrac{10}{3} + 2$

$= \dfrac{12 - 50 + 30}{15} = \dfrac{42 - 50}{15} = \dfrac{-8}{50} = -\dfrac{8}{15}$ Ans.

(b) $I = \int_1^2 \dfrac{3x^3 + 4x^2 - x - 2}{(3x - 2)}\, dx = \int_1^2 \dfrac{(3x - 2)(x^2 + 2x + 1)}{(3x - 2)}\, dx = \int_1^2 (x + 1)^2\, dx = \left[\dfrac{(x + 1)^3}{3}\right]_1^2 = \dfrac{(2 + 1)^3}{3} - \dfrac{(1 + 1)^3}{3} = \dfrac{(3)^3}{3} - \dfrac{(2)^3}{3}$

$$= \dfrac{27}{3} - \dfrac{8}{3} = \dfrac{27 - 8}{3} = \dfrac{19}{3} \qquad \text{Ans.}$$

(c) $I = \int_0^{\frac{\pi}{2}} \dfrac{\sin 2x}{2 - \cos^2 x}\, dx = \int_0^{\frac{\pi}{2}} \dfrac{2 \sin x . \cos x}{2 - (1 - \sin^2 x)}\, dx = \int_0^{\frac{\pi}{2}} \dfrac{2 \sin x . \cos x}{2 - 1 + \sin^2 x}\, dx = \int_0^{\frac{\pi}{2}} \dfrac{2 \sin x . \cos x}{1 + \sin^2 x}\, dx$

Put $1 + \sin^2 x = t \quad \therefore\ 2 \sin x . \cos x\, dx = dt$

$I = \int_0^{\frac{\pi}{2}} \dfrac{dt}{t} = (\log t)_0^{\frac{\pi}{2}} = [\log(1 + \sin^2 x)]_0^{\frac{\pi}{2}} = \log(1 + 1) - \log(1 + 0) = \log 2 - \log 1 = \log 2 \qquad \text{Ans.}$

(d) $I = \int_0^{\log 2} e^{2x}\, dx = \left(\dfrac{e^{2x}}{2}\right)_0^{\log 2} = \dfrac{1}{2}\left(e^{2 \log 2} - e^0\right) = \dfrac{1}{2}\left(e^{\log 4} - e^0\right) = \dfrac{1}{2}(4 - 1) = \dfrac{3}{2} \qquad \text{Ans.}$

(e) $I = \int_{\log 2}^{\log 3} (e^x - 1)\, dx = (e^x - x)_{\log 2}^{\log 3} = \left(e^{\log 3} - \log 3\right) - \left(e^{\log 2} - \log 2\right) = 3 - \log 3 - 2 + \log 2$

$$= 1 + \log 2 - \log 3 \quad \text{or}\quad 1 + \log\left(\dfrac{2}{3}\right) \qquad \text{Ans.}$$

(f) $I = \int_e^{e^2} |x - 4|\, dx$, Put $x - 4 = 0 \quad \therefore\ x = 4 \quad$ then $|x - 4| = \begin{cases} -(x - 4), & \text{if } e < x < 4 \\ (x - 4), & \text{if } 4 < x < e^2 \end{cases}$

$I = \int_e^{e^2} |x - 4|\, dx = \int_e^4 -(x - 4)\, dx + \int_4^{e^2} (x - 4)\, dx = -\int_e^4 x\, dx + \int_e^4 4\, dx + \int_4^{e^2} x\, dx - \int_4^{e^2} 4\, dx = (4x)_e^4 - \left(\dfrac{x^2}{2}\right)_e^4 + \left(\dfrac{x^2}{2}\right)_4^{e^2} - (4x)_4^{e^2}$

$$= (16 - 4e) - \left(\dfrac{16}{2} - \dfrac{e^2}{2}\right) + \left(\dfrac{e^4}{2} - \dfrac{16}{2}\right) - (4e^2 - 16) = 16 - 4e - 8 + \dfrac{e^2}{2} + \dfrac{e^4}{2} - 8 - 4e^2 + 16$$

$$= \dfrac{e^4 + e^2 - 8e^2 - 8e + 32}{2} = \dfrac{e^4 - 7e^2 - 8e + 32}{2} \qquad \text{Ans.}$$

(g) $I = \int_{-2}^2 \dfrac{e^x}{e^x + e^{-x}}\, dx = 2 \qquad \text{Ans.} \quad \left[\text{Do yourself, use property VI} :-\ \int_b^a f(x)\, dx = \int_b^a f(a + b - x)\, dx\right]$

(h) $I = \int_0^{\frac{\pi}{2}} \dfrac{\sin x}{\sin x + \cos x}\, dx = \dfrac{\pi}{4} \qquad \text{Ans.} \quad \left[\text{Do yourself, use property IV} :-\ \int_0^a f(x)\, dx = \int_0^a f(a - x)\, dx\right]$

(i) $I = \int_1^{\sin^2 \theta} \dfrac{\sin^{-1}\sqrt{x}}{\sqrt{1 - x}}\, dx$, Put $\sin^{-1}\sqrt{x} = t \quad \therefore\ \dfrac{1}{\sqrt{1 - x}}\, dx = dt$

$I = \int_1^{\sin^2 \theta} t\, dt = \left(\dfrac{t^2}{2}\right)_1^{\sin^2 \theta} = \left(\dfrac{(\sin^{-1}\sqrt{x})^2}{2}\right)_1^{\sin^2 \theta} = \dfrac{(\sin^{-1}\sqrt{\sin^2 \theta})^2}{2} - \dfrac{(\sin^{-1}\sqrt{1})^2}{2} = \dfrac{(\sin^{-1}(\sin \theta))^2}{2} - \dfrac{(\sin^{-1} 1)^2}{2} = \dfrac{\theta^2}{2} - \dfrac{\left(\frac{\pi}{2}\right)^2}{2}$

$$= \dfrac{\theta^2}{2} - \dfrac{\pi^2}{8} \quad \text{or}\quad \dfrac{1}{2}\left[\theta^2 - \dfrac{\pi^2}{4}\right] \qquad \text{Ans.}$$

(j) $I = \int_0^{\frac{\pi}{2}} \dfrac{dx}{1 + \cos x} = \int_0^{\frac{\pi}{2}} \dfrac{dx}{1 + \cos^2 {x}/{2} - \sin^2 {x}/{2}} = \int_0^{\frac{\pi}{2}} \dfrac{dx}{\cos^2 {x}/{2} + \cos^2 {x}/{2}} = \int_0^{\frac{\pi}{2}} \dfrac{dx}{2 \cos^2 {x}/{2}} = \dfrac{1}{2}\int_0^{\frac{\pi}{2}} \sec^2 {x}/{2}\, dx$

Put $\dfrac{x}{2} = t \quad \therefore\ dx = 2\, dt$

$\therefore\ I = \dfrac{1}{2}\int_0^{\frac{\pi}{2}} 2 \sec^2 t\, dt = \int_0^{\frac{\pi}{2}} \sec^2 t\, dt = (\tan t)_0^{\frac{\pi}{2}} = (\tan {x}/{2})_0^{\frac{\pi}{2}} = \tan\dfrac{\pi}{4} - \tan 0 = 1 - 0 = 1 \qquad \text{Ans.}$

(k) $I = \int_0^{\frac{\pi}{2}} \dfrac{dx}{1 + \tan x} = \int_0^{\frac{\pi}{2}} \dfrac{\cos x\, dx}{\cos x + \sin x} = \dfrac{\pi}{4} \qquad \text{Ans.} \quad \left[\text{Do yourself, use property IV} :-\ \int_0^a f(x)\, dx = \int_0^a f(a - x)\, dx\right]$

(l) $I = \int_0^1 \dfrac{\sqrt{x} + \sqrt[3]{x}}{\sqrt[4]{x}} dx = \int_0^1 \left(\dfrac{\sqrt{x}}{\sqrt[4]{x}} + \dfrac{\sqrt[3]{x}}{\sqrt[4]{x}}\right) dx = \int_0^1 \left(x^{\frac{1}{4}} + x^{\frac{1}{12}}\right) dx = \left[\dfrac{x^{5/4}}{5/4} + \dfrac{x^{13/12}}{13/12}\right]_0^1 = \left[\dfrac{4}{5}x^{5/4} + \dfrac{12}{13}x^{13/12}\right]_0^1 = \dfrac{4}{5} + \dfrac{12}{13} - 0 = \dfrac{4}{5} + \dfrac{12}{13} = \dfrac{52 + 60}{65}$

$\qquad\qquad = \dfrac{112}{65}$ Ans.

(m) $I = \int_0^{\frac{\pi}{2}} \dfrac{\sqrt{\cos x}}{\sqrt{\sin x} + \sqrt{\cos x}} dx = \dfrac{\pi}{4}$ Ans. $\left[\text{Do yourself, use property IV} : - \int_0^a f(x)\, dx = \int_0^a f(a - x)\, dx\right]$

(8) (a) $I = \int_1^2 \dfrac{\sin(1 + \log x)}{x} dx$, Put $1 + \log x = t \quad \therefore \dfrac{1}{x} dx = dt$

$I = \int_1^2 \sin t \, dt = (-\cos t)_1^2 = -[\cos(1 + \log x)]_1^2 = -[\cos(1 + \log 2) - \cos(1 + \log 1)] = \cos 1 - \cos(1 + \log 2)$ Ans.

(b) $I = \int_{-\frac{1}{3}}^{\frac{1}{3}} e^{\cos x} \cdot \sin x \, dx$, Put $\cos x = t \quad \therefore -\sin x \, dx = dt$

$\therefore I = -\int_{-\frac{1}{3}}^{\frac{1}{3}} e^t \, dt = -(e^t)_{-\frac{1}{3}}^{\frac{1}{3}} = -[e^{\cos x}]_{-\frac{1}{3}}^{\frac{1}{3}} = -\left[e^{\cos\left(\frac{1}{3}\right)} - e^{\cos\left(-\frac{1}{3}\right)}\right] = e^{\cos\left(\frac{1}{3}\right)} - e^{\cos\left(\frac{1}{3}\right)} = 0$ Ans.

IInd Method: $- \; I = \int_{-\frac{1}{3}}^{\frac{1}{3}} e^{\cos x} \cdot \sin x \, dx = \int_{-\frac{1}{3}}^{\frac{1}{3}} e^{\cos\left(-\frac{1}{3} + \frac{1}{3} - x\right)} \cdot \sin\left(-\frac{1}{3} + \frac{1}{3} - x\right) dx = -\int_{-\frac{1}{3}}^{\frac{1}{3}} e^{\cos x} \cdot \sin x \, dx = -I$

$\therefore \; I + I = 0 \;$ or $\; 2I = 0 \quad \therefore \; I = 0 \;$ Ans.

IIIrd Method: $- \; f(x) = e^{\cos x} \cdot \sin x, \; f(-x) = -e^{\cos x} \cdot \sin x = -f(x) \quad \therefore \; f(x)$ is odd function , $I = \int_{-\frac{1}{3}}^{\frac{1}{3}} e^{\cos x} \cdot \sin x \, dx = 0$ Ans.

(c) $I = \int_{-\frac{\pi}{4}}^{\frac{\pi}{4}} e^{\sec x} \cdot \sec x \tan x \, dx = 0$ Ans. [Do yourself, $f(x) = e^{\sec x} \cdot \sec x \tan x, \; f(-x) = -f(x)$ it is odd function]

(d) $I = \int_{-5}^5 \dfrac{3x^5 + 2x^4 + 5x^3 + 4x^2 + 2x + 1}{(x^2 + 2)} dx = \int_{-5}^5 \dfrac{-3x^5 + 2x^4 - 5x^3 + 4x^2 - 2x + 1}{(x^2 + 2)} dx = I$

$\left[\text{use property VI} : - \int_b^a f(x)\, dx = \int_b^a f(a + b - x)\, dx\right]$

Adding, $2I = \int_{-5}^5 \left[\dfrac{3x^5 + 2x^4 + 5x^3 + 4x^2 + 2x + 1}{(x^2 + 2)} + \dfrac{-3x^5 + 2x^4 - 5x^3 + 4x^2 - 2x + 1}{(x^2 + 2)}\right] dx = \int_{-5}^5 \dfrac{4x^4 + 8x^2 + 2}{x^2 + 2} dx$

$= \int_{-5}^5 \dfrac{4x^2(x^2 + 2) + 2}{x^2 + 2} dx = \int_{-5}^5 \dfrac{4x^2(x^2 + 2)}{x^2 + 2} dx + \int_{-5}^5 \dfrac{2}{x^2 + 2} dx = \int_{-5}^5 4x^2 \, dx + 2\int_{-5}^5 \dfrac{1}{x^2 + \left(\sqrt{2}\right)^2} dx$

$2I = 4\left(\dfrac{x^3}{3}\right)_{-5}^5 + 2\left[\dfrac{1}{\sqrt{2}} \tan^{-1}\left(\dfrac{x}{\sqrt{2}}\right)\right]_{-5}^5 = \dfrac{4(125 + 125)}{3} + \sqrt{2}\left[\tan^{-1}\left(\dfrac{5}{\sqrt{2}}\right) - \tan^{-1}\left(\dfrac{-5}{\sqrt{2}}\right)\right]$, use formula $\int \dfrac{dx}{x^2 + a^2} = \dfrac{1}{a} \tan^{-1}\left(\dfrac{x}{a}\right)$

$\therefore I = \dfrac{1}{2}\left\{\dfrac{1000}{3} + \sqrt{2}\left[\tan^{-1}\left(\dfrac{5}{\sqrt{2}}\right) - \tan^{-1}\left(\dfrac{-5}{\sqrt{2}}\right)\right]\right\} = \dfrac{500}{3} + \dfrac{1}{\sqrt{2}} \cdot \left[\tan^{-1}\left(\dfrac{5}{\sqrt{2}}\right) - \tan^{-1}\left(\dfrac{-5}{\sqrt{2}}\right)\right]$ Ans.

(e) $I = \int_{-3}^3 \dfrac{5x^3 + x}{x^2 + 1} dx = 0$ Ans. (Do yourself) $\left[\text{use property V} : - \int_{-a}^a f(x)\, dx = \begin{cases} 2\displaystyle\int_0^a f(x)\, dx , & \text{if } f(-x) = f(x) \text{ [Even function]} \\ 0, & \text{if } f(-x) = -f(x) \text{ [Odd function]} \end{cases}\right]$

(f) $I = \int_{-1}^1 \dfrac{x^3 \cos x}{x^2 + 1} dx$, Let $f(x) = \dfrac{x^3 \cos x}{x^2 + 1}$, $f(-x) = -\dfrac{x^3 \cos x}{x^2 + 1} = -f(x)$ use property V.

or $f(-x) = -f(x) \quad \therefore \; f(x)$ is odd function. $\quad \therefore I = \int_{-1}^1 \dfrac{x^3 \cos x}{x^2 + 1} dx = 0$ Ans.

(g) $I = \int_{\frac{\pi}{4}}^{\frac{3\pi}{4}} \frac{dx}{1 - \cos x} = \int_{\frac{\pi}{4}}^{\frac{3\pi}{4}} \frac{dx}{1 - \cos\left(\frac{3\pi}{4} + \frac{\pi}{4} - x\right)} = \int_{\frac{\pi}{4}}^{\frac{3\pi}{4}} \frac{dx}{1 - \cos(\pi - x)} = \int_{\frac{\pi}{4}}^{\frac{3\pi}{4}} \frac{dx}{1 + \cos x} = I$

$$\left[\text{use property VI} : - \int_{b}^{a} f(x)\, dx = \int_{b}^{a} f(a + b - x)\, dx\right]$$

Adding, $2I = \int_{\frac{\pi}{4}}^{\frac{3\pi}{4}} \left(\frac{1}{1 - \cos x} + \frac{1}{1 + \cos x}\right) dx = \int_{\frac{\pi}{4}}^{\frac{3\pi}{4}} \left(\frac{1 + \cos x + 1 - \cos x}{(1 - \cos x)(1 + \cos x)}\right) dx = \int_{\frac{\pi}{4}}^{\frac{3\pi}{4}} \left(\frac{2}{1 - \cos^2 x}\right) dx = \int_{\frac{\pi}{4}}^{\frac{3\pi}{4}} \left(\frac{2}{\sin^2 x}\right) dx$

$$= 2\int_{\frac{\pi}{4}}^{\frac{3\pi}{4}} \text{cosec}^2 x\, dx = 2[-\cot x]_{\frac{\pi}{4}}^{\frac{3\pi}{4}} = -2\left(\cot \frac{3\pi}{4} - \cot \frac{\pi}{4}\right) = -2[-1 - 1] = 4$$

or $I = \frac{4}{2} = 2$ Ans.

(9) (a) $I = \int_{0}^{\frac{\pi}{2}} \frac{\sqrt{\sin x}}{\sqrt{\sin x} + \sqrt{\cos x}}\, dx = \frac{\pi}{4}$ Ans. (Do yourself, use property IV.)

(b) $I = \int_{0}^{1} \frac{e^{1-x}}{e^x + e^{1-x}}\, dx = \frac{1}{2}$ Ans. (Do yourself, use property IV.)

(c) $I = \int_{0}^{\pi} \frac{e^{\cos x}}{e^{\cos x} + e^{-\cos x}}\, dx = \frac{\pi}{2}$ Ans. (Do yourself, use property IV.)

(d) $I = \int_{0}^{\frac{\pi}{2}} \frac{dx}{1 - \tan^2 x} = \int_{0}^{\frac{\pi}{2}} \frac{\cos^2 x\, dx}{\cos^2 x - \sin^2 x} = \frac{\pi}{4}$ Ans. (Do yourself, use property IV.)

(10) (a) $I = \int_{\frac{\pi}{6}}^{\frac{\pi}{3}} \frac{dx}{1 + \sqrt{\tan x}} = \frac{\pi}{12}$ Ans. (Do yourself.) $\left[\text{use property VI} : - \int_{b}^{a} f(x)\, dx = \int_{b}^{a} f(a + b - x)\, dx\right]$

(b) $I = \int_{\sqrt{\log 2}}^{\sqrt{\log 3}} \frac{x \sin x^2}{\sin x^2 + \sin(\log 6 - x^2)}\, dx = \frac{1}{4} \log\left(\frac{3}{2}\right)$ Ans.

(c) $I = \int_{0}^{1} \frac{x^4(1 - x)^4}{1 + x^2}\, dx = \frac{22}{7} - \pi$ Ans.   (d) $I = \int_{e^{-1}}^{e^2} \left|\frac{\log_e x}{x}\right| dx = \frac{5}{2}$ Ans.

(11) (a) $\int_{0}^{\pi} [2 \sin x]\, dx$, Let $y = [2 \sin x]$ or $y = [f(x)]$ where $f(x) = 2 \sin x$

$$y = \begin{cases} 0, & 0 \leq x < \frac{\pi}{6} \\ 1, & \frac{\pi}{6} \leq x < \frac{\pi}{2} \\ 1, & \frac{\pi}{2} \leq x < \frac{5\pi}{6} \\ 0, & \frac{5\pi}{6} < x < \pi \\ 2, & x = \frac{\pi}{2} \end{cases}$$ Draw Graph

or $2 \sin x = 1$ or $\sin x = \frac{1}{2}$ $\therefore x = n\pi \pm (-1)^n . \frac{\pi}{6}$ $\therefore x = \frac{\pi}{6}, \frac{5\pi}{6}$

$\int_{0}^{\pi} [2 \sin x]\, dx = A_1 + A_2 = \left[\frac{\pi}{2} - \frac{\pi}{6}\right] + \left[\frac{5\pi}{6} - \frac{\pi}{2}\right] = \frac{2\pi}{6} + \frac{2\pi}{6} = \frac{4\pi}{6} = \frac{2\pi}{3}$ Ans.

(b) $\int_{0}^{102} [\tan^{-1} x]\, dx$, Let $y = [\tan^{-1} x]$ or $y = [f(x)]$ where $f(x) = \tan^{-1} x$

or $f(x) = 1$ or $\tan^{-1} x = 1$ $\therefore x = \tan 1$

$$y = \begin{cases} 0, & 0 \le x < \tan 1 \\ 1, & \tan 1 \le x \le 102 \end{cases} \qquad \text{Draw Graph}$$

$$\int_0^{102} [\tan^{-1} x] \ dx = \int_0^{\tan 1} [\tan^{-1} x] \ dx + \int_{\tan 1}^{102} [\tan^{-1} x] \ dx = 0 + [x]_{\tan 1}^{102} = [102 - \tan 1] = 102 - \tan 1 \qquad \text{Ans.}$$

(c) $\displaystyle\int_0^{2n\pi} [\sin x + \cos x] \ dx$ (Do yourself)

(d) $\displaystyle\int_0^{\frac{5\pi}{12}} [\tan x] \ dx$, Let $y = [\tan x]$ or $y = [f(x)]$ where $f(x) = \tan x$

$$f(x) = 0,1,2,3 \quad \text{then } x = 0, \tan^{-1} 1, \tan^{-1} 2, \tan^{-1} 3 \quad \text{or } y = \begin{cases} 0, & 0 \le x < \tan^{-1} 1 \\ 1, & \tan^{-1} 1 \le x < \tan^{-1} 2 \\ 2, & \tan^{-1} 2 \le x < \tan^{-1} 3 \\ 3, & \tan^{-1} 3 \le x \le \dfrac{5\pi}{12} \end{cases}$$

$$\int_0^{\frac{5\pi}{12}} [\tan x] \ dx = \int_0^{\tan^{-1} 1} 0 \ dx + \int_{\tan^{-1} 1}^{\tan^{-1} 2} dx + \int_{\tan^{-1} 2}^{\tan^{-1} 3} 2 \ dx + \int_{\tan^{-1} 3}^{\frac{5\pi}{12}} 3 \ dx = (\tan^{-1} 2 - \tan^{-1} 1) + 2(\tan^{-1} 3 - \tan^{-1} 2) + 3\left(\frac{5\pi}{12} - \tan^{-1} 3\right)$$

$$\int_0^{\frac{5\pi}{12}} [\tan x] \ dx = \tan^{-1} 2 - \tan^{-1} 1 + 2\tan^{-1} 3 - 2\tan^{-1} 2 + 3.\frac{5\pi}{12} - 3\tan^{-1} 3 = \frac{5\pi}{4} - \tan^{-1} 1 - \tan^{-1} 2 - \tan^{-1} 3 \qquad \text{Ans.}$$

(e) $\displaystyle\int_0^{n^2} [\sqrt{x}] \ dx, n \in N$

$$I = \int_0^{1^2} 0 \ dx + \int_{1^2}^{2^2} dx + \int_{2^2}^{3^2} 2 \ dx + \cdots \ldots \ldots \ldots \ldots \ldots \ldots + \int_{(n-1)^2}^{n^2} (n-1) \ dx$$
$$= (2^2 - 1^2) + 2(3^2 - 2^2) + 3(4^2 - 3^2) + \cdots \ldots \ldots \ldots \ldots . + (n-1)[n^2 - (n-1)^2]$$

$$I = (2+1) + 2(3+2) + 3(4+3) + \cdots \ldots \ldots . + (n-1)[n^2 - (n-1)^2] = \{1^2 + 2^2 + 3^2 + \cdots \ldots . + (n-1)^2\} + \sum_2^n (n-1)n$$

$$= \frac{1}{6}.N(N+1)(2N+1) + \sum_1^n (n-1)n = \frac{1}{6}(n-1)n(2n-1) + \frac{(n-1)n(n+1)}{3}$$

$$I = \frac{1}{6}n(n-1)[2n - 1 + 2n + 2] = \frac{1}{6}n(n-1)(4n+1) \qquad \text{Ans.}$$

(f) Prove that $\displaystyle\int_0^x [t] \ dt = \frac{[x]([x] - 1)}{2} + [x](x - [x])$

L. H. S $= \displaystyle\int_0^x [t] \ dt$, Let $I = \displaystyle\int_0^x [t] \ dt \quad \because \{x\} = x - [x], \ 0 \le \{x\} < 1 \ or \ x = [x] + \{x\} \ or \ x = n + f, \ 0 \le f < 1 \ n \in I$

$$I = \int_0^{n+f} [x] \ dx = 0 + 1 + 2 + 3 + \cdots \ldots \ldots . + (n-1) + nf = \frac{n(n-1)}{2} + nf \qquad \text{Draw Graph}$$

$$I = \frac{[x]([x] - 1)}{2} + [x](x - [x]) = \text{R. H. S} \quad \text{Proved.}$$

Area

The area between the curve $y = f(x)$, $x -$ axis and two ordinates at the points $x = a$ and $x = b$ $(b > a)$

$$A = \int_a^b f(x) \ dx = \int_a^b y \ dx$$

It represents the shaded and non $-$ shaded area in given below figure $-$

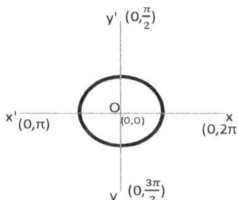

Similarly, if the area be between the curve and y − axis and two abscissas drawn at the points $y = a$ and $y = b$ $(b > a)$

is given by the formula, $\int_a^b x \, dy$

Area between two curves $y = f(x)$ and $y = g(x)$ and the two ordinates drawn at the points $x = a$ and $x = b$ then,

$$A = \int_a^b y_1 \, dx - \int_a^b y_2 \, dx = \text{upper} - \text{lower} = \int_a^b [f(x) - g(x)] \, dx$$

where y_1 and y_2 are ordinates y_1 is the ordinate of $y = f(x)$ which is upper curve and y_2 is the ordinate of $y = g(x)$ which is lower curve.

In the above result we have taken the area to be lying above x − axis and the area to be lying below x − axis then the value of area will

be − ve and hence we take mod. $\quad \therefore \quad A = \left| \int_a^b y \, dx \right|$

similarly, if the area be lying to right and left of y − axis then $\quad \therefore \quad A = \left| \int_a^b x \, dy \right|$

Consider the following graph: −   draw

we shall divide [a, b] into $\quad [a, x_1] + [x_1, x_2] + [x_2, x_3] + [x_3, x_4] + [x_4, x_5] + [x_5, x_6] + [x_6, x_7] + [x_7, b]$

or $\int_a^b f(x) \, dx = \left| \int_a^{x_1} y \, dx \right| + \left| \int_{x_1}^{x_2} y \, dx \right| + \left| \int_{x_2}^{x_3} y \, dx \right| + \left| \int_{x_3}^{x_4} y \, dx \right| + \left| \int_{x_4}^{x_5} y \, dx \right| + \left| \int_{x_5}^{x_6} y \, dx \right| + \left| \int_{x_6}^{x_7} y \, dx \right| + \left| \int_{x_7}^b y \, dx \right|$

where $\left| \int_a^{x_1} y \, dx \right| = -ve, \left| \int_{x_1}^{x_2} y \, dx \right| = +ve, \left| \int_{x_2}^{x_3} y \, dx \right| = -ve, \left| \int_{x_3}^{x_4} y \, dx \right| = +ve, \left| \int_{x_4}^{x_5} y \, dx \right| = -ve,$

$$\left| \int_{x_5}^{x_6} y \, dx \right| = +ve, \quad \left| \int_{x_6}^{x_7} y \, dx \right| = -ve, \left| \int_{x_7}^b y \, dx \right| = +ve$$

From above figure, the area is lying above x − axis as well as below x − axis in the interval [a, b] then we divide the interval [a, b]

into various sub − interval. which the area is lying either above x − axis or below x − axis for below x − axis we take mod. Area of a

sub − interval $[a, x_1], [x_2, x_3], [x_4, x_5]$ and $[x_6, x_7]$ are lying below x − axis then we take mod.

Area bounded by two curves: −

we have shown in (P − 1)that $\quad A = \int_a^b (y_1 - y_2) \, dx = \int_a^b [f(x) - g(x)] \, dx \quad$ (fig P − II) where $y_1 \rightarrow$ Upper and $y_2 \rightarrow$ Lower

The above formula does not change whether the area is above or below x − axis or both.

Draw graph

$A = \int_a^b (x_1 - x_2) \, dy = \int_a^b [f(y) - g(y)] \, dy \quad$ where $x_1 \rightarrow$ Right and $x_2 \rightarrow$ Left

Asymptotes

An asymptote is essentially, a line that a graph approaches, but does not intersect.

An asymptote is a line or curve that approaches a given curve arbitrarily closely, as illustrated in the diagram.

For example in the following graph of $y = \dfrac{1}{x}$, the line approaches the x − axis ($y = 0$) but never touches it.

Draw

The curve has a vertical asymptote at $x = 0$ and a horizontal asymptote at $y = 0$.

Vertical Asymptote

Vertical asymptotes correspond to the 0 of the denominator of a rational function.

Let $f(x) = \dfrac{N(x)}{D(x)}$ be a rational function. the line $x = c$ is a vertical asymptote of the graph of f(x) if $D(x) = 0$ and $N(c) \neq 0$.

Example: − find the vertical asymptote of $f(x) = \dfrac{x^2 + 3x - 2}{x^2 - 1}$.

Solution: − $f(x) = \dfrac{x^2 + 3x - 2}{x^2 - 1}$, $D(x) = x^2 - 1$ and $N(x) = x^2 + 3x - 2$

To be vertical asymptote $D(x) = 0$ and $N(x) \neq 0$ or $D(x) = 0$ or $x^2 - 1 = 0$ ∴ $(x - 1)(x + 1) = 0$ ∴ $x = 1, -1$

$N(x) \neq 0$ or $N(x) = x^2 + 3x - 2$ ∴ $N(1) = 1 + 3 - 2 = 2 \neq 0$ and $N(-1) = 1 - 3 - 2 = -4 \neq 0$

Since $D(1) = 0$ and $N(1) \neq 0$ so, $x = 1$ is vertical asymptote.

Horizontal Asymptote

Let $y = f(x)$ be a function. suppose that $\lim\limits_{x \to \infty} f(x) = L$ or $\lim\limits_{x \to -\infty} f(x) = M$

we refer to the each of the line $y = L$ and $y = M$ as a horizontal asymptote of the function f(x).

Let $f(x) = \dfrac{a_m x^m + a_{m-1} x^{m-1} + \cdots \dots \dots \dots \dots \dots +a_1 x + a_0}{b_n x^n + b_{n-1} x^{n-1} + \cdots \dots \dots \dots \dots \dots \dots +b_1 x + b_0}$, $a_m \neq 0$, $b_n \neq 0$ be a rational function.

\# If $m < n$ then $y = 0$ *is the horizontal asymptote.*

\# If $m = n$ then $y = \dfrac{a_m}{b_m}$ is the horizontal asymptote.

\# If $m > n$ *then there is no horizontal asymptote.*

Example: − find the horizontal asymptote of $f(x) = \dfrac{3x^3 - 2x^2 + 2}{2x^3 + 3x - 5}$.

Solution: − Here $m = 3$, $n = 3$, $a_m = 3$, $b_n = 2$ ∴ This is the case $m = n$ so, the horizontal asymptote is $\boxed{y = \dfrac{a_m}{b_m} = \dfrac{3}{2}}$ Ans.

Example: − find the horizontal asymptote of $f(x) = \dfrac{2x^4 - x^2 + 3}{5x^5 - 20}$.

Solution: − Here $m = 4$, $n = 5$, $a_m = 2$, $b_n = 5$ ∴ This is the case $m < n$ so, the horizontal asymptote is $\boxed{y = 0}$ Ans.

Example: − find the horizontal asymptote of $f(x) = \dfrac{2x^4 - x^3 - 2}{6x^3 + 2x^2 - 5}$.

Solution: − Here $m = 4$, $n = 3$, $a_m = 2$, $b_n = 6$ ∴ This is the case $m > n$ so, there is no horizontal asymptote. Ans.

Symmetry

(a) If the equation of the curve involves even and only even powers of x, then that curve is symmetrical about y − axis.

e.g. Parabola, $x^2 = 4ay$ or $x^2 - 4ay = 0$

Let $f(x) = x^2 - 4ay$, $f(-x) = x^2 - 4ay$ $\therefore$ $f(x) = f(-x)$ Hence this parabola is symmetrical about $y - $ axis.

(b) If the equation of the curve involves even and only even powers of y, then that curve is symmetrical about $x - $ axis.

e.g. Parabola, $y^2 = 4ax$ or $y^2 - 4ax = 0$

Let $f(y) = y^2 - 4ax$, $f(-y) = y^2 - 4ax$ $\therefore$ $f(y) = f(-y)$

Put $y = -y$ in equation $y^2 = 4ax$ then $y^2 = 4ax$ no any change in the equation $y^2 = 4ax$

Hence then that curve is symmetrical about $x - $ axis.

(c) The curve involves even and only even powers of both x and y then it is symmetrical about both the axes.

e.g. $x^2 + y^2 = 1$ and $\dfrac{x^2}{m^2} + \dfrac{y^2}{n^2} = 1$, Put $x = -x$ and $y = -y$ then $x^2 + y^2 = 1$ and $\dfrac{x^2}{m^2} + \dfrac{y^2}{n^2} = 1$

which is no any change in the curve. hence it is symmetrical about both the axes. (x and y)

(d) If x and y be interchanged and the equation of the curve does not change then it is symmetrical about the line $y = x$.

i.e a line through origin making an angle of 45^0 with $x - $ axis.

Point on the curve

Find the point on the curve $x^2 + y^2 + 2x = 3$ ... ... (i)

Put $y = 0$ in equation (i) and find x,

$\therefore$ $x^2 + 0 + 2x = 3$ or $x^2 + 2x - 3 = 0$ or $(x - 1)(x + 3) = 0$ $\therefore$ $x = 1, -3$ $\therefore$ points are $(1,0)$ & $(-3,0)$

Again, Put $x = 0$ in equation (i) and find y, $x^2 + y^2 + 2x = 3$ or $y^2 = 3$ or $y = \pm\sqrt{3}$ $\therefore$ point $= (0, \sqrt{3})$ & $(0, -\sqrt{3})$

Points are $(1,0), (-3,0), (0, \sqrt{3})$ and $(0, -\sqrt{3})$.

Draw graph

Tangents

Tangent at origin $(0,0)$ are obtained by equating to zero (0) the lowest degree terms in the equation of the curve $y^2 = 4ax$, $T_{(0,0)} = 4ax = 0$

i.e $x = 0$ or $y - $ axis and curve $x^2 = 4ay$, $T_{(0,0)} = 4ay = 0$

i.e $y = 0$ or $x - $ axis. curve $x^2 + y^2 = 2axy$, $T_{(0,0)} = 2axy = 0$ i.e $x = 0$ and $y = 0$ i.e both the axes are tangents at the origin.

Tangent at any other point is given by $\dfrac{dy}{dx} = -\dfrac{f_x}{f_y}$ and slope of the tangent at (h, k), $\dfrac{dy}{dx}\Big|_{(h,k)} = -\dfrac{f_x}{f_y}$.

Example: $-$ (1) Find the area between the $x - $ axis, the graph of $y = x$ and $x = 2$.

Solution: $-$ Ist Method: $-$ by elementary geometry $-$ Area $\Delta OAB = \dfrac{OA \times AB}{2}$ or Area $= \dfrac{2 \times 2}{2} = 2$ Ans.

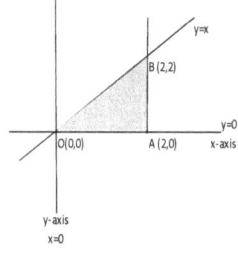

IInd Method: $-$ Area (A) $= \int_0^2 (y_1 - y_2)\, dx = \int_0^2 (x - 0)\, dx = \int_0^2 x\, dx = \left(\dfrac{x^2}{2}\right)_0^2 = \dfrac{4}{2} - 0 = 2$ Ans.

Example: $-$ (2) Find the area bounded by $y = 3x, x = 3, x = 5$ and the $x - $ axis.

Solution: $-$ Ist Method: $-$ find the area of ABCD $= AB\left(\dfrac{AD + BC}{2}\right) = 2\left(\dfrac{9 + 15}{2}\right) = 24$ Ans.

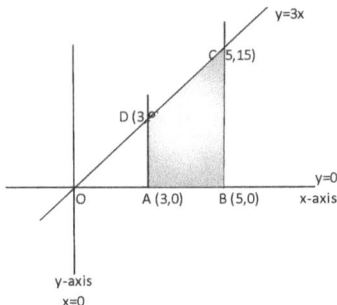

IInd Method: $-$ Area $= \int_3^5 (y_1 - y_2)\, dx = \int_3^5 (3x - 0)\, dx = \int_3^5 3x\, dx = \left(\dfrac{3x^2}{2}\right)_3^5 = \dfrac{75}{2} - \dfrac{27}{2} = \dfrac{75 - 27}{2} = \dfrac{48}{2} = 24$ Ans.

Example: $-$ (3) Find the area under $x = y^2$ from $y = 2$ to $y = 3$ above the $x - $ axis and right sides of $y - $ axis.

Solution: $-$ Area $= \int_2^3 (x_2 - x_1)\, dy = \int_2^3 (y^2 - 0)\, dy = \int_2^3 y^2\, dy = \left(\dfrac{y^3}{3}\right)_2^3 = \dfrac{27}{3} - \dfrac{8}{3} = \dfrac{27 - 8}{3} = \dfrac{19}{3}$ Ans.

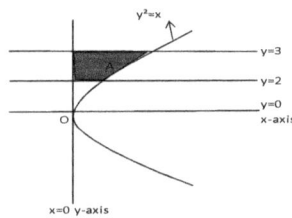

Example: $-$ (4) Find the area between the parabola $y^2 = 4ax$ and $x^2 = 4by$.

Solution: $-$ The two curves meet at point P.

or $y^2 = 4ax \ldots \ldots \ldots$ (i) and $x^2 = 4by \ldots \ldots \ldots \ldots$ (ii)   solving the equation (i) and (ii), we have

From $x^2 = 4by,\ y = \dfrac{x^2}{4b}$ and put value of y in equation (i), we get

or $y^2 = 4ax$ or $\left(\dfrac{x^2}{4b}\right)^2 = 4ax$ or $x^4 = 64ab^2 x$ or $x^4 - 64ab^2 x = 0$ or $x(x^3 - 64ab^2) = 0$

$\therefore\ x = 0$ and $x^3 - 64ab^2 = 0$ or $x^3 = 64ab^2$ $\therefore\ x = 4.\,a^{\frac{1}{3}}.\,b^{\frac{2}{3}}$

Required Area $= \int_0^{4.a^{\frac{1}{3}}.b^{\frac{2}{3}}} (y_1 - y_2)\, dx \ldots \ldots \ldots \ldots \ldots$ (A)

where y_1 is the ordinate of the upper parabola $y^2 = 4ax$ $\therefore\ y = 2.\,\sqrt{a}.\,\sqrt{x}$ or $y_1 = 2.\,\sqrt{a}.\,\sqrt{x}$

and y_2 is the ordinate of the lower parabola $x^2 = 4by$ $\therefore\ y = \dfrac{x^2}{4b}$ or $y_2 = \dfrac{x^2}{4b}$

Putting for y_1 and y_2 in equation (A), we get

Required Area $= \int_0^{4.a^{\frac{1}{3}}.b^{\frac{2}{3}}} (y_1 - y_2)\,dx = \int_0^{4.a^{\frac{1}{3}}.b^{\frac{2}{3}}} \left(2.\sqrt{a}.\sqrt{x} - \frac{x^2}{4b}\right)dx = \int_0^{4.a^{\frac{1}{3}}.b^{\frac{2}{3}}} (2.\sqrt{a}.\sqrt{x})\,dx - \int_0^{4.a^{\frac{1}{3}}.b^{\frac{2}{3}}} \left(\frac{x^2}{4b}\right)dx$

$$= 2\sqrt{a}\int_0^{4.a^{\frac{1}{3}}.b^{\frac{2}{3}}} x^{\frac{1}{2}}\,dx - \frac{1}{4b}\int_0^{4.a^{\frac{1}{3}}.b^{\frac{2}{3}}} x^2\,dx = 2\sqrt{a}.\left(\frac{x^{3/2}}{3/2}\right)_0^{4.a^{\frac{1}{3}}.b^{\frac{2}{3}}} - \frac{1}{4b}.\left(\frac{x^3}{3}\right)_0^{4.a^{\frac{1}{3}}.b^{\frac{2}{3}}}$$

$$A = \frac{4\sqrt{a}}{3}\left[\left(4.a^{\frac{1}{3}}.b^{\frac{2}{3}}\right)^{3/2} - 0\right] - \frac{1}{12b}\left[\left(4.a^{\frac{1}{3}}.b^{\frac{2}{3}}\right)^3 - 0\right] = \frac{4\sqrt{a}}{3}\left[(2)^{2.\frac{3}{2}}.(a)^{\frac{1}{3}.\frac{3}{2}}.(b)^{\frac{2}{3}.\frac{3}{2}}\right] - \frac{1}{12b}\left[4^3.(a)^{\frac{1}{3}.3}.(b)^{\frac{2}{3}.3}\right] = \frac{4\sqrt{a}}{3}\left[8.a^{\frac{1}{2}}.b\right] - \frac{1}{12b}[64ab^2]$$

$$= \frac{32ab}{3} - \frac{16ab}{3} = \frac{32ab - 16ab}{3} = \frac{16ab}{3} \qquad \text{Ans.}$$

Note: – Two parabolas be $y^2 = 4ax$ and $x^2 = 4ay$ then putting $a = b$ the required area is $\frac{16}{3}a^2$ or $\frac{16}{3}b^2$.

Example: – (5) Find the ratio of the area bounded by curves $y^2 = 8x$ and $x^2 = 8y$ is devided by the line $x = 2$.

Solution: – Curves are $y^2 = 8x$ and $x^2 = 8y$, we can write the curves $y^2 = 4.2x$ and $x^2 = 4.2x$

see above question and find the area, $y^2 = 4ax$ and $x^2 = 4ay$ where $a = 2$ and $b = 2$

$$\text{Area (A)} = \frac{16ab}{3} = \frac{16.2.2}{3} = \frac{64}{3}$$

Again, Area $(A_1) = \int_0^2 (y_1 - y_2)\,dx = \int_0^2 \left(2\sqrt{2}.\sqrt{x} - \frac{x^2}{8}\right)dx = 2\sqrt{2}\int_0^2 x^{\frac{1}{2}}\,dx - \frac{1}{8}\int_0^2 x^2\,dx = 2\sqrt{2}\left(\frac{x^{3/2}}{3/2}\right)_0^2 - \frac{1}{8}\left(\frac{x^3}{3}\right)_0^2 = \frac{4\sqrt{2}}{3}(2)^{3/2} - \frac{1}{24}.2^3$

$$= \frac{4\sqrt{2}.2\sqrt{2}}{3} - \frac{8}{24} = \frac{16}{3} - \frac{1}{3} = \frac{16 - 1}{3} = \frac{15}{3} = 5$$

or $A_2 = A - A_1 = \frac{64}{3} - 5 = \frac{64 - 15}{3} = \frac{49}{3}$ \qquad Ratio is $\frac{A_1}{A_2} = \frac{5}{\frac{49}{3}} = \frac{15}{49}$ \qquad Ans.

Example: – (6) Find the area bounded by the curves $y^2 = 12x$ and $x^2 = 16y$.

Solution: – Curves are $y^2 = 12x$ and $x^2 = 16y$, we can write the curves $y^2 = 4.3x$ and $x^2 = 4.4y$

or $y^2 = 4ax$ and $x^2 = 4by$ [see question no. –(4)]

compare the curves, $a = 3$ and $b = 4$ then Area (A) $= \frac{16ab}{3} = \frac{16.3.4}{3} = 64$ \qquad Ans.

Example: – (7) Find the area between the parabola $y^2 = 4ax$ and line $y = mx$.

Solution: – The parabola $y^2 = 4ax \ldots\ldots\ldots\ldots..$ (i) \quad and line $y = mx \ldots\ldots\ldots\ldots\ldots\ldots.$ (ii)

From (i), $\Rightarrow y^2 = 4ax$ or $(mx)^2 = 4ax$ or $m^2x^2 = 4ax$ or $m^2x^2 - 4ax = 0$

or $x(m^2x - 4a) = 0$ \quad $\therefore x = 0$ or $m^2x - 4a = 0$ \quad $\therefore x = \frac{4a}{m^2}$

The required area $A = \int_0^{\frac{4a}{m^2}} (y_1 - y_2)\,dx \ldots\ldots\ldots\ldots\ldots.$ (A)

$\therefore y^2 = 4ax$ or $y = 2\sqrt{a}\sqrt{x}$ or $y_1 = 2\sqrt{a}\sqrt{x}$ (upper curves)

and $y = mx$ or $y_2 = mx$ (Lower curves)

Put y_1 and y_2 in equation (A), we get \quad $\therefore$ Area (A) $= \int_0^{\frac{4a}{m^2}} (y_1 - y_2)\,dx = \int_0^{\frac{4a}{m^2}} \left(2\sqrt{a}\sqrt{x} - mx\right)dx$

$$A = \int_0^{\frac{4a}{m^2}} (2\sqrt{a}\sqrt{x})\, dx + \int_0^{\frac{4a}{m^2}} (mx)\, dx = 2\sqrt{a} \int_0^{\frac{4a}{m^2}} x^{\frac{1}{2}}\, dx + m \int_0^{\frac{4a}{m^2}} x\, dx = 2\sqrt{a}\left(\frac{x^{3/2}}{3/2}\right)_0^{\frac{4a}{m^2}} - m\left(\frac{x^2}{2}\right)_0^{\frac{4a}{m^2}} = \frac{4\sqrt{a}}{3}\left(\frac{4a}{m^2}\right)^{\frac{3}{2}} - \frac{m}{2}\left(\frac{4a}{m^2}\right)^2$$

$$= \frac{4\sqrt{a}}{3m^3} \cdot (4)^{\frac{3}{2}} \cdot a^{\frac{3}{2}} - \frac{16a^2}{2m^3} = \frac{32a\sqrt{a}\sqrt{a}}{3m^3} - \frac{16a^2}{2m^3} = \frac{32a^2}{3m^3} - \frac{8a^2}{m^3} = \frac{32a^2 - 24a^2}{3m^3}$$

$$A = \frac{8a^2}{3m^3} = \frac{8}{3}\left(\frac{a^2}{m^3}\right) \qquad \text{Ans.}$$

Note: $-$ If the curves were $y^2 = 10x$ and $y = 2x$ then $4a = 10$ $\therefore a = \frac{10}{4} = \frac{5}{2}$ and $m = 2$

$$\text{Area (A)} = \frac{8}{3}\left(\frac{a^2}{m^3}\right) = \frac{8}{3} \cdot \frac{\left(\frac{5}{2}\right)^2}{(2)^3} = \frac{8 \times 25}{3 \times 4 \times 8} = \frac{25}{12} \text{ square units} \qquad \text{Ans.}$$

Example: $-$ (8) Find the area cut off the parabola $3y = x^2$ by the straight line $2y = x + 3$.

Solution: $-$ The parabola $3y = x^2 \ldots \ldots \ldots \ldots$ (i) and line $2y = x + 3 \ldots \ldots \ldots \ldots \ldots \ldots$ (ii)

From (ii), $2y = x + 3$ $\therefore y = \frac{x+3}{2}$ Put y in equation (i), we get

$\therefore 3y = x^2$ or $3\left(\frac{x+3}{2}\right) = x^2$ or $3x + 9 = 2x^2$ or $2x^2 - 3x - 9 = 0$ or $2x^2 - 6x + 3x - 9 = 0$

$$\therefore 2x(x - 3) + 3(x - 3) = 0 \quad \text{or} \quad (2x + 3)(x - 3) = 0 \quad \therefore x = -\frac{3}{2}, 3$$

If $x = 3$ then $y = \frac{3+3}{2} = \frac{6}{2} = 3$ and $x = -\frac{3}{2}$ then $y = \frac{-\frac{3}{2}+3}{2} = \frac{-3+6}{4} = \frac{3}{4}$

The Points of intersection are P (3,3) and $Q\left(-\frac{3}{2}, \frac{3}{4}\right)$

$$\text{Area (A)} = \int_{-\frac{3}{2}}^{3} (y_1 - y_2)\, dx = \int_{-\frac{3}{2}}^{3} \left(\frac{x+3}{2} - \frac{x^2}{3}\right) dx = \int_{-\frac{3}{2}}^{3} \left(\frac{3x + 9 - 2x^2}{6}\right) dx = \frac{1}{6}\int_{-\frac{3}{2}}^{3} (3x + 9 - 2x^2)\, dx = \frac{1}{6}\left[3 \cdot \frac{x^2}{2} + 9x - 2 \cdot \frac{x^3}{3}\right]_{-\frac{3}{2}}^{3}$$

$$= \frac{1}{6}\left\{\left[3 \cdot \frac{9}{2} + 9.3 - 2 \cdot \frac{27}{3}\right] - \left[3 \cdot \frac{\left(-\frac{3}{2}\right)^2}{2} + 9. \left(-\frac{3}{2}\right) - 2 \cdot \frac{\left(-\frac{3}{2}\right)^3}{3}\right]\right\}$$

$$A = \frac{1}{6}\left[\frac{27}{2} + 27 - 18\right] - \frac{1}{6}\left[\frac{27}{8} - \frac{27}{2} + \frac{9}{4}\right] = \frac{1}{6}\left[\frac{27 + 18}{2}\right] - \frac{1}{6}\left[\frac{27 - 108 + 18}{8}\right] = \frac{45}{12} + \frac{63}{48} = \frac{135 + 63}{48} = \frac{198}{48} = \frac{33}{8} \text{ sq. units} \qquad \text{Ans.}$$

Example: $-$ (9) Find the area bounded by the curve $x^2 = 2y$ and the straight line $x = 3y - 4$.

Solution: $-$ The curve is $x^2 = 2y$ or $y = \frac{x^2}{2} \ldots \ldots \ldots \ldots$ (i) and line $x = 3y - 4$ or $y = \frac{x+4}{3} \ldots \ldots \ldots$ (ii)

From (i), $\Rightarrow x^2 = 2y$ or $x^2 = 2.\left(\frac{x+4}{3}\right)$ or $3x^2 = 2x + 8$ or $3x^2 - 2x - 8 = 0$

$\therefore x = \frac{-b \pm \sqrt{b^2 - 4ac}}{2a}$, here $a = 3$, $b = -2$ and $c = -8$

then $x = \frac{2 \pm \sqrt{4 + 96}}{6} = \frac{2 \pm 10}{6} = \frac{2 + 10}{6}$ or $\frac{2 - 10}{6} = \frac{12}{6}$ or $\frac{-8}{6}$ $\therefore x = 2, -\frac{4}{3}$

If $x = 2$ then $y = \frac{x+4}{3} = \frac{2+4}{3} = 2$ and $x = -\frac{4}{3}$ then $y = \frac{x+4}{3} = \frac{-\frac{4}{3}+4}{3} = \frac{\frac{-4+12}{3}}{3} = \frac{8}{9}$

Area (A) $= \int_{-\frac{4}{3}}^{2}(y_1 - y_2)\,dx = \int_{-\frac{4}{3}}^{2}\left(\frac{x+4}{3} - \frac{x^2}{2}\right)dx = \int_{-\frac{4}{3}}^{2}\left(\frac{x+4}{3}\right)dx - \int_{-\frac{4}{3}}^{2}\left(\frac{x^2}{2}\right)dx = \frac{1}{3}\left(\frac{x^2}{2} + 4x\right)_{-\frac{4}{3}}^{2} - \frac{1}{2}\left(\frac{x^3}{3}\right)_{-\frac{4}{3}}^{2}$

$= \frac{1}{3}\left[2 + 8 - \frac{8}{9} + \frac{16}{3}\right] - \frac{1}{2}\left[\frac{8}{3} + \frac{64}{81}\right] = \frac{1}{3}\left[\frac{90 - 8 + 48}{9}\right] - \frac{1}{2}\left[\frac{216 + 64}{81}\right] = \frac{130}{27} - \frac{140}{81} = \frac{390 - 140}{81} = \frac{250}{81}$ Ans.

Example: $-$ (10) Find the area bounded by the curves $x = y^2$ and $x = \frac{3}{y^2 - 2}$.

Solution: $-$ The points of intersection are given, $x = y^2$ and $x = \frac{3}{y^2 - 2}$

or $y^2 = \frac{3}{y^2 - 2}$ or $y^4 - 2y^2 = 3$ or $y^4 - 2y^2 - 3 = 0$ or $y^2(y^2 - 3) + 1(y^2 - 3) = 0$

or $(y^2 + 1)(y^2 - 3) = 0$ $\therefore$ $(y^2 + 1) = 0$ it is not possible.   or $(y^2 - 3) = 0$ $\therefore$ $y^2 = 3$ $\therefore$ $y = \pm\sqrt{3}$

If $y = \sqrt{3}$ then $x = y^2 = \left(\sqrt{3}\right)^2 = 3$ and $y = -\sqrt{3}$ then $x = y^2 = \left(-\sqrt{3}\right)^2 = 3$

The points of intersection $P\left(3, \sqrt{3}\right)$ and $Q\left(3, -\sqrt{3}\right)$.

Area (A) $= 2\int_0^{\sqrt{3}}(x_1 - x_2)\,dy = 2\int_0^{\sqrt{3}}\left(\frac{3}{y^2 - 2} - y^2\right)dy = 2\int_0^{\sqrt{3}}\frac{3}{y^2 - 2}\,dy - 2\int_0^{\sqrt{3}}y^2\,dy$ $\left[\text{use formula, } \int\frac{dx}{x^2 - a^2} = \frac{1}{2a}\log\left|\frac{x-a}{x+a}\right|\right]$

$A = 6.\left[\frac{1}{2.\sqrt{2}}\log\left|\frac{y - \sqrt{2}}{y + \sqrt{2}}\right|\right]_0^{\sqrt{3}} - 2\left[\frac{y^3}{3}\right]_0^{\sqrt{3}} = \frac{3}{\sqrt{2}}\left[\log\left|\frac{\sqrt{3} - \sqrt{2}}{\sqrt{3} + \sqrt{2}}\right| - \log\left|\frac{-\sqrt{2}}{\sqrt{2}}\right|\right] - 2\left[\frac{3\sqrt{3}}{3} - 0\right] = \frac{3}{\sqrt{2}}\log\left|\frac{\sqrt{3} - \sqrt{2}}{\sqrt{3} + \sqrt{2}}\right| - 2\sqrt{3}$

$= \frac{3}{\sqrt{2}}\log\left|\frac{\sqrt{3} - \sqrt{2}}{\sqrt{3} + \sqrt{2}} \times \frac{\sqrt{3} - \sqrt{2}}{\sqrt{3} - \sqrt{2}}\right| - 2\sqrt{3} = \frac{3}{\sqrt{2}}\log\left|\frac{\left(\sqrt{3} - \sqrt{2}\right)^2}{3 - 2}\right| - 2\sqrt{3} = \frac{3}{\sqrt{2}}\log\left(\sqrt{3} - \sqrt{2}\right)^2 - 2\sqrt{3}$

$= \frac{3.2}{\sqrt{2}}\log\left(\sqrt{3} - \sqrt{2}\right) - 2\sqrt{3} = 3\sqrt{2}\log(1.732 - 1.414) - 2\sqrt{3} = 3 \times 1.414\log(0.318) - 2 \times 1.732$

$A = 4.242\log(0.318) - 3.464 = 4.242(-0.4975) - 3.464 = -2.110395 - 3.464 = -1.353605$ $[\log(0.318) = -0.4975]$

Area does not negative so Area (A) = 1.35 sq. units (approx.)   Ans.

Example: $-$ (11) The area of the region bounded by the curve $y = 2x + x^2$ and the line $y = 2mx$ equals $\frac{32}{3}$? find the following value of m.

Solution: $-$ The two curves meet at $y = 2x + x^2$ and $y = 2mx$ from, $y = 2x + x^2$ or $2mx = 2x + x^2$ or $x^2 + 2x - 2mx = 0$

or $x^2 + 2x(1 - m) = 0$ or $x[x + 2(1 - m)] = 0$ $\therefore$ $x = 0$ and $x + 2(1 - m) = 0$ $\therefore x = -2(1 - m) = 2m - 2$

If $m > 1$ then $2m - 2$ is positive, Area $(A) = \int_0^{2m-2}(y_1 - y_2)\,dx = \int_0^{2m-2}(2x + x^2 - 2mx)\,dx$

$A = \left[2.\frac{x^2}{2} + \frac{x^3}{3} - 2m.\frac{x^2}{2}\right]_0^{2m-2} = \left[x^2 + \frac{x^3}{3} - mx^2\right]_0^{2m-2} = \left[(2m - 2)^2 + \frac{(2m - 2)^3}{3} - m(2m - 2)^2\right]$

$A = \frac{32}{3} = \left[\frac{3(2m - 2)^2 + (2m - 2)^3 - 3m(2m - 2)^2}{3}\right] = \frac{(2m - 2)^2}{3}[3 + 2m - 2 - 3m] = \frac{(2m - 2)^2}{3}[1 - m]$

or $32 = (2m - 2)^2(1 - m) = -[2(m - 1)]^2(m - 1) = -4(m - 1)^3$ $\therefore$ $-\frac{32}{4} = (m - 1)^3$

or $-8 = (m - 1)^3$ $\therefore$ $(-2)^3 = (m - 1)^3$ or $m - 1 = -2$ $\therefore$ $m = -2 + 1 = -1$

If $m < 1$ then $2m - 2$ is negative, Area $(A) = \int_{2m-2}^0(y_1 - y_2)\,dx = \int_{2m-2}^0(2x + x^2 - 2mx)\,dx$

$$A = \frac{32}{3} = \left[2 \cdot \frac{x^2}{2} + \frac{x^3}{3} - 2m \cdot \frac{x^2}{2}\right]_{2m-2}^{0} = \left[x^2 + \frac{x^3}{3} - mx^2\right]_{2m-2}^{0} = -\left[(2m-2)^2 + \frac{(2m-2)^3}{3} - m(2m-2)^2\right]$$

$$= -\left[\frac{3(2m-2)^2 + (2m-2)^3 - 3m(2m-2)^2}{3}\right] = -\frac{(2m-2)^2}{3}[3 + 2m - 2 - 3m]$$

or $\frac{32}{3} = \frac{(2m-2)^2}{3}[m-1]$ or $32 = [2(m-1)]^2(m-1) = 4(m-1)^3$ or $\frac{32}{4} = (m-1)^3$

or $8 = (m-1)^3$ or $2^3 = (m-1)^3$ or $m-1 = 2$ or $m = 3$ $\therefore$ $m = -1, 3$ Ans.

Example: – (12) The area bounded by the curve $y = f(x)$ and the lines $x = 0, y = 0$ and $x = t$, lies in the interval $\left(\frac{3}{4}, 3\right)$.

Solution: – $-\frac{3}{4} < s < -\frac{1}{2}$ and $\frac{1}{2} < t < \frac{3}{4}$ or $\int_{0}^{\frac{1}{2}}(4x^3 + 3x^2 + 2x + 1)\,dx < area < \int_{0}^{\frac{3}{4}}(4x^3 + 3x^2 + 2x + 1)\,dx$

or $[x^4 + x^3 + x^2 + x]_{0}^{\frac{1}{2}} < area < [x^4 + x^3 + x^2 + x]_{0}^{\frac{3}{4}}$ or $\frac{1}{16} + \frac{1}{8} + \frac{1}{4} + \frac{1}{2} < area < \frac{81}{256} + \frac{27}{64} + \frac{9}{16} + \frac{3}{4}$ $\therefore$ $\frac{15}{16} < area < \frac{525}{256}$ Ans.

Example: – (13) The area enclosed by the curves $y = \sin x + \cos x$ and $y = |\cos x - \sin x|$ over the interval $\left[0, \frac{\pi}{2}\right]$.

Solution: – Given, $y = \sin x + \cos x$ and $y = |\cos x - \sin x|$ or $y_1 = \sin x + \cos x = \sqrt{2}\left[\sin x \cos \frac{\pi}{4} + \cos x \sin \frac{\pi}{4}\right] = \sqrt{2}\sin\left(x + \frac{\pi}{4}\right)$

similarly, $y_2 = |\cos x - \sin x| = \sqrt{2}\left|\sin\left(\frac{\pi}{4} - x\right)\right|$

$\Rightarrow$ Area $(A) = \int_{0}^{\frac{\pi}{4}}[(\sin x + \cos x) - (\cos x - \sin x)]\,dx + \int_{\frac{\pi}{4}}^{\frac{\pi}{2}}[(\sin x + \cos x) - (\sin x - \cos x)]\,dx$

$$A = \int_{0}^{\frac{\pi}{4}} 2\sin x\,dx + \int_{\frac{\pi}{4}}^{\frac{\pi}{2}} 2\cos x\,dx = 2(-\cos x)_{0}^{\frac{\pi}{4}} + 2(\sin x)_{\frac{\pi}{4}}^{\frac{\pi}{2}} = -2\left(\frac{1}{\sqrt{2}} - 1\right) + 2\left(1 - \frac{1}{\sqrt{2}}\right) = 2 - \frac{2}{\sqrt{2}} + 2 - \frac{2}{\sqrt{2}} = 4 - \frac{2}{\sqrt{2}} - \frac{2}{\sqrt{2}} = 4 - \sqrt{2} - \sqrt{2}$$

$$= 4 - 2\sqrt{2} \quad \text{Ans.}$$

Example: – (14) Prove that the area between the parabola $x^2 = 4y$, $x^2 = y$ and $y = 4$, $y = 9$ is $\frac{76}{3}$ sq. units.

Solution: – Given, $x^2 = 4y$ and $x^2 = y$ or $x_1 = 2\sqrt{y}$ and $x_2 = \sqrt{y}$

Area $(A) = 2\int_{4}^{9}(x_1 - x_2)\,dy = 2\int_{4}^{9}(2\sqrt{y} - \sqrt{y})\,dy = 2\int_{4}^{9}\sqrt{y}\,dy = 2\int_{4}^{9} y^{\frac{1}{2}}\,dy = 2\left[\frac{y^{\frac{1}{2}+1}}{\frac{1}{2}+1}\right]_{4}^{9} = 2\left[\frac{y^{\frac{3}{2}}}{\frac{3}{2}}\right]_{4}^{9} = \frac{4}{3}\left[9^{\frac{3}{2}} - 4^{\frac{3}{2}}\right] = \frac{4}{3}\left[(3^2)^{\frac{3}{2}} - (2^2)^{\frac{3}{2}}\right]$

$$= \frac{4}{3}(3^3 - 2^3) = \frac{4}{3}(27 - 8) = \frac{4}{3} \cdot 19 = \frac{76}{3} \text{ sq. units} \quad \text{Proved.}$$

Example: – (15) Find the area bounded by the curve $x = (y+1)(y-1)(y-2)$ lying between the ordinate $y = 0$ and $y = 2$.

Solution: – Given, $x = (y+1)(y-1)(y-2)$ $\therefore$ $x = 0$ then $y = -1, 1, 2$ and $y = 0, x = 2, y = 2, x = 0$

$x = +ve$ for $y > 2$, $x = -ve$ for $y < -1$, $x = +ve$ for $0 < y < 1$, $x = -ve$ for $1 < y < 2$, $x = +ve$ for $-1 < y < 0$

Area $(A) = \int_{0}^{2}|x|\,dy = \int_{0}^{1} x\,dy + \int_{1}^{2}(-x)\,dy = \int_{0}^{1}(y+1)(y-1)(y-2)\,dy - \int_{1}^{2}(y+1)(y-1)(y-2)\,dy$

$$= \int_{0}^{1}(y^3 - 2y^2 - y + 2)\,dy - \int_{1}^{2}(y^3 - 2y^2 - y + 2)\,dy$$

$$A = \left[\frac{y^4}{4} - \frac{2}{3}y^3 - \frac{y^2}{2} + 2y\right]_{0}^{1} - \left[\frac{y^4}{4} - \frac{2}{3}y^3 - \frac{y^2}{2} + 2y\right]_{1}^{2} = \left[\frac{1}{4} - \frac{2}{3} - \frac{1}{2} + 2 - 0\right] - \left[\frac{16}{4} - \frac{16}{3} - \frac{4}{2} + 4 - \frac{1}{4} + \frac{2}{3} + \frac{1}{2} - 2\right]$$

$A = \left[\dfrac{3-8-6+24}{12}\right] - \left[4 - \dfrac{16}{3} - 2 + 4 - \dfrac{1}{4} + \dfrac{2}{3} + \dfrac{1}{2} - 2\right] = \dfrac{13}{12} - \left[4 - \dfrac{16}{3} - \dfrac{1}{4} + \dfrac{2}{3} + \dfrac{1}{2}\right] = \dfrac{13}{12} - \left[\dfrac{48-64-3+8+6}{12}\right] = \dfrac{13}{12} - \dfrac{62-67}{12}$

$= \dfrac{13}{12} + \dfrac{5}{12} = \dfrac{18}{12} = \dfrac{3}{2} = 1\dfrac{1}{2}$ units   Ans.

Example: $-$ (16) Prove that the area bounded by the hyperbola $4x^2 - 9y^2 = a^2$ between the straight lines $x = 2a$

and $x = 3a$ is $\dfrac{a}{3}\left(3\sqrt{35} - 2\sqrt{15}\right) + \dfrac{a^2}{6}\log\left(\dfrac{4+\sqrt{15}}{6+\sqrt{35}}\right)$ sq. units.

Solution: $-$ Given, $4x^2 - 9y^2 = a^2$ or $9y^2 = 4x^2 - a^2$ or $y^2 = \dfrac{4x^2 - a^2}{9}$ or $y = \dfrac{\sqrt{4x^2 - a^2}}{3}$

Area $(A) = \displaystyle\int_{2a}^{3a} \dfrac{\sqrt{4x^2 - a^2}}{3}\,dx = \dfrac{1}{3}\int_{2a}^{3a}\sqrt{4x^2 - a^2}\,dx = \dfrac{1}{3}\int_{2a}^{3a}\sqrt{(2x)^2 - a^2}\,dx = \dfrac{1}{3}\left[\dfrac{2x}{2}\sqrt{4x^2 - a^2} - \dfrac{a^2}{2}\log\left(2x + \sqrt{4x^2 - a^2}\right)\right]_{2a}^{3a}$

$= \dfrac{1}{3}\left[x\sqrt{4x^2 - a^2} - \dfrac{a^2}{2}\log\left(2x + \sqrt{4x^2 - a^2}\right)\right]_{2a}^{3a}$ $\left[\text{use formula,}\quad \displaystyle\int \sqrt{x^2 - a^2}\,dx = \dfrac{x}{2}\sqrt{x^2 - a^2} - \dfrac{a^2}{2}\log\left(x + \sqrt{x^2 - a^2}\right)\right]$

$A = \dfrac{1}{3}\left\{3a\sqrt{4(3a)^2 - a^2} - \dfrac{a^2}{2}\log\left(2.3a + \sqrt{4(3a)^2 - a^2}\right) - \left[2a\sqrt{4(2a)^2 - a^2} - \dfrac{a^2}{2}\log\left(2.2a + \sqrt{4(2a)^2 - a^2}\right)\right]\right\}$

$A = \dfrac{1}{3}\left[3a\sqrt{36a^2 - a^2} - \dfrac{a^2}{2}\log\left(6a + \sqrt{36a^2 - a^2}\right) - 2a\sqrt{16a^2 - a^2} + \dfrac{a^2}{2}\log\left(4a + \sqrt{16a^2 - a^2}\right)\right]$

$A = \dfrac{1}{3}\left[3a\sqrt{35a^2} - \dfrac{a^2}{2}\log\left(6a + \sqrt{35a^2}\right) - 2a\sqrt{15a^2} + \dfrac{a^2}{2}\log\left(4a + \sqrt{15a^2}\right)\right]$

$A = \dfrac{1}{3}\left[3\sqrt{35}a - 2\sqrt{15}a + \dfrac{a^2}{2}\log(4a + \sqrt{15}a) - \dfrac{a^2}{2}\log(6a + \sqrt{35}a)\right] = \dfrac{a}{3}\left(3\sqrt{35} - 2\sqrt{15}\right) + \dfrac{a^2}{6}\log\left(\dfrac{4a + \sqrt{15}a}{6a + \sqrt{35}a}\right)$

$A = \dfrac{a}{3}\left(3\sqrt{35} - 2\sqrt{15}\right) + \dfrac{a^2}{6}\log\left(\dfrac{4+\sqrt{15}}{6+\sqrt{35}}\right)$ sq. units Proved.

Example: $-$ (17) Prove that the area common to the parabolas $y = 3x^2$ and $y = x^2 + 2$ is $\dfrac{8}{3}$ sq. units.

Solution: $-$ Given parabolas $y = 3x^2 \ldots \ldots \ldots$ (i) and $y = x^2 + 2 \ldots \ldots \ldots \ldots$ (ii)

solve equation (i) and (ii), Put $y = 3x^2$ in equation $y = x^2 + 2$ $\therefore$ $3x^2 = x^2 + 2$ or $2x^2 = 2$ $\therefore$ $x^2 = 1$ $\therefore$ $x = \pm 1$

Area $(A) = \displaystyle\int_{-1}^{1}(y_1 - y_2)\,dx = \int_{-1}^{1}(x^2 + 2 - 3x^2)\,dx = \int_{-1}^{1}(2 - 2x^2)\,dx = \left(2x - \dfrac{2x^3}{3}\right)_{-1}^{1} = \left(2 - \dfrac{2}{3}\right) - \left(-2 + \dfrac{2}{3}\right) = 2 - \dfrac{2}{3} + 2 - \dfrac{2}{3} = 4 - \dfrac{2}{3} - \dfrac{2}{3}$

$= \dfrac{12 - 2 - 2}{3} = \dfrac{8}{3}$ sq. units Proved.

Example: $-$ (18) Prove that the area bounded by the parabola $2x = y^2$ and the line $x = 4y$ is $\dfrac{128}{3}$ sq. units.

Solution: $-$ Given parabola $2x = y^2$ and line $x = 4y$

solve, $2x = y^2$ and $x = 4y$ Put $x = 4y$ in $2x = y^2$ or $2.4y = y^2$ or $y^2 - 8y = 0$ or $y(y - 8) = 0$ $\therefore$ $y = 0, 8$

Area $(A) = \displaystyle\int_{0}^{8}\left(4y - \dfrac{y^2}{2}\right)dy = \left(\dfrac{4y^2}{2} - \dfrac{y^3}{6}\right)_{0}^{8} = \dfrac{256}{2} - \dfrac{512}{6} = 128 - \dfrac{256}{3} = \dfrac{384 - 256}{3} = \dfrac{128}{3}$ sq. units. Proved.

(19) Find the area bounded by the curve $y = 3x + x^2$ and the straight line $y = x$.

Solution: $-$ Given curve $y = 3x + x^2$ and line $y = x$ Put $y = x$ in the curve $y = 3x + x^2$ and

find the point, $x = 3x + x^2$ or $x^2 + 3x - x = 0$ or $x^2 + 2x = 0$ or $x(x + 2) = 0$ $\therefore$ $x = 0, -2$

$$\text{Area (A)} = \int_{-2}^{0} (y_1 - y_2)\, dx = \int_{-2}^{0} (x - 3x - x^2)\, dx = \int_{-2}^{0} (-2x - x^2)\, dx = \left(\frac{-2x^2}{2}\right)_{-2}^{0} - \left(\frac{x^3}{3}\right)_{-2}^{0} = (0+4) - \left(0 + \frac{8}{3}\right) = 4 - \frac{8}{3} = \frac{12-8}{3}$$

$$= \frac{4}{3} \text{ sq. units} \quad \text{Ans.}$$

(20) Find the area in the plane bounded by the curves $x = y + 1$ and $(x-2)^2 = 2(y-1)$.

Solution: – Given, $x = y + 1$ or $y = x - 1 \ldots \ldots \ldots$ (i) and $(x-2)^2 = 2(y-1)$ or $y = \dfrac{(x-2)^2 + 2}{2} \ldots \ldots \ldots$ (ii)

solve the equation (i) & (ii), or $(x-2)^2 = 2(x-1-1)$ or $(x-2)^2 = 2(x-2)$

or $x^2 - 4x + 4 = 2x - 4$ or $x^2 - 6x + 8 = 0$ or $x^2 - 4x - 2x + 8 = 0$ or $x(x-4) - 2(x-4) = 0$ or $(x-2)(x-4) = 0$ $\therefore x = 2, 4$

$$\text{Area (A)} = \int_{2}^{4} \left[\frac{(x-2)^2 + 2}{2} - (x-1)\right] dx = \int_{2}^{4} \left[\frac{(x-2)^2 + 2 - 2x + 2}{2}\right] dx$$

$$A = \frac{1}{2}\int_{2}^{4}(x^2 - 4x + 4 - 2x + 4)\, dx = \frac{1}{2}\int_{2}^{4}(x^2 - 6x + 8)\, dx = \frac{1}{2}\left[\left(\frac{x^3}{3}\right)_{2}^{4} - \left(\frac{6x^2}{2}\right)_{2}^{4} + (8x)_{2}^{4}\right] = \frac{1}{2}\left[\left(\frac{64}{3} - \frac{8}{3}\right) - 3(16-4) + (32-16)\right]$$

$$= \frac{1}{2}\left[\left(\frac{64-8}{3}\right) - 36 + 16\right] = \frac{1}{2}\left[\frac{56}{3} - 20\right] = \frac{1}{2}\left[\frac{56-60}{3}\right] = \frac{1}{2}\left[\frac{-2}{3}\right] = -\frac{1}{3} \quad \text{(Area is not negative)}$$

$$\text{Area (A)} = \frac{1}{3} \text{ sq. units} \quad \text{Ans.}$$

(21) Find the area bounded by the parabola $x = y^2 - 6$ and the straight line $x = -y$.

Solution: – Put $x = -y$ in the parabola $x = y^2 - 6$, we get

or $-y = y^2 - 6$ or $y^2 + y - 6 = 0$ or $y^2 + 3y - 2y - 6 = 0$ or $y(y+3) - 2(y+3) = 0$ or $(y-2)(y+3) = 0$ $\therefore y = -3, 2$

$$\text{Area (A)} = \int_{-3}^{2} (x_1 - x_2)\, dy = \int_{-3}^{2} (-y - y^2 + 6)\, dy = \left[-\frac{y^2}{2} - \frac{y^3}{3} + 6y\right]_{-3}^{2} = \left[-2 - \frac{8}{3} + 12\right] - \left[-\frac{9}{2} + 9 - 18\right] = \left[10 - \frac{8}{3}\right] - \left[-9 - \frac{9}{2}\right]$$

$$= \left(\frac{30-8}{3}\right) - \left(\frac{-18-9}{2}\right) = \frac{22}{3} + \frac{27}{2} = \frac{44+81}{6} = \frac{125}{6} \text{ sq. units} \quad \text{Ans.}$$

(22) The area of the region bounded by the curve $x^2 = y + 1$ and $x + y = 1$.

Solution: – Given, $x^2 = y + 1 \ldots \ldots \ldots$ (i) and $x + y = 1 \ldots \ldots \ldots$ (ii)

solve equation (i) & (ii), or $x^2 = y + 1$ or $y = x^2 - 1$ and $y = 1 - x$

Put $y = 1 - x$ in equation (i), we get or $x^2 = 1 - x + 1 = 2 - x$ or $x^2 + x - 2 = 0$

or $x^2 + 2x - x - 2 = 0$ or $x(x+2) - 1(x+2) = 0$ or $(x-1)(x+2) = 0$ $\therefore x = -2, 1$

$$\text{Area (A)} = \int_{-2}^{1} (y_1 - y_2)\, dx = \int_{-2}^{1} [(1-x) - (x^2-1)]\, dx = \int_{-2}^{1} (1 - x - x^2 + 1)\, dx = \int_{-2}^{1} (2 - x - x^2)\, dx = \left[2x - \frac{x^2}{2} - \frac{x^3}{3}\right]_{-2}^{1}$$

$$= \left[2 - \frac{1}{2} - \frac{1}{3}\right] - \left[-4 - \frac{4}{2} + \frac{8}{3}\right] = \left(\frac{12-3-2}{6}\right) - \left(\frac{-24-12+16}{6}\right) = \frac{7}{6} + \frac{20}{6} = \frac{7+20}{6} = \frac{27}{6} = \frac{9}{2} \text{ sq. units} \quad \text{Ans.}$$

Differential Equation

Definitions: – **(a) Differential equations** – consider the following equations

$$\therefore \frac{d^3y}{dx^3} + 2\frac{d^2y}{dx^2} - 2\frac{dy}{dx} + y = \cos 2x \dots\dots\dots\dots (i)$$

$$\therefore \frac{dy}{dx} = \frac{\sqrt{x}}{\sqrt{1+y^2}} \dots\dots\dots (ii), \quad xy\frac{dy}{dx} = \frac{x^2-1}{1-y} \dots\dots\dots (iii)$$

$$\therefore \ x\frac{\partial z}{\partial \theta} + y\frac{\partial z}{\partial y} + z = 0 \dots\dots\dots (iv) \quad \therefore \ \frac{\partial^2 y}{\partial t^2} = b^2\frac{\partial^2 y}{\partial x^2} \dots\dots\dots (v)$$

$$\therefore \ \frac{\left[1+\left(\frac{dy}{dx}\right)^2\right]^{\frac{1}{2}}}{\frac{d^2y}{dx^2}} = \rho \ \dots\dots\dots\dots (vi)$$

All the above equation are called differential equations.

(b) ordinary and partial differential equations: –

\# ordinary differential equations are those which involve only one independent variable.

e.g. $\frac{d^2y}{dx^2} + 2\frac{dy}{dx} + y = \sin x \dots\dots\dots (i)$ and $\frac{dy}{dx} = \frac{x+1}{\sqrt{1+y^2}} \dots\dots\dots (ii)$

Both of equations (i) & (ii) involve only one independent variable x.

\# Partial differential equations are those which involve two or more than two independent variable.

e.g. $\frac{\partial^2 y}{\partial t^2} = a^2\frac{\partial^2 y}{\partial x^2} + b\frac{\partial y}{\partial x} \dots\dots\dots (iii)$ $\quad \therefore \ x\frac{\partial z}{\partial \theta} + y\frac{\partial z}{\partial y} + z = 0 \dots\dots\dots (iv)$

Both of equations (iii) & (iv), (θ, y) and (t, x) are independent variable respectively.

(c) order and degree of differential equations: – The order of differential equation is defined

to be the order of the highest derivative or differential coefficient occurring in it.

Example: – Equation $\frac{dy}{dx} = \frac{\sqrt{x+1}}{\sqrt{1+y^3}}$ and $xy\frac{dy}{dx} = \frac{x^2-1}{y-1}$ are of first order.

Equation $\frac{\left[1-\left(\frac{dy}{dx}\right)^3\right]^{\frac{2}{3}}}{\frac{d^2y}{dx^2}} = \rho$ and $\frac{d^3y}{dx^3} + 3\frac{d^2y}{dx^2} - \frac{dy}{dx} + y = \cos 2x$ are of second and third order respectively.

Equation $\frac{\left[1+\left(\frac{dy}{dx}\right)^2\right]^{\frac{3}{2}}}{\frac{d^2y}{dx^2}} = \rho$ or $\left[1+\left(\frac{dy}{dx}\right)^2\right]^3 = \rho^2\left(\frac{d^2y}{dx^2}\right)^2$

Hence the equation is of second order and second degree.

Example: – (1) $y = Ae^{-3x} - Be^{3x}$, Eliminating the arbitrary constant A and B. $\therefore \ \frac{dy}{dx} = -3Ae^{-3x} - 3Be^{3x} \dots\dots\dots\dots (i)$

Again, Differentiate $\frac{d^2y}{dx^2} = 9Ae^{-3x} - 9Be^{3x} = 9(Ae^{-3x} - Be^{3x}) = 9y$

$\therefore \dfrac{d^2y}{dx^2} - 9y = 0$ it is differential equation. hence the equation is of second order.

Example: – (2) $y = A\cos 3x - B\sin 3x$ Differentiate, $\dfrac{dy}{dx} = -3A\sin 3x - 3B\cos 3x$

$\therefore \dfrac{d^2y}{dx^2} = -9A\cos 3x + 9B\sin 3x$ $\therefore \dfrac{d^2y}{dx^2} = -9(A\cos 3x - B\sin 3x) = -9y$

$\therefore \dfrac{d^2y}{dx^2} + 9y = 0$, hence the differential equation is of second order.

Example: – (3) (a) Prove that $Ax + By^2 = 1$ is the solution of $y\dfrac{d^2y}{dx^2} + \left(\dfrac{dy}{dx}\right)^2 = 0$.

Solution: – Given, $Ax + By^2 = 1$ $\therefore Ax + By^2 - 1 = 0$ Differentiating, $A + 2By\dfrac{dy}{dx} = 0$ or $2By\dfrac{dy}{dx} = -A$ or $y\dfrac{dy}{dx} = -\dfrac{A}{2B}$

Again Differentiating, $y\dfrac{d^2y}{dx^2} + \dfrac{dy}{dx} \cdot \dfrac{dy}{dx} = 0$ $\therefore y\dfrac{d^2y}{dx^2} + \left(\dfrac{dy}{dx}\right)^2 = 0$ Proved.

(b) Prove that $By^2 = 1 + Ax^2$ is the solution of $y\dfrac{d^2y}{dx^2} = \dfrac{y}{x} \cdot \dfrac{dy}{dx} - \left(\dfrac{dy}{dx}\right)^2$.

Solution: – $By^2 = 1 + Ax^2$ Differentiating, $2By\dfrac{dy}{dx} = 2Ax$ $\therefore \dfrac{dy}{dx} = \dfrac{2Ax}{2By} = \dfrac{Ax}{By}$

Again Differentiating, $\dfrac{d^2y}{dx^2} = \dfrac{A}{B}\left[\dfrac{y.1 - x.\dfrac{dy}{dx}}{y^2}\right]$ or $By^2\dfrac{d^2y}{dx^2} = Ay - Ax\dfrac{dy}{dx}$ …………………(A)

Put $\dfrac{dy}{dx} = \dfrac{Ax}{By}$ $\therefore By = \dfrac{Ax}{\dfrac{dy}{dx}}$ in equation (A), we get

or $\dfrac{Ax}{\dfrac{dy}{dx}} \cdot y\dfrac{d^2y}{dx^2} = Ay - Ax\dfrac{dy}{dx}$ or $Axy\dfrac{d^2y}{dx^2} = Ay\dfrac{dy}{dx} - Ax\left(\dfrac{dy}{dx}\right)^2$ or $y\dfrac{d^2y}{dx^2} = \dfrac{y}{x} \cdot \dfrac{dy}{dx} - \left(\dfrac{dy}{dx}\right)^2$ Proved.

Example: – (4)(a) Prove that $y = e^x + \log x$ is the solution of $\dfrac{d^2y}{dx^2} - \dfrac{dy}{dx} + \dfrac{(x+1)}{x^2} = 0$.

Solution: – Given, $y = e^x + \log x$ $\therefore \dfrac{dy}{dx} = e^x + \dfrac{1}{x}$ $\therefore \dfrac{d^2y}{dx^2} = e^x - \dfrac{1}{x^2} = \dfrac{dy}{dx} - \dfrac{1}{x} - \dfrac{1}{x^2} = \dfrac{dy}{dx} - \left(\dfrac{x+1}{x^2}\right)$ $\therefore \dfrac{d^2y}{dx^2} - \dfrac{dy}{dx} + \left(\dfrac{x+1}{x^2}\right) = 0$ Proved.

(b) Prove that $ye^x + \log x = 0$ is the solution of $\dfrac{d^2y}{dx^2} + \dfrac{dy}{dx} = \dfrac{(x+1)}{x^2e^x}$ or $x^2e^x\dfrac{d^2y}{dx^2} + x^2e^x\dfrac{dy}{dx} = (x+1)$.

Solution: – Given, $ye^x + \log x = 0$ Differentiating, $ye^x + e^x\dfrac{dy}{dx} + \dfrac{1}{x} = 0$ $\therefore e^x\left(y + \dfrac{dy}{dx}\right) = -\dfrac{1}{x}$ or $y + \dfrac{dy}{dx} = -\dfrac{1}{xe^x}$ $\therefore \dfrac{dy}{dx} = -y - \dfrac{1}{xe^x}$

Again Differentiating, $\dfrac{d^2y}{dx^2} = -\dfrac{dy}{dx} - \left[\dfrac{0 - (xe^x + e^x)}{x^2e^{2x}}\right] = -\dfrac{dy}{dx} + \dfrac{e^x(x+1)}{x^2e^{2x}} = -\dfrac{dy}{dx} + \dfrac{(x+1)}{x^2e^x}$

$\therefore \dfrac{d^2y}{dx^2} + \dfrac{dy}{dx} = \dfrac{(x+1)}{x^2e^x}$ or $x^2e^x\dfrac{d^2y}{dx^2} + x^2e^x\dfrac{dy}{dx} = (x+1)$ Proved.

(5) Find the Order, Degree, Linear and Non – linear of the differential equation: –

(a) $y'' + 3y' + 4y = 0$ or $\dfrac{d^2y}{dx^2} + 3\dfrac{dy}{dx} + 4y = 0$ Order = 2, Degree = 1 and Linear differential equation.

(b) $x^2y'' + xy' + 3y = 5x$ or $x^2\dfrac{d^2y}{dx^2} + x\dfrac{dy}{dx} + 3y = 5x$ Order = 2, Degree = 1 and Linear.

(c) $(y')^2 + 3xy' + 3y = 0$ Ans: $-$ Order $= 1$, Degree $= 1$ and Non $-$ linear.

(d) $\sqrt{1 + 3x^2}\ dx + \sqrt{1 + 3y^2}\ dy = 0 \Rightarrow \dfrac{dy}{dx} = -\dfrac{\sqrt{1 + 3x^2}}{\sqrt{1 + 3y^2}}$ Order $= 1$, Degree $= 1$ and Non $-$ linear.

(6) Find the differential equation: $-$ (a) $y = Ae^{Px}$ $\therefore$ $\dfrac{dy}{dx} = APe^{Px} = P(Ae^{Px}) = Py$, Put $y = Ae^{Px}$ or $\dfrac{dy}{dx} = Py$ $\therefore$ $\dfrac{dy}{dx} - Py = 0$ Ans.

(b) $y = A\sin Px - B\cos Px$ Differentiating, $\dfrac{dy}{dx} = AP\cos Px + BP\sin Px$

Again Differentiating, $\dfrac{d^2y}{dx^2} = -AP^2\sin Px + BP^2\cos Px = -P^2(A\sin Px - B\cos Px) = -P^2y$

or $\dfrac{d^2y}{dx^2} = -P^2y$ $\therefore$ $\dfrac{d^2y}{dx^2} + P^2y = 0$ Ans.   [put $y = A\sin Px - B\cos Px$]

(c) $y = A\sin(Px - q)$ Differentiating, $\dfrac{dy}{dx} = AP\cos(Px - q)$

Again Differentiatin $\dfrac{d^2y}{dx^2} = -AP^2\sin(Px - q) = -P^2[A\sin(Px - q)] = -P^2y$ $\therefore$ $\dfrac{d^2y}{dx^2} + P^2y = 0$ Ans.   [put $y = A\sin(Px - q)$]

(d) $y = \tan(Px + q)$ Ans: $-$ $y'' + P^2y = 0$ or $\dfrac{d^2y}{dx^2} + P^2y = 0$ (Do yourself)

(7) Find the differential equation: $-$ (a) $y = a\sin x + b\cos x$, eliminate the constant a, b from the relation.

Solution: $-$ Given, $y = a\sin x + b\cos x$ Differentiating, $\dfrac{dy}{dx} = a\cos x - b\sin x$

Again Differentiating, $\dfrac{d^2y}{dx^2} = -a\sin x - b\cos x = -(a\sin x + b\cos x) = -y$ $\therefore$ $\dfrac{d^2y}{dx^2} + y = 0$ $\therefore$ $y'' + y = 0$ Ans.

(b) Prove that $y = e^x\log x + x$ is the solution of $x^2\left[\dfrac{d^2y}{dx^2} - \dfrac{dy}{dx} + 1\right] = e^x(x - 1)$.

Solution: $-$ Given, $y = e^x\log x + x$ Diff. $\dfrac{dy}{dx} = e^x.\dfrac{1}{x} + \log x.e^x + 1 = \dfrac{e^x}{x} + e^x\log x + 1$ or $\dfrac{dy}{dx} = \dfrac{e^x}{x} + e^x\log x + 1$

Again Differentiating, $\dfrac{d^2y}{dx^2} = \dfrac{xe^x - e^x}{x^2} + e^x.\dfrac{1}{x} + \log x.e^x = \dfrac{e^x(x - 1)}{x^2} + \left(\dfrac{e^x}{x} + e^x\log x\right)$

put $\dfrac{dy}{dx} = \dfrac{e^x}{x} + e^x\log x + 1$ or $\dfrac{e^x}{x} + e^x\log x = \dfrac{dy}{dx} - 1$ or $\dfrac{d^2y}{dx^2} = \dfrac{e^x(x - 1)}{x^2} + \dfrac{dy}{dx} - 1$

or $\dfrac{d^2y}{dx^2} - \dfrac{dy}{dx} + 1 = \dfrac{e^x(x - 1)}{x^2}$ $\therefore$ $x^2\left[\dfrac{d^2y}{dx^2} - \dfrac{dy}{dx} + 1\right] = e^x(x - 1)$ Proved.

(8) (a) Find the differential equation of the family of curves $y = Ae^{2x} + Be^{3x}$.

Solution: $-$ Given, $y = Ae^{2x} + Be^{3x}$ ... ... (i)

Differentiating, $y' = 2Ae^{2x} + 3Be^{3x}$ ... ... (ii) and $y'' = 4Ae^{2x} + 9Be^{3x}$ ... ... (iii)

Eliminating A and B from the above three equation, we get

or $\begin{vmatrix} e^{2x} & e^{3x} & -y \\ 2e^{2x} & 3e^{3x} & -y' \\ 4e^{2x} & 9e^{3x} & -y'' \end{vmatrix} = 0$ or $(-e^{2x}.e^{3x})\begin{vmatrix} 1 & 1 & y \\ 2 & 3 & y' \\ 4 & 9 & y'' \end{vmatrix} = 0$

or $-e^{2x}.e^{3x}[1(3y'' - 9y') - 1(4y' - 2y'') + y(18 - 12)] = 0$ or $3y'' - 9y' - 4y' + 2y'' + 6y = 0$

or $5y'' - 13y' + 6y = 0$ $\therefore$ $5\dfrac{d^2y}{dx^2} - 13\dfrac{dy}{dx} + 6y = 0$ Ans.

(b) Find the differential equation of the family of curves $y = Ae^{3x} + Be^{-5x} + Ce^x$ where A, B and C are arbitrary constant.

Solution: – Do yourself.

Ordinary differential equations of the first order and first degree: –

An ordinary differential equation of the first order and first degree is of the form: –

$$M + N\frac{dy}{dx} = 0 \quad \text{or} \quad M\,dx + N\,dy = 0$$

where M and N are functions of x and y or constant.

Some special cases: –

Case I: – Variable separable: – Equations which are capable of being put in the form

$f(x)dx + \phi(y)dy = 0$ will be termed as variable separable.

Hence complete general solution of $f(x)dx + \phi(y)dy = 0$ will be $\int f(x)\,dx + \int \phi(y)\,dy = a$ (constant)

Example: – Solve, $\dfrac{dy}{dx} = \dfrac{1+x^2}{1+y^2}$ or $(1+y^2)dy = (1+x^2)dx$, Integrating both sides

or $\int (1+y^2)dy = \int (1+x^2)dx$ or $y + \dfrac{y^3}{3} = x + \dfrac{x^3}{3} + c$ or $\dfrac{3y + y^3}{3} = \dfrac{3x + x^3 + 3c}{3}$

or $3y + y^3 = 3x + x^3 + 3c$ or $y^3 - x^3 + 3(y - x) = k$ Ans. (where $3c = k$)

Example: – Solve, $\dfrac{dy}{dx} = \dfrac{\sqrt{1-y^2}}{\sqrt{1-x^2}}$ or $\dfrac{dy}{\sqrt{1-y^2}} = \dfrac{dx}{\sqrt{1-x^2}}$ Integrating both of sides, $\int \dfrac{dy}{\sqrt{1-y^2}} = \int \dfrac{dx}{\sqrt{1-x^2}}$

or $\sin^{-1} y = \sin^{-1} x + c$ or $\sin^{-1} y - \sin^{-1} x = c$ or $\sin^{-1}\left[y\sqrt{1-x^2} - x\sqrt{1-y^2}\right] = c$

use formula, $\int \dfrac{dx}{\sqrt{a^2 - x^2}} = \sin^{-1}\left(\dfrac{x}{a}\right)$ and $\sin^{-1} x \pm \sin^{-1} y = \sin^{-1}\left[x\sqrt{1-y^2} \pm y\sqrt{1-x^2}\right]$

or $\sin^{-1}\left[y\sqrt{1-x^2} - x\sqrt{1-y^2}\right] = c$ or $y\sqrt{1-x^2} - x\sqrt{1-y^2} = \sin c = k$ (say) $\therefore$ $y\sqrt{1-x^2} - x\sqrt{1-y^2} = k$ Ans.

Remember the following result

(a) $d\left(\dfrac{x}{y}\right) = \dfrac{y.\,dx - x.\,dy}{y^2}$ (b) $d\left(\dfrac{y}{x}\right) = \dfrac{x.\,dy - y.\,dx}{x^2}$ (c) $d(x^2 + y^2) = 2(x.\,dx + y.\,dy)$ (d) $d(x^2 - y^2) = 2(x.\,dx - y.\,dy)$

Case II: – Exact differential equations: – $M\,dx + N\,dy = 0$ where M and N are functions of x and y if $\dfrac{\partial M}{\partial y} = \dfrac{\partial N}{\partial x}$

then the equation is exact and its solution is $\int M\,dx + \int N\,dy = c$ where $\int M\,dx \to y - \text{constant}$ and $\int N\,dy \to \text{free form of } x$

Example: – Solve, $(x^2 + 2xy + 3y^2)\,dx + (y^2 + 6xy + x^2)\,dy = 0$

Solution: – Let $M = x^2 + 2xy + 3y^2$ and $N = y^2 + 6xy + x^2$

or $\dfrac{\partial M}{\partial y} = 6y + 2x$ and $\dfrac{\partial N}{\partial x} = 2x + 6y$ since $\dfrac{\partial M}{\partial y} = \dfrac{\partial N}{\partial x}$ the equation is exact.

Its solution is $\int (x^2 + 2xy + 3y^2)\,dx + \int N\,dy = c$ where $\int N\,dy \to \text{free form of } x$

---

or $\dfrac{x^3}{3} + x^2y + 3y^2x + \displaystyle\int y^2\,dy = c$ or $\dfrac{x^3}{3} + x^2y + 3y^2x + \dfrac{y^3}{3} = c$ or $x^3 + 3x^2y + 9y^2x + y^3 = 3c$

$$\therefore\quad x^3 + 3x^2y + 9y^2x + y^3 = k \quad \text{where } k = 3c \quad \text{Ans.}$$

Example: $-$ Solve, $y(1 + e^x)\,dx + (x + e^x)\,dy = 0$

Solution: $-$ Given, $y(1 + e^x)\,dx + (x + e^x)\,dy = 0$ Here $M = y(1 + e^x)$ and $N = x + e^x$

or $\dfrac{\partial M}{\partial y} = (1 + e^x)$ and $\dfrac{\partial N}{\partial x} = 1 + e^x$ $\quad\therefore\quad \dfrac{\partial M}{\partial y} = \dfrac{\partial N}{\partial x}$ the equation is exact.

Its solution is $\quad \displaystyle\int M\,dx + \int N\,dy = c \quad$ or $\displaystyle\int y(1 + e^x)\,dx + \int dy = c$

or $y\displaystyle\int dx + y\int e^x\,dx + \int dy = c \quad$ or $y.x + y.e^x + y = c$ or $y(x + e^x + 1) = c$ $\therefore$ $y(1 + x + e^x) = c$ Ans.

Case III: $-$ **Homogeneous equations:** $-$ when $\dfrac{dy}{dx} = \dfrac{x+y}{x-y}$ whose N^r and D^r both are homogeneous function of x and y of the same

degree then the differential equation is called to be homogeneous equation. $\dfrac{dy}{dx} = \dfrac{f(x,y)}{\phi(x,y)}$ where f and ϕ are both homogeneous function

of x and y of the same degree.

Method of solution: $-$ Let $\dfrac{dy}{dx} = \dfrac{x+y}{x-y}$ it is homogeneous function. $\quad\therefore\quad \dfrac{dy}{dx} = \dfrac{x\left(1 + \frac{y}{x}\right)}{x\left(1 - \frac{y}{x}\right)}$ $\quad$ Put $y = vx$ $\quad\therefore\quad \dfrac{dy}{dx} = v + x\dfrac{dv}{dx}$

Now, substitute the value of $\dfrac{dy}{dx}$ and y in the given equation

or $\dfrac{dy}{dx} = \dfrac{\left(1 + \frac{y}{x}\right)}{\left(1 - \frac{y}{x}\right)}$ $\quad$ or $v + x\dfrac{dv}{dx} = \dfrac{1+v}{1-v}$ $\quad$ or $x\dfrac{dv}{dx} = \dfrac{1+v}{1-v} - v = \dfrac{1 + v - v + v^2}{1-v} = \dfrac{1+v^2}{1-v}$

or $\dfrac{1-v}{1+v^2}\,dv = \dfrac{dx}{x}$ $\quad$ or $\dfrac{1}{1+v^2}\,dv - \dfrac{v}{1+v^2}\,dv = \dfrac{dx}{x}$

Integrating, $\displaystyle\int \dfrac{1}{1+v^2}\,dv - \int \dfrac{v}{1+v^2}\,dv = \int \dfrac{dx}{x}$ $\quad$ or $\tan^{-1}v - \dfrac{1}{2}\log(1+v^2) = \log x + \log c$

or $\tan^{-1}\left(\dfrac{y}{x}\right) - \log\sqrt{1 + \left(\dfrac{y}{x}\right)^2} = \log x + \log c$ or $\tan^{-1}\left(\dfrac{y}{x}\right) = \log\left(\sqrt{1 + \left(\dfrac{y}{x}\right)^2}\right) + \log x.c$

or $\tan^{-1}\left(\dfrac{y}{x}\right) = \log\left(\sqrt{\dfrac{x^2+y^2}{x^2}}\right) + \log x.c$ or $\tan^{-1}\left(\dfrac{y}{x}\right) = \log\left\{x.c.\dfrac{\sqrt{x^2+y^2}}{x}\right\} = \log\left(\sqrt{x^2+y^2}.c\right)$ Ans.

Example: $-$ Solve, $x\dfrac{dy}{dx} = x\sin\left(\dfrac{x+y}{x}\right) + y$

Solution: $-$ Given, $x\dfrac{dy}{dx} = x\sin\left(\dfrac{x+y}{x}\right) + y$ or $\dfrac{dy}{dx} = \sin\left[1 + \dfrac{y}{x}\right] + \dfrac{y}{x}$ $\quad$ Put $v = \dfrac{y}{x}$ or $y = vx$ $\therefore$ $\dfrac{dy}{dx} = v + x\dfrac{dv}{dx}$

or $v + x\dfrac{dv}{dx} = \sin(1+v) + v$ $\quad$ or $x\dfrac{dv}{dx} = \sin(1+v) + v - v = \sin(1+v)$ $\quad$ or $\dfrac{dv}{\sin(1+v)} = \dfrac{dx}{x}$

Integrating, $\displaystyle\int \dfrac{dv}{\sin(1+v)} = \int \dfrac{dx}{x}$ $\quad$ or $\displaystyle\int \operatorname{cosec}(1+v)\,dv = \int \dfrac{dx}{x}$ $\quad$ or $\log\left(\dfrac{1+v}{2}\right) = \log x + \log c$

or $\log\left(\dfrac{1+\frac{y}{x}}{2}\right) = \log x.c$ or $\log\left(\dfrac{x+y}{2x}\right) - \log x.c = 0$ or $\log\left(\dfrac{x+y}{2x^2c}\right) = 0$ Ans. $\left[\int \operatorname{cosec} x\, dx = \log\left(\tan \frac{x}{2}\right)\right]$

Example: $-$ Solve, $\dfrac{dx}{dy} = e^{-\frac{x}{2}} + \dfrac{x}{y}$

Solution: $-$ Given, $\dfrac{dx}{dy} = e^{-\frac{x}{2}} + \dfrac{x}{y}$ Put $v = \dfrac{x}{y}$ or $x = yv$ $\therefore \dfrac{dx}{dy} = v + y\dfrac{dv}{dy}$

or $v + y\dfrac{dv}{dy} = e^{-v} + v$ or $y\dfrac{dv}{dy} = e^{-v}$ or $\dfrac{dv}{e^{-v}} = \dfrac{dy}{y}$ or $e^v\, dv = \dfrac{dy}{y}$

Integrating, $\int e^v\, dv = \int \dfrac{dy}{y}$ or $e^v = \log y + \log k$ or $e^v = \log(y.k)$ put $v = \dfrac{x}{y}$

or $e^v = \log(y.k)$ $\therefore$ $e^{\frac{x}{y}} = \log(y.k)$ Ans.

Example: $-$ Solve, $x\dfrac{dy}{dx} = y + x\operatorname{cosec}\left(\dfrac{y}{x}\right)$

Solution: $-$ Given, $x\dfrac{dy}{dx} = y + x\operatorname{cosec}\left(\dfrac{y}{x}\right)$ Put $v = \dfrac{y}{x}$ or $y = vx$ $\therefore \dfrac{dy}{dx} = v + x\dfrac{dv}{dx}$

or $x\dfrac{dy}{dx} = y + x\operatorname{cosec}\left(\dfrac{y}{x}\right)$ or $\dfrac{dy}{dx} = \dfrac{y}{x} + \operatorname{cosec}\left(\dfrac{y}{x}\right)$ or $v + x\dfrac{dv}{dx} = v + \operatorname{cosec} v$ or $x\dfrac{dv}{dx} = \operatorname{cosec} v$

or $\dfrac{dv}{\operatorname{cosec} v} = \dfrac{dx}{x}$ Integrating, $\int \dfrac{dv}{\operatorname{cosec} v} = \int \dfrac{dx}{x}$ or $\int \sin v\, dv = \int \dfrac{dx}{x}$

or $-\cos v = \log x + \log k$ or $-\cos v = \log(x.k)$ Put $v = \dfrac{y}{x}$ $\therefore$ $\cos\left(\dfrac{y}{x}\right) + \log(x.k) = 0$ Ans.

Example: $-$ Solve, $(x+y)\dfrac{dy}{dx} = y$

Solution: $-$ Given, $(x+y)\dfrac{dy}{dx} = y$ or $\dfrac{dy}{dx} = \dfrac{y}{x+y} = \dfrac{y}{x}.\dfrac{1}{\left(1+\frac{y}{x}\right)}$ Put $y = vx$ $\therefore \dfrac{dy}{dx} = v + x\dfrac{dv}{dx}$

or $v + x\dfrac{dv}{dx} = v.\dfrac{1}{1+v} = \dfrac{v}{1+v}$ or $x\dfrac{dv}{dx} = \dfrac{v}{1+v} - v = \dfrac{v - v - v^2}{1+v} = -\dfrac{v^2}{1+v}$ or $\dfrac{1+v}{v^2}\, dv = -\dfrac{dx}{x}$

Integrating, $\int \dfrac{1+v}{v^2}\, dv = -\int \dfrac{dx}{x}$ or $\int\left(\dfrac{1}{v^2} + \dfrac{v}{v^2}\right) dv = -\int \dfrac{dx}{x}$ or $\int \dfrac{1}{v^2}\, dv + \int \dfrac{dv}{v} = -\int \dfrac{dx}{x}$

or $-\dfrac{1}{v} + \log v = -\log x - \log k$ or $-\dfrac{1}{v} = -[\log x + \log v + \log k]$ or $\dfrac{1}{v} = \log|x.v.k|$ Put $v = \dfrac{y}{x}$

or $\dfrac{1}{\frac{y}{x}} = \log\left|x.\dfrac{y}{x}.k\right| = \log|y.k|$ $\therefore$ $\dfrac{x}{y} = \log|y.k|$ $\therefore$ $|y.k| = e^{\frac{x}{y}}$ Ans.

Example: $-$ Solve, $y\dfrac{dx}{dy} = x + ye^{\frac{x}{y}}$

Solution: $-$ Given, $y\dfrac{dx}{dy} = x + ye^{\frac{x}{y}}$ or $\dfrac{dx}{dy} = \dfrac{y\left(\frac{x}{y} + e^{\frac{x}{y}}\right)}{y}$ Put $v = \dfrac{x}{y}$ or $x = vy$ $\therefore \dfrac{dx}{dy} = v + y\dfrac{dv}{dy}$

or $\dfrac{dx}{dy} = \dfrac{y\left(\frac{x}{y} + e^{\frac{x}{y}}\right)}{y} = \dfrac{x}{y} + e^{\frac{x}{y}}$ or $v + y\dfrac{dv}{dy} = v + e^v$ or $y\dfrac{dv}{dy} = e^v$ or $\dfrac{dv}{e^v} = \dfrac{dy}{y}$

Integrating, $\int \dfrac{dv}{e^v} = \int \dfrac{dy}{y}$ or $\int e^{-v}\, dv = \int \dfrac{dy}{y}$ or $-e^{-v} = \log y + \log k = \log|y.k|$ put $v = \dfrac{x}{y}$

or $e^{-v} = -\log|y.k|$ or $e^{-\frac{x}{y}} = -\log|y.k|$ $\therefore$ $e^{-\frac{x}{y}} = \log\left(\frac{1}{|y.k|}\right)$ Ans.

(I). Equations reduciable to homogeneous form: $-$ Consider the following equations $\dfrac{dy}{dx} = \dfrac{ax + by + c}{a'x + b'y + c'}$ ... . (i)

where $\dfrac{a}{a'} \neq \dfrac{b}{b'}$ it is not a homogeneous equation.

Method of solution: $-$ Put $x = \alpha + h$ and $y = \beta + k$ where α and β are variables but h and k are constant.   then, $\dfrac{dy}{dx} = \dfrac{d\beta}{d\alpha}$

and the above equation becomes

from equation (i), $\dfrac{dy}{dx} = \dfrac{ax + by + c}{a'x + b'y + c'}$ Put value x, y and $\dfrac{dy}{dx}$ we have

or $\dfrac{d\beta}{d\alpha} = \dfrac{a(\alpha + h) + b(\beta + k) + c}{a'(\alpha + h) + b'(\beta + k) + c'} = \dfrac{a\alpha + b\beta + (ah + bk + c)}{a'\alpha + b'\beta + (a'h + b'k + c')}$

Now, choose h and k such that $ah + bk + c = 0$ or $a'h + b'k + c' = 0$ ... ... (A)

Then, $\dfrac{d\beta}{d\alpha} = \dfrac{a\alpha + b\beta}{a'\alpha + b'\beta}$ which is a homogeneous equation in α and β.

Put $\beta = v\alpha$ $\therefore$ $\dfrac{d\beta}{d\alpha} = v + \alpha\dfrac{dv}{d\alpha}$ or $\dfrac{d\beta}{d\alpha} = \dfrac{a\alpha + b\beta}{a'\alpha + b'\beta}$ or $v + \alpha\dfrac{dv}{d\alpha} = \dfrac{\alpha\left(a + b.\frac{\beta}{\alpha}\right)}{\alpha\left(a' + b'.\frac{\beta}{\alpha}\right)} = \dfrac{a + bv}{a' + b'v}$

or $\alpha\dfrac{dv}{d\alpha} = \dfrac{a + bv}{a' + b'v} - v = \dfrac{a + bv - a'v - b'v^2}{a' + b'v}$ or $\dfrac{a' + b'v}{a + bv - a'v - b'v^2} dv = \dfrac{d\alpha}{\alpha}$

In the end put $v = \dfrac{\beta}{\alpha}$ and $\alpha = x - h$, $\beta = y - k$ where h and k are determined from equation (A).

(II). If $\dfrac{dy}{dx} = \dfrac{ax + by + c}{a'x + b'y + c'}$ and $\dfrac{a}{a'} = \dfrac{b}{b'} = m$ (say)

Method of solution: $-$ $\dfrac{dy}{dx} = \dfrac{ax + by + c}{a'x + b'y + c'} = \dfrac{ax + by + c}{\frac{1}{m}(ax + by) + c'}$

Put $ax + by = v$ $\therefore$ $a + b\dfrac{dy}{dx} = \dfrac{dv}{dx}$

or $\dfrac{dy}{dx} = \dfrac{1}{b}\left(\dfrac{dv}{dx} - a\right)$ or $\dfrac{dy}{dx} = \dfrac{ax + by + c}{\frac{1}{m}(ax + by) + c'}$ or $\dfrac{1}{b}\left(\dfrac{dv}{dx} - a\right) = \dfrac{v + c}{\frac{v}{m} + c'} = \dfrac{m(v + c)}{v + mc'}$

Now the variables can be separated.

(III). Linear Differential Equations: $-$ A differential equation of the form $\dfrac{dy}{dx} + Py = Q$ where P and Q are functions of x (not of y)

Similarly, $\dfrac{dx}{dy} + Px = Q$ where P and Q are functions of y (not of x)

Method of solution: $-$ The equation in the form $\dfrac{dy}{dx} + Py = Q$

Integrate P with respect to x and make the integrand the power of e to get I.F (Integrating factor I. F).

i.e $I.F = e^{\int P\, dx}$ The required solution as $\boxed{y \times I.F = \int Q.(I.F)\, dx + c}$ formula

Similarly, $\dfrac{dx}{dy} + Px = Q$ then the required solution as $\boxed{x \times I.F = \int Q(I.F)\ dy + c}$ formula

Example: – Solve, $(x + 3y + 1)\ dy = (2x + y + 3)\ dx$

Solution: – Given, $(x + 3y + 1)\ dy = (2x + y + 3)\ dx$ or $\dfrac{dy}{dx} = \dfrac{2x + y + 3}{x + 3y + 1}$

Put $x = \alpha + h$ and $y = \beta + k$ $\quad \therefore \dfrac{dy}{dx} = \dfrac{d\beta}{d\alpha}$

or $\dfrac{d\beta}{d\alpha} = \dfrac{2(\alpha + h) + (\beta + k) + 3}{(\alpha + h) + 3(\beta + k) + 1} = \dfrac{2\alpha + \beta + (2h + k + 3)}{\alpha + 3\beta + (h + 3k + 1)}$ ... ... (A)

Now, choose h and k such that $\therefore\ 2h + k + 3 = 0$ ... ... (i) and $h + 3k + 1 = 0$ ... ... (ii)

Multiplying, (i) × 1 & (ii) × 2, we have $\therefore\ 2h + k + 3 = 0$ ... ... (iii) and $2h + 6k + 2 = 0$ ... ... (iv)

Substracting, $2h + k + 3 - 2h - 6k - 2 = 0$ $\quad \therefore\ -5k + 1 = 0$ or $5k = 1$ $\therefore\ k = \dfrac{1}{5}$

Put value of k in equation (i), we have or $2h + k + 3 = 0$ or $2h = -3 - \dfrac{1}{5} = \dfrac{-15 - 1}{5} = -\dfrac{16}{5}$ $\therefore\ h = -\dfrac{8}{5}$

From (A), $\dfrac{d\beta}{d\alpha} = \dfrac{2\alpha + \beta + (2h + k + 3)}{\alpha + 3\beta + (h + 3k + 1)} = \dfrac{2\alpha + \beta}{\alpha + 3\beta} = \dfrac{\alpha\left(2 + \dfrac{\beta}{\alpha}\right)}{\alpha\left(1 + 3.\dfrac{\beta}{\alpha}\right)} = \dfrac{2 + \dfrac{\beta}{\alpha}}{1 + 3.\dfrac{\beta}{\alpha}}$

which is homogeneous equation.

Put $\beta = v\alpha$ $\therefore\ \dfrac{d\beta}{d\alpha} = v + \alpha\dfrac{dv}{d\alpha}$ or $\dfrac{d\beta}{d\alpha} = \dfrac{2 + \dfrac{\beta}{\alpha}}{1 + 3.\dfrac{\beta}{\alpha}}$ or $v + \alpha\dfrac{dv}{d\alpha} = \dfrac{2 + v}{1 + 3v}$ or $\alpha\dfrac{dv}{d\alpha} = \dfrac{2 + v}{1 + 3v} - v$

or $\alpha\dfrac{dv}{d\alpha} = \dfrac{2 + v - v - 3v^2}{1 + 3v} = \dfrac{2 - 3v^2}{1 + 3v}$ or $\dfrac{1 + 3v}{2 - 3v^2}\ dv = \dfrac{d\alpha}{\alpha}$

Integrating, $\int \dfrac{1 + 3v}{2 - 3v^2}\ dv = \int \dfrac{d\alpha}{\alpha}$ or $\int \dfrac{1}{2 - 3v^2}\ dv + \int \dfrac{3v}{2 - 3v^2}\ dv = \int \dfrac{d\alpha}{\alpha}$ ... ... (B)

Let $I_1 = \int \dfrac{1}{2 - 3v^2}\ dv = \int \dfrac{1}{\left(\sqrt{2}\right)^2 - \left(\sqrt{3}v\right)^2}\ dv = \dfrac{1}{2.\sqrt{2}} \log\left|\dfrac{\sqrt{2} + \sqrt{3}v}{\sqrt{2} - \sqrt{3}v}\right|$ using formula, $\int \dfrac{dx}{a^2 - x^2} = \dfrac{1}{2a} \log\left|\dfrac{a + x}{a - x}\right|$, $x < a$

and $I_2 = \int \dfrac{3v}{2 - 3v^2}\ dv$ Let $2 - 3v^2 = z$ or $-6v\ dv = dz$ $\therefore\ 2.3v\ dv = -dz$ $\therefore\ 3v\ dv = -\dfrac{dz}{2}$

$\therefore\ I_2 = -\dfrac{1}{2}\int \dfrac{dz}{z} = -\dfrac{1}{2} \log|z| = -\dfrac{1}{2} \log|2 - 3v^2|$

Put value of I_1 and I_2 in equation (B), we get

or $\int \dfrac{1}{2 - 3v^2}\ dv + \int \dfrac{3v}{2 - 3v^2}\ dv = \int \dfrac{d\alpha}{\alpha}$ or $\dfrac{1}{2.\sqrt{2}} \log\left|\dfrac{\sqrt{2} + \sqrt{3}v}{\sqrt{2} - \sqrt{3}v}\right| - \dfrac{1}{2} \log|2 - 3v^2| = \log \alpha + k$

Put $v = \dfrac{\beta}{\alpha}$ and $x = \alpha + h$ $\therefore\ \alpha = x - h$, $y = \beta + k$ $\therefore\ \beta = y - k$

or $\dfrac{1}{2.\sqrt{2}} \log\left|\dfrac{\sqrt{2} + \sqrt{3}.\dfrac{\beta}{\alpha}}{\sqrt{2} - \sqrt{3}.\dfrac{\beta}{\alpha}}\right| - \dfrac{1}{2} \log\left|2 - 3\left(\dfrac{\beta}{\alpha}\right)^2\right| = \log \alpha + k$ or $\dfrac{1}{2\sqrt{2}} \log\left|\dfrac{\sqrt{2}\alpha + \sqrt{3}\beta}{\sqrt{2}\alpha - \sqrt{3}\beta}\right| - \dfrac{1}{2} \log\left|\dfrac{2\alpha^2 - 3\beta^2}{\alpha^2}\right| = \log \alpha + k$

or $\dfrac{1}{2\sqrt{2}} \log\left|\dfrac{\sqrt{2}(x - h) + \sqrt{3}(y - k)}{\sqrt{2}(x - h) - \sqrt{3}(y - k)}\right| - \dfrac{1}{2} \log\left|\dfrac{2(x - h)^2 - 3(y - k)^2}{(x - h)^2}\right| = \log(x - h) + k$

$$\text{Put } \alpha = x - h = x + \frac{8}{5} \quad \text{and} \quad \beta = y - k = y - \frac{1}{5}$$

or $\quad \dfrac{1}{2\sqrt{2}} \log \left| \dfrac{\sqrt{2}\left(x+\frac{8}{5}\right)+\sqrt{3}\left(y-\frac{1}{5}\right)}{\sqrt{2}\left(x+\frac{8}{5}\right)-\sqrt{3}\left(y-\frac{1}{5}\right)} \right| - \dfrac{1}{2} \log \left| \dfrac{2\left(x+\frac{8}{5}\right)^2 - 3\left(y-\frac{1}{5}\right)^2}{\left(x+\frac{8}{5}\right)^2} \right| = \log\left(x+\frac{8}{5}\right) + k$

or $\quad \dfrac{1}{2\sqrt{2}} \log \left| \dfrac{\sqrt{2}(5x+8)+\sqrt{3}(5y-1)}{\sqrt{2}(5x+8)-\sqrt{3}(5y-1)} \right| - \dfrac{1}{2} \log \left| \dfrac{2(5x+8)^2 - 3(5y-1)^2}{(5x+8)^2} \right| = \log\left(\dfrac{5x+8}{5}\right) + k \quad$ **Ans.**

Example: – Solve, $\dfrac{dy}{dx} + \dfrac{y}{x} = \dfrac{\sin x}{x}$

Solution: – Given, $\dfrac{dy}{dx} + \dfrac{y}{x} = \dfrac{\sin x}{x}$ The equation in the form $\dfrac{dy}{dx} + Py = Q$

where $P = \dfrac{1}{x}$ and $Q = \dfrac{\sin x}{x}$ $\quad \therefore$ $\text{I.F} = e^{\int P\, dx} = e^{\int \frac{1}{x} dx} = e^{\log x} = x$

The required equation is – $\quad y \times \text{I.F} = \displaystyle\int (\text{I.F}).Q\ dx + c$

or $\quad y.x = \displaystyle\int x.\dfrac{\sin x}{x}\ dx + c = \int \sin x\, dx + c = -\cos x + c \quad$ or $\quad xy = -\cos x + c \quad \therefore \ xy + \cos x = c \quad$ **Ans.**

Example: – Solve, $\dfrac{dx}{dy} - \dfrac{x}{y} = y\cos y$

Solution: – Given, $\dfrac{dx}{dy} - \dfrac{x}{y} = y\cos y \quad$ form of $\dfrac{dx}{dy} + Px = Q \quad$ where $P = -\dfrac{1}{y}$ and $Q = y\cos y$

$\therefore \ \text{I.F} = e^{\int P\, dy} = e^{-\int \frac{1}{y} dy} = e^{-\log y} = e^{\log \frac{1}{y}} = \dfrac{1}{y}$

The required solution is – $\quad x \times \text{I.F} = \displaystyle\int \text{I.F} \times Q\ dy + c$

or $\quad x.\dfrac{1}{y} = \displaystyle\int \dfrac{1}{y}.y\cos y\ dy + c = \int \cos y\ dy + c = \sin y + c$

or $\quad \dfrac{x}{y} = \sin y + c \quad$ or $\quad x = y\sin y + yc \quad$ or $\quad x - y\sin y = yc \quad \therefore \ c = \dfrac{x - y\sin y}{y} \quad$ **Ans.**

Example: – Solve, $\dfrac{dy}{dx} = \dfrac{x+y+1}{x+y+3}$

Solution: – Given, $\dfrac{dy}{dx} = \dfrac{x+y+1}{x+y+3} \quad$ here $a = a'$, $b = b' \quad \therefore \ \dfrac{dy}{dx} = \dfrac{(x+y)+1}{(x+y)+3}$

Put $x + y = v \quad \therefore \ 1 + \dfrac{dy}{dx} = \dfrac{dv}{dx} \quad \therefore \ \dfrac{dy}{dx} = \dfrac{dv}{dx} - 1$

or $\quad \dfrac{dy}{dx} = \dfrac{(x+y)+1}{(x+y)+3} \quad$ or $\quad \dfrac{dv}{dx} - 1 = \dfrac{v+1}{v+3} \quad$ or $\quad \dfrac{dv}{dx} = \dfrac{v+1}{v+3} + 1 = \dfrac{v+1+v+3}{v+3} = \dfrac{2v+4}{v+3} \quad$ or $\quad \dfrac{v+3}{2v+4}\, dv = dx$

Integrating both of sides, $\displaystyle\int \dfrac{v+3}{2v+4}\, dv = \int dx \quad$ or $\quad \displaystyle\int \dfrac{v}{(2v+4)}\, dv + \int \dfrac{3}{(2v+4)}\, dv = \int dx$

or $\quad \displaystyle\int \dfrac{v}{2(v+2)}\, dv + \int \dfrac{3}{2(v+2)}\, dv = \int dx \quad$ or $\quad \dfrac{1}{2}\displaystyle\int \dfrac{v}{v+2}\, dv + \dfrac{3}{2}\int \dfrac{dv}{v+2} = \int dx$

or $\quad \dfrac{1}{2}\displaystyle\int \left(1 - \dfrac{2}{v+2}\right) dv + \dfrac{3}{2}\int \dfrac{dv}{v+2} = \int dx \quad$ or $\quad \dfrac{1}{2}\displaystyle\int dv - \dfrac{1}{2}.2\int \dfrac{dv}{v+2} + \dfrac{3}{2}\int \dfrac{dv}{v+2} = \int dx$

or $\frac{1}{2}v - \log|v + 2| + \frac{3}{2}\log|v + 2| = x + c$ or $\frac{v - 2\log|v + 2| + 3\log|v + 2|}{2} = x + c$ Put $v = x + y$

or $x + y + \log|x + y + 2| = 2x + 2c$ $\therefore$ $y - x + \log|x + y + 2| = k$ where $k = 2c$ Ans.

Solved Example

(1) Solve the following differential equations:— (a) $(x + y)dx - 2x\,dy = 0$ (b) $\frac{dy}{dx} = \frac{yx^3 - xy^3}{x^4}$ (c) $\frac{dy}{dx} = \frac{y}{x} + \cos\left(\frac{y}{x}\right)$

(d) $(x^2 + 2x + 1)dy - (x^2 + 3x + 2)dx = 0$ (e) $dy + \log(x + y)\,dx = \log(2x + y)\,dx$ (f) $e^{\frac{dy}{dx}} = 2^x + 2$

Solution:— (a) $(x + y)dx - 2x\,dy = 0$ or $(x + y)dx = 2x\,dy$ or $2\frac{dy}{dx} = \frac{x + y}{x} = 1 + \frac{y}{x}$ .... (i)

Put $y = vx$ $\therefore$ $\frac{dy}{dx} = v + x\frac{dv}{dx}$ or $y = vx$ $\therefore$ $v = \frac{y}{x}$

Put value of $\frac{dy}{dx}$ in equation (i), we get $2\left(v + x\frac{dv}{dx}\right) = 1 + v$ or $2v + 2x\frac{dv}{dx} = 1 + v$

or $2x\frac{dv}{dx} = 1 + v - 2v$ or $2x\frac{dv}{dx} = 1 - v$ or $\frac{2\,dv}{1 - v} = \frac{dx}{x}$

Integrating, $2\int\frac{dv}{1 - v} = \int\frac{dx}{x}$ or $2\log|1 - v| = \log|x| + \log k$ or $\log|1 - v|^2 = \log|xk|$

or $|1 - v|^2 = |xk|$ or $\left|1 - \frac{y}{x}\right|^2 = |xk|$ or $\left|\frac{x - y}{x}\right|^2 = |xk|$ or $|x - y|^2 = |x^3 k|$ Ans.

(b) $\frac{dy}{dx} = \frac{yx^3 - xy^3}{x^4}$ or $\frac{dy}{dx} = \frac{x^4\left[\frac{y}{x} - \left(\frac{y}{x}\right)^3\right]}{x^4}$ or $\frac{dy}{dx} = \frac{y}{x} - \left(\frac{y}{x}\right)^3$ .... (i)

Put $v = \frac{y}{x}$ or $y = vx$ $\therefore$ $\frac{dy}{dx} = v + x\frac{dv}{dx}$

Put value of $\frac{dy}{dx}$ in equation (i), we get $v + x\frac{dv}{dx} = v - v^3$ or $x\frac{dv}{dx} = -v^3$ or $\frac{dv}{v^3} = -\frac{dx}{x}$

Integrating, $\int\frac{dv}{v^3} = -\int\frac{dx}{x}$ or $\int v^{-3}\,dv = -\int\frac{dx}{x}$ or $\frac{v^{-3+1}}{-3 + 1} = -\log x + \log k$

or $-\frac{1}{2}v^{-2} = -(\log x - \log k)$ or $\frac{1}{2v^2} = \log\left(\frac{x}{k}\right)$ or $1 = 2v^2\log\left(\frac{x}{k}\right)$ or $2\left(\frac{y}{x}\right)^2 \cdot \log\left(\frac{x}{k}\right) = 1$

or $\frac{2y^2}{x^2}\cdot\log\left(\frac{x}{k}\right) = 1$ or $2y^2\cdot\log\left(\frac{x}{k}\right) = x^2$ Ans.

(c) $\frac{dy}{dx} = \frac{y}{x} + \cos\left(\frac{y}{x}\right)$ .... (i) Put $y = vx$ $\therefore$ $\frac{dy}{dx} = v + x\frac{dv}{dx}$

Put value of $\frac{dy}{dx}$ in equation (i), we get $v + x\frac{dv}{dx} = v + \cos v$ or $x\frac{dv}{dx} = \cos v$ or $\frac{dv}{\cos v} = \frac{dx}{x}$

Integrating, $\int\frac{dv}{\cos v} = \int\frac{dx}{x}$ or $\int\sec v\,dv = \int\frac{dx}{x}$ or $\int\frac{\sec v.\tan v}{\tan v}\,dv = \int\frac{dx}{x}$

or $\int\frac{\sec v.\tan v}{\sqrt{\sec^2 v - 1}}\,dv = \int\frac{dx}{x}$ Put $\sec v = t$ $\therefore$ $\sec v.\tan v\,dv = dt$

or $\int\frac{dt}{\sqrt{t^2 - 1}} = \int\frac{dx}{x}$ or $\log\left|t + \sqrt{t^2 - 1}\right| = \log|x| + \log|k|$ using formula, $\int\frac{dx}{\sqrt{x^2 - a^2}} = \log\left|x + \sqrt{x^2 - a^2}\right|$

or $\log\left|\sec v + \sqrt{\sec^2 v - 1}\right| = \log|xk|$ or $\log\left|\sec\left(\frac{y}{x}\right) + \sqrt{\sec^2\left(\frac{y}{x}\right) - 1}\right| - \log|x| = \log|k|$ Ans.

(d) $(x^2 + 2x + 1)dy - (x^2 + 3x + 2)dx = 0$ $\quad \therefore \dfrac{dy}{dx} = \dfrac{x^2 + 3x + 2}{x^2 + 2x + 1} = \dfrac{(x+1)(x+2)}{(x+1)^2} = \dfrac{x+2}{x+1}$

or $dy = \dfrac{x+2}{x+1}\, dx$ $\quad$ Integrating, $\displaystyle\int dy = \int \dfrac{x+2}{x+1}\, dx$ or $\displaystyle\int dy = \int \left(1 + \dfrac{1}{x+1}\right) dx$

or $y = x + \log(x+1) + \log c$ $\quad \therefore y - x = \log(x+1) + k$ $\quad$ Ans. $(\log c = k)$

(f) $e^{\frac{dy}{dx}} = 2^x + 2 = 2^{x+1}$ $\quad$ or $\log e^{\frac{dy}{dx}} = \log(2^{x+1}) = \log_e 2^{x+1}$ $\quad$ or $\dfrac{dy}{dx} = (x+1)\log 2$ $\quad$ or $dy = (x+1)\log 2 \; dx$

Intgerating, $\displaystyle\int dy = \int (x+1)\log 2 \; dx = \log 2 \int (x+1)\,dx$ or $y = \log 2 \left[\dfrac{x^2}{2} + x\right] + c$

or $y = \dfrac{\log 2 \,(x^2 + 2x) + 2c}{2}$ $\quad$ or $2y = \log 2\,(x^2 + 2x) + 2c$ $\therefore 2y = (x^2 + 2x)\log 2 + k$ Ans. $(k = 2c)$

(2) (a) Solve, $y^3 \dfrac{dx}{dy} = x^3 + x^2\sqrt{x^2 - y^2}$

Solution: $-$ Given, $y^3 \dfrac{dx}{dy} = x^3 + x^2\sqrt{x^2 - y^2} = x^3 + x^2\sqrt{y^2\left(\dfrac{x^2}{y^2} - 1\right)} = x^3 + x^2 y\sqrt{\left(\dfrac{x}{y}\right)^2 - 1}$

or $\dfrac{dx}{dy} = \left(\dfrac{x}{y}\right)^3 + \left(\dfrac{x}{y}\right)^2\sqrt{\left(\dfrac{x}{y}\right)^2 - 1}$ $\quad$ Put $x = yv$ $\quad$ or $v = \dfrac{x}{y}$ $\quad \therefore \dfrac{dx}{dy} = v + y\dfrac{dv}{dy}$

or $v + y\dfrac{dv}{dy} = v^3 + v^2\sqrt{v^2 - 1}$ $\quad$ or $y\dfrac{dv}{dy} = v^3 + v^2\sqrt{v^2 - 1} - v = v\left[(v^2 - 1) + v\sqrt{v^2 - 1}\right]$

or $y\dfrac{dv}{dy} = v\sqrt{v^2 - 1}\left[\sqrt{v^2 - 1} + v\right]$ $\quad$ or $\dfrac{dv}{v\sqrt{v^2 - 1}\left[\sqrt{v^2 - 1} + v\right]} = \dfrac{dy}{y}$ or $\dfrac{(\sqrt{v^2 - 1} - v)}{v\sqrt{v^2 - 1}(v^2 - 1 + v^2)}\,dv = \dfrac{dy}{y}$ or $\dfrac{(\sqrt{v^2 - 1} - v)}{-v\sqrt{v^2 - 1}}\,dv = \dfrac{dy}{y}$

Integrating both of sides, $\quad -\displaystyle\int \dfrac{(\sqrt{v^2 - 1} - v)}{v\sqrt{v^2 - 1}}\,dv = \int \dfrac{dy}{y}$ $\quad$ or $\displaystyle\int \dfrac{(v - \sqrt{v^2 - 1})}{v\sqrt{v^2 - 1}}\,dv = \int \dfrac{dy}{y}$

or $\displaystyle\int \dfrac{dv}{\sqrt{v^2 - 1}} - \int \dfrac{dv}{v} = \int \dfrac{dy}{y}$ $\quad$ or $\log\left|v + \sqrt{v^2 - 1}\right| - \log v = \log y + \log a$

or $\log\left(\dfrac{v + \sqrt{v^2 - 1}}{v}\right) = \log ay$ $\quad$ or $\dfrac{v + \sqrt{v^2 - 1}}{v} = ay$ $\quad$ or $ay = \dfrac{\dfrac{x}{y} + \sqrt{\left(\dfrac{x}{y}\right)^2 - 1}}{\dfrac{x}{y}}$ $\quad$ or $\dfrac{x + \sqrt{x^2 - y^2}}{x} = ay$

or $axy = x + \sqrt{x^2 - y^2}$ $\quad \therefore ax = \dfrac{1}{y}\left(x + \sqrt{x^2 - y^2}\right)$ $\quad$ Ans. $\quad$ using formula, $\displaystyle\int \dfrac{dx}{\sqrt{x^2 - a^2}} = \log\left|x + \sqrt{x^2 - a^2}\right|$

(b) Solve, $\dfrac{dy}{dx} = \dfrac{x - 2y + 3}{2x - 4y + 5}$

Solution: $-$ Given, $\dfrac{dy}{dx} = \dfrac{x - 2y + 3}{2x - 4y + 5} = \dfrac{(x - 2y) + 3}{2(x - 2y) + 5}$ ....(i) $\quad$ Put $x - 2y = v$ $\quad$ or $1 - 2\dfrac{dy}{dx} = \dfrac{dv}{dx}$ $\quad$ or $\dfrac{dy}{dx} = \dfrac{1 - \dfrac{dv}{dx}}{2}$

From (i), $\Rightarrow \dfrac{1 - \dfrac{dv}{dx}}{2} = \dfrac{v + 3}{2v + 5}$ $\quad$ or $1 - \dfrac{dv}{dx} = \dfrac{2v + 6}{2v + 5}$ $\quad$ or $\dfrac{dv}{dx} = 1 - \dfrac{2v + 6}{2v + 5} = \dfrac{2v + 5 - 2v - 6}{2v + 5} = \dfrac{-1}{2v + 5}$ $\quad$ or $(2v + 5)\,dv = -dx$

Integrating, $\displaystyle\int (2v + 5)\,dv = -\int dx$ $\quad$ or $2.\dfrac{v^2}{2} + 5v = -x + c$ $\quad$ or $v^2 + 5v = -x + c$

or $(x - 2y)^2 + 5(x - 2y) = -x + c$ $\quad$ or $x^2 - 4xy + 4y^2 + 5x - 10y + x = c$

or $x^2 + 4y^2 - 4xy + 6x - 10y = c$ $\quad \therefore x^2 - 4xy + 6x - 10y + 4y^2 = c$ $\quad$ Ans.

(c) Solve, $\dfrac{dy}{dx} = \dfrac{3(x+y)+1}{x+y+3}$

Solution: – Given, $\dfrac{dy}{dx} = \dfrac{3(x+y)+1}{x+y+3}$ .... (i)   Put $x+y = v$ $\therefore\ 1+\dfrac{dy}{dx} = \dfrac{dv}{dx}$ $\therefore\ \dfrac{dy}{dx} = \dfrac{dv}{dx} - 1$

Put value $\dfrac{dy}{dx}$ in equation (i), we get $\Rightarrow\ \dfrac{dv}{dx} - 1 = \dfrac{3v+1}{v+3}$

or $\dfrac{dv}{dx} = \dfrac{3v+1}{v+3} + 1 = \dfrac{3v+1+v+3}{v+3} = \dfrac{4v+4}{v+3}$ or $\dfrac{dv}{dx} = \dfrac{4(v+1)}{v+3}$ or $\dfrac{v+3}{4(v+1)}\,dv = dx$

Integrating, $\displaystyle\int \dfrac{v+3}{4(v+1)}\,dv = \int dx$ or $\dfrac{1}{4}\displaystyle\int \dfrac{v+3}{v+1}\,dv = \int dx$ or $\dfrac{1}{4}\displaystyle\int \dfrac{(v+1)+2}{v+1}\,dv = \int dx$

or $\dfrac{1}{4}\displaystyle\int \dfrac{v+1}{v+1}\,dv + \dfrac{1}{4}\int \dfrac{2}{v+1}\,dv = \int dx$ or $\dfrac{1}{4}\displaystyle\int dv + \dfrac{1}{2}\int \dfrac{dv}{v+1} = \int dx$

or $\dfrac{1}{4}v + \dfrac{1}{2}\log|v+1| = x+c$ or $\dfrac{v+2\log|v+1|}{4} = x+c$ or $v+2\log|v+1| = 4x+4c$

Put $v = x+y$ or $x+y+2\log|x+y+1| = 4x+4c$ or $x+y+2\log|x+y+1| - 4x = 4c$

or $y-3x+2\log|x+y+1| = 4c$ $\therefore\ y-3x+2\log|x+y+1| = k$ where $4c = k$ Ans.

(d) solve, $\dfrac{dy}{dx} = \dfrac{2x+3y-1}{2x+3y+3}$ (Do yourself, solve same as above question)

Ans: – $x+3y - \dfrac{4}{5}(10x+15y+3) + \dfrac{12}{5}\log|10x+15y+3| = c$

(3) (a) Solve, $\cot^{-1}\left[\dfrac{x}{y} + \cos^2\left(\dfrac{x}{y}\right)\right] = \theta$

Solution: – Given, $\cot^{-1}\left[\dfrac{x}{y} + \cos^2\left(\dfrac{x}{y}\right)\right] = \theta$ or $\cot\theta = \dfrac{x}{y} + \cos^2\left(\dfrac{x}{y}\right)$ $\therefore\ \tan\theta = \dfrac{dy}{dx}$ $\therefore\ \cot\theta = \dfrac{dx}{dy}$

or $\dfrac{dx}{dy} = \dfrac{x}{y} + \cos^2\left(\dfrac{x}{y}\right)$ Put $v = \dfrac{x}{y}$ or $x = vy$ $\therefore\ \dfrac{dx}{dy} = v + y\dfrac{dv}{dy}$

or $v + y\dfrac{dv}{dy} = v + \cos^2 v$ or $y\dfrac{dv}{dy} = \cos^2 v$ or $\dfrac{dv}{\cos^2 v} = \dfrac{dy}{y}$

Integrating, $\displaystyle\int \dfrac{dv}{\cos^2 v} = \int \dfrac{dy}{y}$ or $\displaystyle\int \sec^2 v\,dv = \int \dfrac{dy}{y}$ or $\tan v = \log y + \log a = \log ay$ Put $v = \dfrac{x}{y}$

or $\tan\left(\dfrac{x}{y}\right) = \log ay$ or $\dfrac{x}{y} = \tan^{-1}(\log ay)$ $\therefore\ x = y\tan^{-1}(\log ay)$ Ans.

(b) Solve, $y\,dx = \left[x + y \cdot \dfrac{f\left(\frac{x}{y}\right)}{f'\left(\frac{x}{y}\right)}\right] dy$

Solution: – (Do yourself.)   Given, $y\,dx = \left[x + y \cdot \dfrac{f\left(\frac{x}{y}\right)}{f'\left(\frac{x}{y}\right)}\right] dy$ Put $v = \dfrac{x}{y}$ $\therefore\ f\left(\dfrac{x}{y}\right) = ky$ Ans.

(c) Solve, $\dfrac{dx}{dy} = e^{-\frac{x}{y}} + \dfrac{x}{y}$

Solution: – Given, $\dfrac{dx}{dy} = e^{\frac{x}{y}} + \dfrac{x}{y}$ Put $v = \dfrac{x}{y}$ or $x = yv$ $\therefore\ \dfrac{dx}{dy} = v + y\dfrac{dv}{dy}$

or $v + y\dfrac{dv}{dy} = e^{-v} + v$ or $y\dfrac{dv}{dy} = e^{-v}$ or $\dfrac{dv}{e^{-v}} = \dfrac{dy}{y}$ or $e^v\,dv = \dfrac{dy}{y}$

Integrating, $\int e^v \, dv = \int \dfrac{dy}{y}$ or $e^v = \log y + \log a = \log|ay|$ or $e^{\frac{x}{y}} = \log|ay|$ Ans. $\left(\text{put } v = \dfrac{x}{y}\right)$

(d) Solve, $xy\dfrac{dx}{dy} = x^2 + y^2 e^{\left(\frac{x^2}{y^2}\right)}$

Solution:− Given, $xy\dfrac{dx}{dy} = x^2 + y^2 e^{\left(\frac{x^2}{y^2}\right)}$ or $\dfrac{dx}{dy} = \dfrac{x}{y} + \dfrac{y}{x}e^{\left(\frac{x^2}{y^2}\right)}$

Put $v = \dfrac{x}{y}$ or $x = yv$ $\therefore \dfrac{dx}{dy} = v + y\dfrac{dv}{dy}$ or $v + y\dfrac{dv}{dy} = v + \dfrac{e^{v^2}}{v}$ or $y\dfrac{dv}{dy} = \dfrac{e^{v^2}}{v}$ or $\dfrac{v}{e^{v^2}}\,dv = \dfrac{dy}{y}$

Integrating, $\int \dfrac{v}{e^{v^2}}\,dv = \int \dfrac{dy}{y}$ Let $v^2 = t$ $\therefore 2v\,dv = dt$ or $v\,dv = \dfrac{dt}{2}$

or $\dfrac{1}{2}\int \dfrac{dt}{e^t} = \int \dfrac{dy}{y}$ or $\dfrac{1}{2}\int e^{-t}\,dt = \int \dfrac{dy}{y}$ or $-\dfrac{1}{2}e^{-t} = \log y + \log k = \log|ky|$

or $e^{-t} = -2\log|ky| = \log\left(\dfrac{1}{ky}\right)^2$ Put $t = v^2$ and $v = \dfrac{x}{y}$ or $e^{-v^2} = \log\left(\dfrac{1}{ky}\right)^2$ $\therefore e^{-\left(\frac{x}{y}\right)^2} = \log\left(\dfrac{1}{ky}\right)^2$ Ans.

(4) (a) Solve, $\dfrac{dy}{dx} = \dfrac{x+y+1}{x-y+5}$

Solution:− Given, $\dfrac{dy}{dx} = \dfrac{x+y+1}{x-y+5}$ Put $x = \alpha + h$, $y = \beta + k$ $\therefore \dfrac{dy}{dx} = \dfrac{d\beta}{d\alpha}$

$\therefore \dfrac{d\beta}{d\alpha} = \dfrac{\alpha+h+\beta+k+1}{\alpha+h-\beta-k+5} = \dfrac{\alpha+\beta+(h+k+1)}{\alpha-\beta+(h-k+5)}$ choose h and k such that

or $h+k+1 = 0 \ldots\ldots\ldots.(i)$ and $h-k+5 = 0$ $\therefore k = h+5 \ldots\ldots\ldots (ii)$

Put $k = h+5$ in equation (i), we get $\therefore h+h+5+1 = 0$ or $2h = -6$ $\therefore h = -3$ from (ii), $k = h+5 = -3+5 = 2$

or $\dfrac{d\beta}{d\alpha} = \dfrac{\alpha+\beta}{\alpha-\beta} = \dfrac{\alpha\left(1+\frac{\beta}{\alpha}\right)}{\alpha\left(1-\frac{\beta}{\alpha}\right)} = \dfrac{1+\frac{\beta}{\alpha}}{1-\frac{\beta}{\alpha}}$ Put $\beta = v\alpha$ or $v = \dfrac{\beta}{\alpha}$ $\therefore \dfrac{d\beta}{d\alpha} = v + \alpha\dfrac{dv}{d\alpha}$

or $v + \alpha\dfrac{dv}{d\alpha} = \dfrac{1+v}{1-v}$ or $\alpha\dfrac{dv}{d\alpha} = \dfrac{1+v}{1-v} - v = \dfrac{1+v-v+v^2}{1-v} = \dfrac{1+v^2}{1-v}$ or $\dfrac{1-v}{1+v^2}\,dv = \dfrac{d\alpha}{\alpha}$

Integrating, $\int \dfrac{1-v}{1+v^2}\,dv = \int \dfrac{d\alpha}{\alpha}$ or $\int \dfrac{1}{1+v^2}\,dv - \int \dfrac{v}{1+v^2}\,dv = \int \dfrac{d\alpha}{\alpha}$

Put IInd integral $1+v^2 = z$ $\therefore 2v\,dv = dz$ or $v\,dv = \dfrac{dz}{2}$

or $\int \dfrac{1}{1+v^2}\,dv - \dfrac{1}{2}\int \dfrac{dz}{z} = \int \dfrac{d\alpha}{\alpha}$ or $\tan^{-1}v - \dfrac{1}{2}\log|z| = \log|\alpha| + k$

or $\tan^{-1}v - \dfrac{1}{2}\log|1+v^2| - \log|\alpha| = k$ or $\tan^{-1}\left(\dfrac{\beta}{\alpha}\right) - \log\left(\sqrt{1+\left(\dfrac{\beta}{\alpha}\right)^2}\right) - \log|\alpha| = k$

Put $x = \alpha + h$, $y = \beta + k$ $\therefore \alpha = x+3$, $\beta = y-2$

or $\tan^{-1}\left(\dfrac{y-k}{x-h}\right) = \log\left(\dfrac{\sqrt{\alpha^2+\beta^2}}{|\alpha|}\right) + \log|\alpha| + k$ or $\tan^{-1}\left(\dfrac{y-2}{x+3}\right) = \log\left(\sqrt{\alpha^2+\beta^2}\right) - \log|\alpha| + \log|\alpha| + k$

or $\tan^{-1}\left(\dfrac{y-2}{x+3}\right) = \log\left(\sqrt{(x-h)^2+(y-k)^2}\right) + k$ or $\tan^{-1}\left(\dfrac{y-2}{x+3}\right) - \log\left(\sqrt{(x+3)^2+(y-2)^2}\right) = k$

---

or $\tan^{-1}\left(\dfrac{y-2}{x+3}\right) - \log\left(\sqrt{x^2 + 6x + 9 + y^2 - 4y + 4}\right) = k$ or $\tan^{-1}\left(\dfrac{y-2}{x+3}\right) - \log\left(\sqrt{x^2 + y^2 + 6x - 4y + 13}\right) = k$

$\therefore\ \tan^{-1}\left(\dfrac{y-2}{x+3}\right) - \log\left(\sqrt{x^2 + y^2 + 6x - 4y + 13}\right) = k$ Ans.

(b) Solve, $\dfrac{dy}{dx} = \dfrac{2x - y + 1}{x + 2y - 5}$

Solution:– Given, $\dfrac{dy}{dx} = \dfrac{2x - y + 1}{x + 2y - 5}$ Put $x = \alpha + h$, $y = \beta + k$ $\therefore\ \dfrac{dy}{dx} = \dfrac{d\beta}{d\alpha}$

$\therefore\ \dfrac{d\beta}{d\alpha} = \dfrac{2(\alpha + h) - (\beta + k) + 1}{(\alpha + h) + 2(\beta + k) - 5} = \dfrac{2\alpha - \beta + (2h - k + 1)}{\alpha + 2\beta + (h + 2k - 5)}$

choose h and k such that $\therefore\ 2h - k + 1 = 0 \ldots\ldots\ldots$ (i)   and $h + 2k - 5 = 0 \ldots\ldots\ldots\ldots\ldots$ (ii)

solve (i) & (ii) and find h and k, $\therefore\ 2h - k + 1 = 0$ or $k = 2h + 1$

put value of k in equation (ii), we get $h + 2k - 5 = 0$ or $h + 2(2h + 1) = 5$ or $h + 4h + 2 = 5$

or $5h = 5 - 2 = 3$ $\therefore\ h = \dfrac{3}{5}$ from (i), $2h - k + 1 = 0$ or $k = 2.\dfrac{3}{5} + 1 = \dfrac{6 + 5}{5} = \dfrac{11}{5}$

or $\dfrac{d\beta}{d\alpha} = \dfrac{2\alpha - \beta}{\alpha + 2\beta} = \dfrac{\alpha\left(2 - \dfrac{\beta}{\alpha}\right)}{\alpha\left(1 + 2.\dfrac{\beta}{\alpha}\right)} = \dfrac{2 - \dfrac{\beta}{\alpha}}{1 + 2.\dfrac{\beta}{\alpha}}$ Put $\beta = v\alpha$ or $v = \dfrac{\beta}{\alpha}$ $\therefore\ \dfrac{d\beta}{d\alpha} = v + \alpha\dfrac{dv}{d\alpha}$

or $v + \alpha\dfrac{dv}{d\alpha} = \dfrac{2 - v}{1 + 2v}$ or $\alpha\dfrac{dv}{d\alpha} = \dfrac{2 - v}{1 + 2v} - v = \dfrac{2 - v - v - 2v^2}{1 + 2v} = \dfrac{2 - 2v - 2v^2}{1 + 2v}$

or $\dfrac{1 + 2v}{2 - 2v - 2v^2}\,dv = \dfrac{d\alpha}{\alpha}$

Integrating, $\displaystyle\int \dfrac{1 + 2v}{2 - 2v - 2v^2}\,dv = \int \dfrac{d\alpha}{\alpha}$ Let $2 - 2v - 2v^2 = z$ $\therefore\ -2(1 + 2v)\,dv = dz$

or $(1 + 2v)\,dv = -\dfrac{dz}{2}$ or $-\dfrac{1}{2}\displaystyle\int \dfrac{dz}{z} = \int \dfrac{d\alpha}{\alpha}$ or $-\dfrac{1}{2}\log|z| = \log|\alpha| + k$ or $\log\left|\dfrac{1}{z}\right| = 2\log|\alpha| + 2k$ using formula, $\displaystyle\int \dfrac{f'(x)}{f(x)}\,dx = \log|f(x)|$

or $\log\left|\dfrac{1}{2 - 2v - 2v^2}\right| = \log|\alpha|^2 + 2k$ or $\log\left|\dfrac{1}{2 - 2.\dfrac{\beta}{\alpha} - 2\left(\dfrac{\beta}{\alpha}\right)^2}\right| = \log|\alpha|^2 + 2k$ or $\log\left|\dfrac{\alpha^2}{2\alpha^2 - 2\alpha\beta - 2\beta^2}\right| = \log|\alpha|^2 + 2k$

or $\log\left|\dfrac{(x - h)^2}{2(x - h)^2 - 2(x - h)(y - k) - 2(y - k)^2}\right| = \log|x - h|^2 + 2k$

or $\log(x - h)^2 - \log[2(x - h)^2 - 2(x - h)(y - k) - 2(y - k)^2] = \log|x - h|^2 + 2k$

Put $h = \dfrac{3}{5}$ and $k = \dfrac{11}{5}$

or $\log\left(x - \dfrac{3}{5}\right)^2 - \log\left[2\left(x - \dfrac{3}{5}\right)^2 - 2\left(x - \dfrac{3}{5}\right)\left(y - \dfrac{11}{5}\right) - 2\left(y - \dfrac{11}{5}\right)^2\right] = \log\left(x - \dfrac{3}{5}\right)^2 + 2k$

or $-\log\left[\dfrac{2(5x - 3)^2 - 2(5x - 3)(5y - 11) - 2(5y - 11)^2}{25}\right] = 2k$

$\therefore\ -\log\left\{\dfrac{2}{25}\left[(5x - 3)^2 - (5x - 3)(5y - 11) - (5y - 11)^2\right]\right\} = 2k$ Ans.

(5) (a) Solve, $\dfrac{dy}{dx} + \dfrac{y}{x} = e^x$

Solution: − Given, $\dfrac{dy}{dx} + \dfrac{y}{x} = e^x$ form of $\dfrac{dy}{dx} + Py = Q$ here $P = \dfrac{1}{x}$ and $Q = e^x$

$I.F = e^{\int P \, dx} = e^{\int \frac{1}{x} dx} = e^{\log x} = x$

The required solution is $y \times I.F = \int I.F \times Q \, dx$ or $y.x = \int xe^x \, dx = x \int e^x \, dx - \int \left[\dfrac{d(x)}{dx}. \int e^x \, dx\right] dx$

or $xy = xe^x - \int e^x \, dx$ or $xy = xe^x - e^x + c$ $\boxed{\therefore \;\; xy = e^x(x-1) + c}$ Ans.

(b) Solve, $\dfrac{1}{\sin x}. \dfrac{dy}{dx} + y \cos x = \cos x$

Solution: − Given, $\dfrac{1}{\sin x}. \dfrac{dy}{dx} + y \cos x = \cos x$ or $\dfrac{dy}{dx} + y \sin x \cos x = \sin x \cos x$

Here $P = \sin x \cos x$ and $Q = \sin x \cos x$ $\therefore$ $I.F = e^{\int P \, dx} = e^{\int \sin x \cos x \, dx} = e^{\int z \, dz} = e^{\frac{z^2}{2}} = e^{\frac{\sin^2 x}{2}}$

The required solution is $y \times I.F = \int I.F \times Q \, dx$ or $y. e^{\frac{\sin^2 x}{2}} = \int e^{\frac{\sin^2 x}{2}}. \sin x \cos x \, dx$

Put $\dfrac{\sin^2 x}{2} = t$ or $\sin^2 x = 2t$ $\therefore$ $2 \sin x \cos x \, dx = 2 \, dt$ or $\sin x \cos x \, dx = dt$

or $y. e^{\frac{\sin^2 x}{2}} = \int e^t \, dt$ or $y. e^{\frac{\sin^2 x}{2}} = e^t + c$ or $y. e^{\frac{\sin^2 x}{2}} = e^{\frac{\sin^2 x}{2}} + c$ or $y = 1 + c e^{-\frac{\sin^2 x}{2}}$

$\therefore$ $y = 1 + k$ Ans.   where $k = ce^{-\frac{\sin^2 x}{2}}$

(c) Solve, $\dfrac{dx}{dy} - \dfrac{xy}{1 + y^2} = y + y^3$

Solution: − Given, $\dfrac{dx}{dy} - \dfrac{xy}{1 + y^2} = y + y^3$ or $\dfrac{dx}{dy} - \dfrac{y}{1 + y^2} x = y + y^3$

Here $P = -\dfrac{y}{1 + y^2}$ and $Q = y + y^3$ $\therefore$ $I.F = e^{\int P \, dy} = e^{-\int \frac{y}{1+y^2} dy}$

Put $1 + y^2 = z$ $\therefore$ $2y \, dy = dz$ or $y \, dy = \dfrac{dz}{2}$

$\therefore$ $I.F = e^{-\frac{1}{2}\int \frac{dz}{z}} = e^{-\frac{1}{2}\log z} = e^{\log\left(\frac{1}{\sqrt{z}}\right)} = e^{\log\left(\frac{1}{\sqrt{1+y^2}}\right)} = \dfrac{1}{\sqrt{1 + y^2}}$

The required solution is − $x \times I.F = \int I.F \times Q \, dy$

or $x. \dfrac{1}{\sqrt{1 + y^2}} = \int \dfrac{1}{\sqrt{1 + y^2}}. (y + y^3) \, dy = \int \dfrac{y(1 + y^2)}{\sqrt{1 + y^2}} \, dy$ or $\dfrac{x}{\sqrt{1 + y^2}} = \int y\sqrt{1 + y^2} \, dy$

Put $1 + y^2 = t^2$ or $y = \sqrt{t^2 - 1}$ $\therefore$ $2y \, dy = 2t \, dt$ or $y \, dy = t \, dt$

or $\dfrac{x}{\sqrt{1 + y^2}} = \int t^2 \, dt$ or $\dfrac{x}{\sqrt{1 + y^2}} = \dfrac{t^3}{3} + c$ or $\dfrac{x}{\sqrt{1 + y^2}} = \dfrac{(1 + y^2)\sqrt{1 + y^2}}{3} + c = \dfrac{(1 + y^2)\sqrt{1 + y^2} + 3c}{3}$

or $3x = (1 + y^2)\sqrt{1 + y^2}. \sqrt{1 + y^2} + 3c\sqrt{1 + y^2}$ $\therefore$ $3x = (1 + y^2)^2 + 3c\sqrt{1 + y^2}$ Ans.

(6) (a) Solve, $\dfrac{dy}{dx} + y \cot x = y^2 \cos^3 x$

Solution: – Given, $\dfrac{dy}{dx} + y \cot x = y^2 \cos^3 x$ or $\dfrac{1}{y^2}\dfrac{dy}{dx} + \dfrac{\cot x}{y} = \cos^3 x$ Divide by y^2

Let $\dfrac{1}{y} = z$ $\therefore -\dfrac{1}{y^2}\dfrac{dy}{dx} = \dfrac{dz}{dx}$ or $\dfrac{1}{y^2}\dfrac{dy}{dx} = -\dfrac{dz}{dx}$

Now, the original equation is $-\dfrac{dz}{dx} + z \cot x = \cos^3 x$ or $\dfrac{dz}{dx} - z \cot x = \cos^3 x$

Here $P = -\cot x$ and $Q = \cos^3 x$

$\therefore$ I.F $= e^{\int P\,dx} = e^{-\int \cot x\,dx} = e^{-\log(\sin x)} = e^{\log(\sin x)^{-1}} = (\sin x)^{-1} = \dfrac{1}{\sin x}$

The required solution is – $z \times$ I.F $= \displaystyle\int (\text{I.F} \times Q)\,dx$

or $z.\dfrac{1}{\sin x} = \displaystyle\int \dfrac{1}{\sin x}.\cos^3 x\,dx = \int \dfrac{\cos^2 x.\cos x}{\sin x}\,dx$ or $\dfrac{z}{\sin x} = \displaystyle\int \dfrac{(1 - \sin^2 x).\cos x}{\sin x}\,dx$

Put $\sin x = t$ $\therefore \cos x\,dx = dt$ also put $z = \dfrac{1}{y}$

or $\dfrac{1}{y \sin x} = \displaystyle\int \dfrac{1 - t^2}{t}\,dt = \int \dfrac{dt}{t} - \int \dfrac{t^2}{t}\,dt = \log t - \int t\,dt = \log t - \dfrac{t^2}{2} + c = \log(\sin x) - \dfrac{\sin^2 x}{2} + c$

or $\dfrac{1}{y \sin x} = \dfrac{2\log(\sin x) - \sin^2 x + 2c}{2}$ or $2 = y\sin x\,[2\log(\sin x) - \sin^2 x + 2c]$ Ans.

(b) Solve, $\dfrac{dy}{dx}(x^2 y + xy) = \dfrac{1}{y}$

Solution: – Given, $\dfrac{dy}{dx}(x^2 y + xy) = \dfrac{1}{y}$ or $\dfrac{dy}{dx} = \dfrac{1}{y(x^2 y + xy)}$ or $\dfrac{dx}{dy} = y(x^2 y + xy)$

or $\dfrac{dx}{dy} = x^2 y^2 + xy^2$ or $\dfrac{dx}{dy} - xy^2 = x^2 y^2$ or $\dfrac{1}{x^2}\dfrac{dx}{dy} - \dfrac{y^2}{x} = y^2$

Put $\dfrac{1}{x} = z$ or $-\dfrac{1}{x^2}\dfrac{dx}{dy} = \dfrac{dz}{dy}$ or $\dfrac{1}{x^2}\dfrac{dx}{dy} = -\dfrac{dz}{dy}$

Now, the original equation is $-\dfrac{dz}{dy} - zy^2 = y^2$ or $\dfrac{dz}{dy} + zy^2 = -y^2$

$\therefore$ I.F $= e^{\int P\,dy} = e^{\int y^2\,dy} = e^{\frac{y^3}{3}}$

The required solution is – $z.e^{\frac{y^3}{3}} = -\displaystyle\int e^{\frac{y^3}{3}}.y^2\,dy$

Let $\dfrac{y^3}{3} = t$ or $3y^2 dy = 3dt$ or $y^2 dy = dt$ or $\dfrac{e^{\frac{y^3}{3}}}{x} = -\displaystyle\int e^t\,dt = -e^t - c$ or $e^{\frac{y^3}{3}} = -xe^{\frac{y^3}{3}} - cx$ or $1 = -x - cxe^{\frac{-y^3}{3}}$

or $x + 1 = -cxe^{\frac{-y^3}{3}}$ Ans.

IInd Method: – $\dfrac{dz}{dy} + zy^2 = -y^2$ or $\dfrac{dz}{dy} = -y^2 - zy^2 = -y^2(z + 1)$ or $\dfrac{dz}{z+1} = -y^2 dy$

Integrating, $\displaystyle\int \dfrac{dz}{z+1} = -\int y^2\,dy$ or $\log|z + 1| = -\dfrac{y^3}{3} - c$ or $\log\left(\dfrac{1+x}{x}\right) = -\left(\dfrac{y^3}{3} + c\right)$ Ans.

(c) Solve, $(xy^2 + xy)dx = dy$

Solution: − Given, $(xy^2 + xy)dx = dy$ or $\dfrac{dy}{dx} = xy^2 + xy$ or $\dfrac{dy}{dx} - xy = xy^2$ or $\dfrac{1}{y^2}\dfrac{dy}{dx} - \dfrac{x}{y} = x$

Let $\dfrac{1}{y} = z$ $\therefore$ $-\dfrac{1}{y^2}\dfrac{dy}{dx} = \dfrac{dz}{dx}$ or $\dfrac{1}{y^2}\dfrac{dy}{dx} = -\dfrac{dz}{dx}$

Now, the original equation is $-\dfrac{dz}{dx} - zx = x$ or $\dfrac{dz}{dx} + zx = -x$ $\therefore$ $I.F = e^{\int x\,dx} = e^{\frac{x^2}{2}}$

The required solution is − $z \times I.F = \int (I.F \times Q)\,dx$ or $z.e^{\frac{x^2}{2}} = -\int e^{\frac{x^2}{2}}.x\,dx$

Put $\dfrac{x^2}{2} = t$ or $x\,dx = dt$ or $z.e^{\frac{x^2}{2}} = -\int e^t\,dt = -e^t - c$ or $\dfrac{e^{\frac{x^2}{2}}}{y} = -e^{\frac{x^2}{2}} - c$

or $1 = -\left(y + cye^{-\frac{x^2}{2}}\right)$ or $1 + y = -cye^{-\frac{x^2}{2}}$ Ans.

(d) Solve, $\dfrac{dy}{dx} = xy + x^5y^4$

Solution: − Given, $\dfrac{dy}{dx} = xy + x^5y^4$ or $\dfrac{dy}{dx} - xy = x^5y^4$ Divide by y^4

or $\dfrac{1}{y^4}\dfrac{dy}{dx} - \dfrac{x}{y^3} = x^5$ Let $\dfrac{1}{y^3} = z$ $\therefore$ $-\dfrac{3y^2}{y^6}\dfrac{dy}{dx} = \dfrac{dz}{dx}$ or $\dfrac{1}{y^4}\dfrac{dy}{dx} = -\dfrac{1}{3}.\dfrac{dz}{dx}$

The original equation is $-\dfrac{1}{3}.\dfrac{dz}{dx} - zx = x^5$ or $\dfrac{dz}{dx} + 3xz = -3x^5$ $\therefore$ $I.F = e^{\int 3x\,dx} = e^{\frac{3x^2}{2}}$

The required solution is − $z \times I.F = \int (I.F \times Q)\,dx$

or $z.e^{\frac{3x^2}{2}} = -3\int e^{\frac{3x^2}{2}}.x^5\,dx = -3\int e^{\frac{3x^2}{2}}.x^4.x\,dx$ Let $\dfrac{3x^2}{2} = t$ or $x^2 = \dfrac{2t}{3}$ or $6x\,dx = 2\,dt$ $\therefore$ $3x\,dx = dt$

or $\dfrac{e^{\frac{3x^2}{2}}}{y^3} = -\int e^t.\left(\dfrac{2t}{3}\right)^2 dt$ or $\dfrac{e^{\frac{3x^2}{2}}}{y^3} = -\dfrac{4}{9}\int t^2.e^t\,dt$ (use integration by part formula)

or $\dfrac{e^{\frac{3x^2}{2}}}{y^3} = -\dfrac{4}{9}.e^t[t^2 - 2t + 2] - c$ or $\dfrac{e^{\frac{3x^2}{2}}}{y^3} = -\dfrac{4}{9}.e^{\frac{3x^2}{2}}\left[\left(\dfrac{3x^2}{2}\right)^2 - 2\left(\dfrac{3x^2}{2}\right) + 2\right] - c$ Ans.

(8) (a) Solve, $\dfrac{dy}{dx} + \dfrac{y}{x} = \dfrac{\sqrt{y}}{x\sqrt{x}}$

Solution: − Given, $\dfrac{dy}{dx} + \dfrac{y}{x} = \dfrac{\sqrt{y}}{x\sqrt{x}}$ Divide by $\sqrt{y}$, $\therefore$ $\dfrac{1}{\sqrt{y}}\dfrac{dy}{dx} + \dfrac{y}{x\sqrt{y}} = \dfrac{1}{x\sqrt{x}}$

or $\dfrac{1}{\sqrt{y}}\dfrac{dy}{dx} + \dfrac{\sqrt{y}}{x} = \dfrac{1}{x\sqrt{x}}$ Let $\sqrt{y} = z$ $\therefore$ $\dfrac{1}{2\sqrt{y}}\dfrac{dy}{dx} = \dfrac{dz}{dx}$ or $\dfrac{1}{\sqrt{y}}\dfrac{dy}{dx} = 2\dfrac{dz}{dx}$

The original equation is $2\dfrac{dz}{dx} + \dfrac{z}{x} = \dfrac{1}{x\sqrt{x}}$ or $\dfrac{dz}{dx} + \dfrac{z}{2x} = \dfrac{1}{2x\sqrt{x}}$

$\therefore$ $I.F = e^{\int \frac{dx}{2x}} = e^{\frac{1}{2}\int \frac{dx}{x}} = e^{\frac{1}{2}\log x} = e^{\log \sqrt{x}} = \sqrt{x}$

The required solution is − $z \times I.F = \int (I.F \times Q)\,dx$

or $z.\sqrt{x} = \int \sqrt{x}.\dfrac{1}{2x\sqrt{x}}\,dx = \dfrac{1}{2}\int \dfrac{dx}{x} = \dfrac{1}{2}\log x + \log k$ or $\sqrt{y}.\sqrt{x} = \log \sqrt{x} + \log k = \log(\sqrt{x}.k)$ or $\sqrt{xy} = \log(\sqrt{x}.k)$ Ans.

(b) Solve, $\dfrac{dy}{dx} + \dfrac{1}{1+x} = \dfrac{e^{-y}}{x}$

Solution: − Given, $\dfrac{dy}{dx} + \dfrac{1}{1+x} = \dfrac{e^{-y}}{x}$ Divide by e^{-y}, we get ∴ $\dfrac{1}{e^{-y}}\cdot\dfrac{dy}{dx} + \dfrac{1}{e^{-y}(1+x)} = \dfrac{1}{x}$ or $e^y\dfrac{dy}{dx} + \dfrac{e^y}{1+x} = \dfrac{1}{x}$

Let $e^y = z$ Differentiate, $e^y\dfrac{dy}{dx} = \dfrac{dz}{dx}$

The original equation is $\dfrac{dz}{dx} + \dfrac{z}{1+x} = \dfrac{1}{x}$ ∴ I.F $= e^{\int P\,dx} = e^{\int \frac{dx}{1+x}} = e^{\log(1+x)} = 1+x$

The required solution is $z \times I.F = \int (I.F \times Q)\,dx$ or $z.(1+x) = \int \dfrac{1+x}{x}\,dx = \int \dfrac{dx}{x} + \int dx$

or $e^y(1+x) = \log x + x + k$ or $-x - \log x + e^y(1+x) = k$ Ans.

(c) Solve, $x^2\dfrac{dy}{dx} - xy = y^3 \cos x$

Solution: − Given, $x^2\dfrac{dy}{dx} - xy = y^3 \cos x$ Divid by $x^2 y^3$, we get $\dfrac{1}{y^3}\dfrac{dy}{dx} - \dfrac{1}{xy^2} = \dfrac{\cos x}{x^2}$

Let $\dfrac{1}{y^2} = z$ or $-\dfrac{2}{y^3}\dfrac{dy}{dx} = \dfrac{dz}{dx}$ or $\dfrac{1}{y^3}\dfrac{dy}{dx} = -\dfrac{1}{2}\cdot\dfrac{dz}{dx}$

The original equation is $-\dfrac{1}{2}\cdot\dfrac{dz}{dx} - \dfrac{z}{x} = \dfrac{\cos x}{x^2}$ or $\dfrac{dz}{dx} + \dfrac{2z}{x} = -\dfrac{\cos x}{x^2}$ ∴ I.F $= e^{2\int\frac{dx}{x}} = e^{2\log x} = e^{\log x^2} = x^2$

The required solution is $z \times I.F = \int (I.F \times Q)\,dx$

or $z.x^2 = -\int \dfrac{\cos x}{x^2}.x^2\,dx = -\int \cos x\,dx = -\sin x - c$ ∴ $\dfrac{x^2}{y} = -\sin x - c$ or $x^2 = -y\sin x - cy$ Ans.

(d) Solve, $\dfrac{1}{y}\cdot\dfrac{dy}{dx} + \dfrac{1}{1+x} = \dfrac{x}{y^3}$

Solution: − Given, $\dfrac{1}{y}\cdot\dfrac{dy}{dx} + \dfrac{1}{1+x} = \dfrac{x}{y^3}$ Multiplying by y^3, we get $y^2\dfrac{dy}{dx} + \dfrac{y^3}{1+x} = x$

Let $y^3 = z$ Diff. $3y^2\dfrac{dy}{dx} = \dfrac{dz}{dx}$ ∴ $y^2\dfrac{dy}{dx} = \dfrac{1}{3}\cdot\dfrac{dz}{dx}$

The original equation is $\dfrac{1}{3}\cdot\dfrac{dz}{dx} + \dfrac{z}{1+x} = x$ or $\dfrac{dz}{dx} + \dfrac{3z}{1+x} = 3x$ ∴ I.F $= e^{\int\frac{3}{1+x}dx} = e^{3\log(1+x)} = e^{\log(1+x)^3} = (1+x)^3$

The required solution is $z \times I.F = \int (I.F \times Q)\,dx$ or $z(1+x)^3 = \int x.(1+x)^3\,dx$

Put $1+x = t$ ∴ $dx = dt$

or $(1+x)^3.y^3 = \int (t-1).t^3\,dt = \int t^4\,dt - \int t^3\,dt = \dfrac{t^5}{5} - \dfrac{t^4}{4} + c = \dfrac{(1+x)^5}{5} - \dfrac{(1+x)^4}{4} + c$

or $y^3 = \dfrac{(1+x)^2}{5} - \dfrac{1+x}{4} + \dfrac{c}{(1+x)^3}$ Ans.

Exercise − A11

(1) Find the differential equation: − (a) $y = (c_1 + c_2)e^x + \cos(x + c_3)$ where c_1, c_2 and c_3 are constant.

(b) $y = c_1 \sin(x + c_2) - c_3 e^{x+c_4}$ where c_1, c_2, c_3 and c_4 are constant.   (c) $xy = Ae^{3x} + Be^{-3x}$ where A and B are constant.

(2) (a) Prove that the differential equation of the family of parabola $y^2 = 4ax$ is $2x\dfrac{d^2y}{dx^2} + \dfrac{dy}{dx} = 0$ or $x\dfrac{d^2y}{dx^2} - y = 0$.

(b) Prove that the differential equation of the family of parabola $x^2 = 4ay$ is $x\dfrac{dy}{dx} - 2y = 0$ or $x^2\dfrac{d^2y}{dx^2} - 2y = 0$.

(c) Prove that the differential equation of the circle $x^2 + y^2 = a^2$ is $y\dfrac{dy}{dx} + x = 0$.

(3) (a) Eliminate the constants a, b from the relation $y = a\cos x + b\sin x + x\cos x$.

(b) Eliminate the constants c_1, c_2 from the relation $y = c_1 e^x \sin x + c_2 e^x \cos x$.

(4) (a) Prove that $y = ax^2 + bx + c$ is a solution of $x^2\dfrac{d^2y}{dx^2} - 2x\dfrac{dy}{dx} + 2y = 0$.

(b) Prove that $u = \dfrac{A}{v} + Bv$ is a solution of $v^2\dfrac{d^2u}{dv^2} + v\dfrac{du}{dv} - u = 0$.

(c) Find the differential equation corresponding to the family of curves $xy = (x + k)^2$ where k is an arbitrary constant.

(5) (a) Find the differential equation of the family of the curves $y = Ae^x + Be^{3x}$.

(b) Find the differential equation of the curves $y = ae^x + be^{-2x} + ce^{3x}$ where a, b and c are arbitrary constant.

(c) Find the differential equation of the curves $y = Ae^{5x} + Be^{-7x}$.

(6) Find the differential equation of the family of the curves

(a) $x^2 + y^2 + 2ax + c = 0$ (b) $x^2 + y^2 + 2by + c = 0$ (c) $\dfrac{x^2}{a^2 + m} + \dfrac{y^2}{b^2} = 1$ where m is parameter.

(7) (a) $(1 + x^2)^2\dfrac{d^2y}{dx^2} + 2x(1 + x^2)\dfrac{dy}{dx} - 3y = 0$ it being given that $x = \tan\theta$.

(b) $(1 - x^2)\dfrac{d^2y}{dx^2} - x\dfrac{dy}{dx} + 2y = 0$ it being given that $x = \sin\theta$.

(c) $(1 - e^{2x})\dfrac{d^2y}{dx^2} - \dfrac{e^{2x}}{\sqrt{1 - e^{2x}}}\cdot\dfrac{dy}{dx} = 0$ it being given that $x = \log(\sin\theta)$.

(8) (a) $x[1 - (\log x)^2]\cdot\dfrac{d^2y}{dx^2} + [1 + \log x - (\log x)^2]\cdot\dfrac{dy}{dx} - 3y = 0$ it being given that $x = e^{\cos\theta}$.

(b) $\sin x\cdot\dfrac{d^2y}{dx^2} - \cos x\cdot\dfrac{dy}{dx} + 2y\sin^3 x = 3\sin^5 x$ it being given that $\theta = \cos x$.

(c) If $x = \cos\theta$, $y = \cos k\theta$ then show that $(1 - x^2)\dfrac{d^2y}{dx^2} - x\dfrac{dy}{dx} + k^2y = 0$.

Answer

(1) (a) $\dfrac{d^4y}{dx^4} = y$ or $y^{iv} - y = 0$ or $\dfrac{d^4y}{dx^4} - y = 0$ (b) $\dfrac{d^4y}{dx^4} = y$ or $\dfrac{d^4y}{dx^4} - y = 0$ (c) $xy'' + 2y' = 9xy$ or $x\dfrac{d^2y}{dx^2} + 2\dfrac{dy}{dx} = 9xy$

(2) (a) $y^2 = 4ax$ Differentiating, $\therefore\ 2y\dfrac{dy}{dx} = 4a$ or $y\dfrac{dy}{dx} = 2a$ $\therefore\ \dfrac{dy}{dx} = \dfrac{2a}{y}$ or $\dfrac{dy}{dx} = \dfrac{2ay}{y^2} = \dfrac{2ay}{4ax} = \dfrac{y}{2x}$

or $2x\dfrac{dy}{dx} = y$ $\therefore\ 2x\dfrac{dy}{dx} - y = 0$ proved.

Again Differentiating, $y\dfrac{d^2y}{dx^2} + \dfrac{dy}{dx}\cdot\dfrac{dy}{dx} = 0$ or $y\dfrac{d^2y}{dx^2} + \dfrac{2a}{y}\cdot\dfrac{dy}{dx} = 0$

or $\dfrac{d^2y}{dx^2} + \dfrac{2a}{y^2} \cdot \dfrac{dy}{dx} = 0$ or $\dfrac{d^2y}{dx^2} + \dfrac{2a}{4ax} \cdot \dfrac{dy}{dx} = 0$ or $\dfrac{d^2y}{dx^2} + \dfrac{1}{2x} \cdot \dfrac{dy}{dx} = 0$ or $2x\dfrac{d^2y}{dx^2} + \dfrac{dy}{dx} = 0$ Proved.

(b) and (c) Do yourself.   (see above question)

(3) (a) $\dfrac{d^2y}{dx^2} + y = -2\sin x$ (b) $\dfrac{d^2y}{dx^2} - 2\dfrac{dy}{dx} + 2y = 0$

(4) (a) $y = ax^2 + bx + c$ Diff. $\dfrac{dy}{dx} = 2ax + b$ Again Diff. $\dfrac{d^2y}{dx^2} = 2a$

or $x^2\dfrac{d^2y}{dx^2} - 2x\dfrac{dy}{dx} + 2y = 0$ or $x^2 . 2a - 2x(2ax + b) + 2(ax^2 + bx + c) = 0$

or $2ax^2 - 4ax^2 - 2bx + 2ax^2 + 2bx + 2c = 0$ or $2c = 0$ $\therefore$ $c = 0$ Ans.

(b) Do yourself.   (c) Do yourself, $\left(x\dfrac{dy}{dx} + y\right)^2 = 4xy$ or $x^2\left(\dfrac{dy}{dx}\right)^2 + 2xy\dfrac{dy}{dx} + y^2 = 4xy$ Ans.

(5) (a) $2\dfrac{d^2y}{dx^2} - 5\dfrac{dy}{dx} + 3y = 0$ (b) $5\dfrac{d^3y}{dx^3} - 18\dfrac{d^2y}{dx^2} + 19\dfrac{dy}{dx} - 6y = 0$ (c) $6\dfrac{d^2y}{dx^2} - 37\dfrac{dy}{dx} + 35y = 0$

(6) (a) $y\dfrac{d^2y}{dx^2} + \left(\dfrac{dy}{dx}\right)^2 + 1 = 0$ (b) $x\dfrac{d^2y}{dx^2} = \dfrac{dy}{dx}\left[\left(\dfrac{dy}{dx}\right)^2 + 1\right]$ (c) $xy\dfrac{dy}{dx} - y^2 + b^2 = 0$

(7) (a) $\dfrac{d^2y}{d\theta^2} - 3y = 0$ (b) $\dfrac{d^2y}{d\theta^2} + 2y = 0$ (c) $\sin\theta\dfrac{d^2y}{d\theta^2} + \cos\theta\dfrac{dy}{d\theta} + y = 0$

(8) (a) $\dfrac{1}{x} \cdot \dfrac{d^2y}{d\theta^2} - 3y = 0$ (b) $\dfrac{d^2y}{d\theta^2} + 2y = 3(1 - \theta^2)$

Exercise – A12

(1) solve the following differential equation: –   (a) $(2 + x)\, dy - (3 - y)\, dx = 0$

(b) $(e^y + 1)x\, dx = e^y(x + 1)\, dy$ (c) $\cot x\, dy + \cot y\, dx = 0$ (d) $\mathrm{cosec}^2 x . \cot y\, dx = \mathrm{cosec}^2 y . \cot x\, dy$

(2) (a) $x\sin^2 y\, dx - y\sin^2 x\, dy = 0$ (b) $\dfrac{dy}{dx} = \cos(5x + 3y)$ (c) $\cos(x + y)\, dy = \sin(x + y)\, dx$ (d) $e^x(y + 1)dy = e^y\, dx$

(3) (a) $y\, dx + x\, dy = x\, dx$ (b) $x\, dy - y\, dx = xy\, dx$ (c) $\dfrac{dx}{dy} - \sqrt{\dfrac{1 - x^2}{1 - y^2}} = 0$ (d) $\log\left(\dfrac{dy}{dx}\right) = x - y$

(4) (a) $\dfrac{dy}{dx} - \dfrac{xy - x}{xy - y} = 0$ (b) $\dfrac{dy}{dx} + \dfrac{e^x \tan y}{(e^x + 1)\sec^2 y} = 0$ (c) $\dfrac{y}{1 - x} \cdot \dfrac{dy}{dx} = \dfrac{1 - y^2}{1 - x^2}$ (d) $\sin^2 x\dfrac{dy}{dx} = \sin 2x . \cos^2 y$

(5) (a) $\dfrac{dy}{dx} = (x + y + 1)^2$ (b) $\dfrac{dy}{dx} = (4x + y)^2$ (c) $\dfrac{dy}{dx} - 1 = e^{x-y}$ (d) $\dfrac{dy}{dx} + \sin(x + y) = \sin(x - y)$

(6) (a) $\sin^{-1}\left(\dfrac{dy}{dx}\right) = 3x + y$ (b) $\dfrac{dy}{dx} = \cot(y - x) + 1$ (c) $\dfrac{dy}{dx} - 1 = x\sec(x - y)$ (d) $(3x + y - 1)^2 . \dfrac{dy}{dx} = a^2$

(7) (a) $(x - y + 1)\, dy = (3x - 3y + 5)\, dx$ (b) $(2x - y - 3)\, dy + (4x - 2y + 5)\, dx = 0$

(c) $\cot y . \dfrac{dy}{dx} = \cos(x + y) + \cos(x - y)$ (d) $x\,\mathrm{cosec}^2 y\, dy = (x + 1)\cot y\, dx$

(8) (a) $(x + y)^2 . \dfrac{dy}{dx} = 4$ (b) $(y + 5)\cot x . \dfrac{dy}{dx} = -y\,\mathrm{cosec}^2 x$ (c) $\log\left(\dfrac{dy}{dx}\right) = 3x + 4y$ (d) $y - (1 + x)\dfrac{dy}{dx} = (1 + x^2)\dfrac{dy}{dx}$

(9) (a) $2xb^2 + 2ya^2 . \dfrac{dy}{dx} = 2x$ (b) $\dfrac{dy}{dx} + x^2 e^{x+by} = e^{ax+by}$ (c) $\dfrac{dy}{dx} = \sin(10x + 6y) . \cos(10x + 6y)$ (d) $\dfrac{dy}{dx} = \dfrac{x + 1}{y}$

(10) (a) $\dfrac{dy}{dx} + 1 = x\sin(x+y)$ (b) $x\csc^2 y + (x+3)\cot y \cdot \dfrac{dx}{dy} = 0$ (c) $x - y \cdot \dfrac{dx}{dy} = x^2 - \dfrac{dx}{dy}$ (d) $y\cos y \cdot \dfrac{dy}{dx} = x(2\log x - 1)$

(11) (a) $y\,dx - x\,dy = \sqrt{x^2 + y^2}\,dy$ (b) $y\,dx - x\,dy = \sqrt{x^2 - y^2}\,dy$ (c) $x\,dy + y\,dx = xy\,dx$ (d) $y\,dx - x\,dy = \dfrac{x}{y}\,dy$

(12) (a) $\dfrac{y - x \cdot \dfrac{dy}{dx}}{\sqrt{x^2 + y^2}} = \dfrac{x^2 + y^2}{x}$ (b) $-x\,dy - y\,dx = (x^2 y^2 + xy)dx$ (c) $dx + \dfrac{(x+1)}{y}\,dy = y(x+1)\,dx$ (d) $(x+y)\dfrac{dy}{dx} = \dfrac{y}{x} + (x-y)\dfrac{dy}{dx}$

(13) (a) $\dfrac{\cos y}{\sin x} \cdot \dfrac{dy}{dx} = \dfrac{\sqrt{1 + \sin^2 y}}{\sqrt{1 + \cos^2 x}}$ (b) $e^{y+1} \cdot x\dfrac{dy}{dx} = \dfrac{x+1}{y+1}$ (c) $(x-1)y\,dx = (y+2)x\,dy$

(d) $(y+1) \cdot \cos(y+1)\,dy = (x+2) \cdot \sin(x+2)\,dx$ (e) $e^{y-x} \cdot \dfrac{dy}{dx} = \dfrac{e^y + 1}{e^x + 1}$

Answer

(1) (a) $(2+x)(3-y) = \dfrac{1}{k}$ (b) $(x+1)(e^y + 1) = ke^x$ (c) $\cos x \cos y = \dfrac{1}{c}$ or $c(\cos x \cos y) = 1$ (d) $\tan y - \tan x = c$

(2) (a) $\log\left(\dfrac{\sin x}{\sin y}\right) = x\cot x - y\cot y - c$ or $\sin x \sin y \cdot \log\left(\dfrac{\sin x}{\sin y}\right) = x\cos x \sin y - y\cos y \sin x + k$

(b) $\dfrac{1}{2}\tan^{-1}\left[\dfrac{\sqrt{7}}{2}\tan(5x + 3y)\right] + c$ (c) $e^{x+c} \cdot \sqrt{2 + 2\tan(x+y) + \tan^2(x+y)} = 1 + \tan(x+y)$ (d) $e^{x-y}(y-2) = 1 + k$

(3) (a) $x^2 - 2xy + 2c = 0$ (b) $k = 1 - x\log x$ or $x\log x = 1 - k$ (c) $\sin^{-1} y - \sin^{-1} x = c$ (d) $e^y - e^x = c$

(4) (a) $y - x + \log\left(\dfrac{y-1}{x-1}\right) = c$ or $\dfrac{y-1}{x-1} = e^{x-y+c}$ or $y = (x-1) \cdot e^{x-y+c} + 1$

(b) $\sin y\,(e^x + 1) = k\cos y$ [where $k = \log c$]   (c) $(1+x)^2(1-y^2) = \dfrac{1}{c^2}$ (d) $c \cdot \sin^2 x = e^{\tan y}$ or $\dfrac{e^{\tan y}}{\sin^2 x} = c$

(5) (a) $\tan^{-1}(x+y+1) = x + c$ or $x + y + 1 = \tan(x+c)$ or $x + y = \tan(x+c) - 1$

(b) $\tan^{-1}\left(\dfrac{4x+y}{2}\right) = 2(x+c)$ (c) $(x+c)e^{x-y} = 1$ (d) $\log(\sec x + \tan x) = -2\sin x + c$

(6) (a) $\dfrac{\sqrt{6}}{2}\tan^{-1}\left[\dfrac{3\tan\left(\dfrac{3x+y}{2}\right) + 1}{2\sqrt{2}}\right] = x + c$ (b) $\cos(y-x)\,e^{x+c} = 1$

(c) $2\sin(x-y) + x^2 = 2c$ (d) $y - 3c = 1 + a\tan^{-1}\left[\dfrac{\sqrt{3}}{a}(3x+y-1)\right]$

(7) (a) $3x - y - \log(x-y+2) = c$ (b) $x - y - 8\log(8x - 4y - 1) = c$

(c) $\log\left(\tan\dfrac{y}{2}\right) = 2\sin x + c$ (d) $\log(x\cot y) = -(x+c)$ or $x\cot y = e^{-(x+c)}$

(8) (a) $y - 2\tan^{-1}\left(\dfrac{x+y}{2}\right) = c$ (b) $y^5 \cdot e^y = k\cot x$ or $y^5 = \cot x \cdot e^{k-y}$ where $k = \log c$

(c) $-\dfrac{1}{4}e^{-4y} = \dfrac{1}{3}e^{3x} + c$ or $3e^{-4y} + 4e^{3x} = k$ where $k = 4c$ (d) $\log y = \dfrac{2}{\sqrt{7}}\tan^{-1}\left(\dfrac{2x+1}{\sqrt{7}}\right) + c$

(9) (a) $a^2 y^2 - (1-b^2)x^2 = c$ (b) $\dfrac{e^{-by}}{b} = e^x(x^2 - 2x + 2) - \dfrac{e^{ax}}{a} - c$

(c) $\dfrac{2\sqrt{5}}{\sqrt{91}}\tan^{-1}\left[\dfrac{10\tan\left(\dfrac{10x+6y}{2}\right) + 3}{\sqrt{91}}\right] = x + c$ (d) $y^2 - x^2 - 2x = 2c$ or $y^2 - x^2 - 2x = 1 + 2c$

(10) (a) $\log\left[\tan\left(\dfrac{x+y}{2}\right)\right] = \dfrac{x^2}{2} + c$ (b) $\log\left(\dfrac{\cot y}{x^3}\right) = x + c$

(c) $x(1-y) = (x-1)c$ or $x - xy = (x-1)c$ (d) $y\sin y - \cos y = x^2(\log x - 1) + c$

(11) (a) $x + \sqrt{x^2 + y^2} = y^2 k$ or $x = y\sin(\log ky)$ (b) $x + \sqrt{x^2 - y^2} = y^2 k$ or $x = y\cos(\log ky)$

(c) $\log(xy) = x + k$ or $xy = e^{x+k}$ (d) $\log\left(\dfrac{x}{y}\right) + \dfrac{1}{y} = c$

(12) (a) $x\sqrt{x^2 + y^2} + y = k$ where $k = -c\sqrt{x^2 + y^2}$ (b) $-x + \log\left(1 + \dfrac{1}{xy}\right) = c$

(c) $\log(x+1) + \dfrac{1}{y(x+1)} = -k$ where $k = \log c$ (d) $y = \log\sqrt{x} + \log k$ or $y = \log(k\sqrt{x})$

(13) (a) $\left(\sin y + \sqrt{1 + \sin^2 y}\right)\left(\cos x + \sqrt{1 + \cos^2 x}\right) = k$ (b) $ye^{y+1} - x - \log x = k$

(c) $y - x + \log xy^2 = k$ (d) $t\sin t + \cos t + z\cos z - \sin z = k$ where $t = y + 1, z = x + 2$ (e) $e^y + 1 = (e^x + 1)e^k$

Exercise − A13

Solve the following differential equations: −

(1) (a) $(x - y)\dfrac{dy}{dx} = y$ (b) $(x^3 + y^3)\dfrac{dy}{dx} = x^2 y$ (c) $(x + y)\dfrac{dy}{dx} = y$ (d) $\dfrac{dy}{dx} = \dfrac{y}{x} + \dfrac{1}{\log\left(1 + \dfrac{y}{x}\right)}$

(2) (a) $(x - y)\dfrac{dy}{dx} = x + y$ (b) $\dfrac{dy}{dx} = \dfrac{3y - x}{3x - y}$ (c) $(y^2 - x^2)\,dx = xy\,dy$ (d) $xy^2\,dy = (x^3 + y^3)\,dx$

(3) (a) $(x^2 + y^2)\,dy = \dfrac{y^3}{x}\,dx$ (b) $x + y\dfrac{dy}{dx} = \sqrt{x^2 + y^2}$ (c) $(xy + x^2)\dfrac{dx}{dy} = (y^2 + xy)$ (d) $xy + y^2\dfrac{dx}{dy} = x^2 + x^2\dfrac{dx}{dy}$

(4) (a) $xy\dfrac{dy}{dx} = y^2 + x^2\sec\left(\dfrac{y}{x}\right)$ (b) $\dfrac{y}{x}\cdot\dfrac{dy}{dx} = \left(\dfrac{y}{x}\right)^2 + \left(\dfrac{x^2 + y^2}{x^2}\right)$ (c) $x\sqrt{1 + \dfrac{x}{y}}\cdot\dfrac{dy}{dx} = \sqrt{1 + \dfrac{y}{x}} + y\sqrt{1 + \dfrac{x}{y}}$

(d) $(x^2 + y^2)\dfrac{dy}{dx} = xy$ (e) $\dfrac{dy}{dx} = \dfrac{x - y}{x - y + 1}$

(5) (a) $y\dfrac{dx}{dy} = x - y\sin^2\left(\dfrac{x}{y}\right)$ (b) $3xy^2\dfrac{dy}{dx} = x^3 + y^3$ (c) $2x\dfrac{dy}{dx} = \log_y y - \log_x y + 2y$ (d) $\dfrac{dy}{dx} = \dfrac{y(x - y)}{x^2}$

(6) (a) $2x\dfrac{dy}{dx} = 2y - \sqrt{x^2 - y^2}$ (b) $y\dfrac{dx}{dy} = x - \sqrt{x^2 + y^2}$ (c) $x\,dy = (\sqrt{xy} + y)\,dx$ (d) $\left(y + \sqrt{x^2 + xy + y^2}\right)dx = x\,dy$

(7) (a) $y^2\,dx = (x^2 + xy + y^2)\,dy$ (b) $x^2 y\dfrac{dy}{dx} = x(y^2 + xy + x^2)$ (c) $(x^2 + 2xy + y^2)\dfrac{dy}{dx} = x^2 - y^2$

(8) (a) $\dfrac{dy}{dx} - \dfrac{y}{x} = \cos^2\left(\dfrac{y}{x}\right)$ (b) $\dfrac{y}{x}\cdot\dfrac{dy}{dx} = \dfrac{x^2 + y^2}{x^2} + \cos\left(\dfrac{2y^2}{x^2}\right)$ (c) $\dfrac{dy}{dx} + \dfrac{y^2}{x^2} = 1 + \dfrac{y}{x}$ (d) $\cos\left(\dfrac{y}{x}\right)\dfrac{dy}{dx} = \dfrac{y}{x}\cos\left(\dfrac{y}{x}\right) + \sin\left(\dfrac{y}{x}\right)$

(9) (a) $\dfrac{y}{x}\cot\left(\dfrac{y}{x}\right) + \dfrac{y^2}{x^2} = \dfrac{y}{x}\cdot\dfrac{dy}{dx}$ (b) $y^2\dfrac{dx}{dy} = \dfrac{y(x + y)}{3}$ (c) $x\dfrac{dy}{dx} + y\log x = y\log y$ (d) $(x + 2y)\,dy - y\,dx = x\,dx$

(10) (a) $\dfrac{dy}{dx} = \dfrac{3x + 5y + 1}{x + y + 2}$ (b) $\dfrac{dy}{dx} = \dfrac{2x + y + 1}{x + 2y + 3}$ (c) $\dfrac{dy}{dx} = \dfrac{3y - x + 5}{x - y + 7}$ (d) $\dfrac{dy}{dx} = \dfrac{x - 4y + 1}{4x - 3y + 5}$

(11) (a) $\dfrac{dx}{dy} = \dfrac{x - y + 1}{x + y + 3}$ (b) $\dfrac{dx}{dy} = \dfrac{x + 2y + 1}{x + y}$ (c) $\dfrac{dy}{dx} = \dfrac{10x + 3y + 15}{5x + 6y + 30}$ (d) $\dfrac{dy}{dx} = \dfrac{x - y}{2y - 3x + 5}$

(12) (a) $\dfrac{dy}{dx} = \dfrac{7x + 5y + 14}{5x - 3y + 10}$ (b) $\dfrac{dy}{dx} = \dfrac{10x + 7y + 20}{9x + 5y + 30}$

(1) (a) $xc = e^{-\frac{x}{y}}$ or $xe^{\frac{x}{y}} = \frac{1}{c}$ (b) $\frac{-x^3 + 2y^3}{2xy^2} = \log x + k$ (c) $x = ce^{\frac{x}{y}}$ (d) $(x+y)\left[\log\left(\frac{x+y}{x}\right) - 1\right] = x\log x + k$

(2) (a) Hint: $-\tan^{-1}\left(\frac{y}{x}\right) - \frac{1}{2}\log\left(\frac{x^2 + y^2}{x^2}\right) = \log x + \log c$ or $2\tan^{-1}\left(\frac{y}{x}\right) - \log\left(\frac{x^2 + y^2}{x^2}\right) = 2\log x + 2\log c$

or $2\tan^{-1}\left(\frac{y}{x}\right) = \log\left(\frac{x^2 + y^2}{x^2}\right) + \log x^2 + k$ or $2\tan^{-1}\left(\frac{y}{x}\right) = \log\left(\frac{x^2 + y^2}{x^2}\cdot x^2\right) + k$

or $2\tan^{-1}\left(\frac{y}{x}\right) - \log(x^2 + y^2) = k$ Ans.   where $k = 2\log c$

(b) $\log\left(\frac{x}{x+y}\right) = x + c$ (c) $\frac{y^2}{2x^2} + \log x = k$ (where $k = \log c$)   (d) $\frac{y^3}{3x^3} - \log x = k$

(3) (a) $2x^2.\log y + y^2 = 2x^2k$ where $k = \log c$ (b) $\log\left(\frac{x - \sqrt{x^2 + y^2}}{x^2}\right) = k$

(c) $\frac{x^2}{2y^2} + \log\left(\frac{x}{y}\right) - \log y = k$ (d) $x^2 + 2xy = e^{2k}$ where $k = \log c$

(4) (a) $y\sin\left(\frac{y}{x}\right) + x\cos\left(\frac{y}{x}\right) = x\log x + xk$ where $k = \log c$ (b) $\sqrt{x^2 + y^2} = x^2k$ or $x^2 + y^2 = x^4k^2$ (c) $2\sqrt{\frac{y}{x}} - \log x = k$

(d) $\log x - \frac{x^2}{2y^2} = k$ (e) $x^2 + y^2 - 2y(x + 1) = k$

(5) (a) $ye^{-\cot\left(\frac{x}{y}\right)} = k$ (b) $x^3 - 2y^3 = \frac{x}{k^2}$ (c) $y = xe^{\sqrt{x}.k}$ or $\log\left(\frac{y}{x}\right) = k\sqrt{x}$ (d) $k = xe^{-\frac{x}{y}}$

(6) (a) $2\sin^{-1}\left(\frac{y}{x}\right) + \log x = k$ (b) $x + \sqrt{x^2 + y^2} = e^k$ where $k = \log c$ (c) $2\sqrt{y} = \sqrt{x}(\log x + k)$ (d) $2y + x + 2\sqrt{x^2 + xy + y^2} = 2x^2k$

(7) (a) $\tan^{-1}\left(\frac{x}{y}\right) = \log y + k$ or $x = y\tan(\log ky)$ (b) $y - x\log(x + y) = k$ (c) $\log\left(\frac{1}{\sqrt{x^2 - 2xy - y^2}}\right) = k$

(8) (a) $\tan\left(\frac{y}{x}\right) - \log x = k$ (b) $\frac{1}{4}\tan\left(\frac{y^2}{x^2}\right) = \log(k.x)$ (c) $\sin^{-1}\left(\frac{y}{x}\right) - \log x = k$ (d) $\sin\left(\frac{y}{x}\right) = kx$

(9) (a) $\sec\left(\frac{y}{x}\right) = kx$ (b) $-\frac{3}{2}\log(y - 2x) - \frac{1}{2}\log y = k$ (c) $y = xe^{1+xc}$ (d) $\frac{1}{2}\log\left(\frac{x + \sqrt{2}y}{x - \sqrt{2}y}\right) - \frac{1}{4}\log\left(\frac{x^2 - 2y^2}{x^2}\right) = \log(xk)$

(10) (a) $-\frac{3}{2\sqrt{7}}\log\left|\frac{\sqrt{7} + (v - 2)}{\sqrt{7} - (v - 2)}\right| - \frac{1}{2}\log|3 + 4v - v^2| = \log|xk|$

Put $v = \frac{\beta}{\alpha}$ and $\alpha = x - h$, $\beta = y - k$ value of $k = \frac{5}{2}, h = -\frac{9}{2}$

(b) $\frac{1}{4}\log\left|\frac{3x + 3y + 4}{3x - 3y - 6}\right| - \frac{1}{2}\log\frac{2}{3}|3x^2 - 3y^2 - 2x - 10y - 8| = k$ where $k = \log c$

(c) $\frac{1}{\sqrt{2}}\log\left|\frac{(x + y + 19) - \sqrt{2}(x + 13)}{(x + y + 19) + \sqrt{2}(x + 13)}\right| - \frac{1}{2}\log|y^2 - x^2 + 2xy + 38y - 14x + 23| = k$ where $k = \log c$

(d) $-\frac{1}{2}\log\left|\frac{3(13y + 1)^2 - 8(13y + 1)(13x + 17) + (13x + 17)^2}{169}\right| = k$ where $k = \log c$

(11) (a) $-\frac{1}{2}\log|x^2 + y^2 + 4x + 2y + 5| - \tan^{-1}\left(\frac{x + 2}{y + 1}\right) = k$ where $k = \log c$

(b) $\dfrac{1}{2\sqrt{2}}\log\left|\dfrac{\sqrt{2}(y+1)+(x-1)}{\sqrt{2}(y+1)-(x-1)}\right|-\dfrac{1}{2}\log|2y^2-x^2+4y+2x+1|=k$ where $k=\log c$

(c) $-\dfrac{1}{2}\log|3y^2-5x^2+xy+30y+5x+75|-\dfrac{2}{\sqrt{183}}\log\left|\dfrac{12\sqrt{3}(y+5)+x(2\sqrt{3}-\sqrt{61})}{12\sqrt{3}(y+5)+x(2\sqrt{3}+\sqrt{61})}\right|=k$

(d) $\dfrac{1}{\sqrt{3}}\log\left|\dfrac{2y-x(1+\sqrt{3})}{2y-x(1-\sqrt{3})-10}\right|-\dfrac{1}{2}\log|x^2-2y^2+2xy-20x+10y+25|=k$ where $k=\log c$

(12) (a) $\dfrac{5}{\sqrt{21}}\tan^{-1}\left[\dfrac{\sqrt{3}y}{\sqrt{7}(x+2)}\right]-\dfrac{1}{2}\log|7x^2+3y^2+28x+28|=k$ where $k=\log c$

(b) $-\dfrac{1}{2}\log|5\beta^2+2\alpha\beta-10\alpha^2|-\dfrac{4\sqrt{5}}{5\sqrt{51}}\log\left|\dfrac{\sqrt{5}(5\beta+\alpha)-5\sqrt{51}\alpha}{\sqrt{5}(5\beta+\alpha)+5\sqrt{51}\alpha}\right|=k$ where $k=\log c$

Put $\alpha=x-h,\ \beta=y-k$ and $h=-\dfrac{110}{13},\ k=\dfrac{120}{13}$

Exercise – A15

Solve the following differential equations: –

(1) (a) $\dfrac{dy}{dx}+\dfrac{y}{x}=x^2$ (b) $\dfrac{dy}{dx}=1+\dfrac{y}{x}$ (c) $\dfrac{dy}{dx}+\dfrac{y}{x}=\log x$ (d) $\dfrac{dy}{dx}=(e^x+1)+y$

(2) (a) $\dfrac{dy}{dx}+y\sin x=\sin x\cos x$ (b) $\dfrac{dx}{dy}+\dfrac{xy}{\sqrt{1+y^2}}=y+y^3$ (c) $\dfrac{dy}{dx}=xe^{2x}-y$ (d) $\dfrac{dy}{dx}+\dfrac{y}{x}=\sqrt{1-x^2}$

(3) (a) $\dfrac{dx}{dy}-\dfrac{2x}{y}=\log y$ (b) $(x+1)\,dy-(y+1)\,dx$ (c) $\dfrac{dy}{dx}=\dfrac{y}{1+x}+\dfrac{x}{\sqrt{1+x}}$ (d) $\dfrac{dy}{dx}+y\tan x=\dfrac{\sin x}{\cos^2 x}$

(4) (a) $\dfrac{dy}{dx}+\dfrac{y}{x}\log y=y(\log y)^2$ (b) $x\dfrac{dy}{dx}+\dfrac{xy}{1+x}=\sqrt{1-x}$ (c) $(1+y)\dfrac{dy}{dx}+\dfrac{y}{x}=e^{x-y}$ (d) $\dfrac{dy}{dx}+\dfrac{y}{x\log x}=\dfrac{\log x}{x}$

(5) (a) $\dfrac{dy}{dx}-y\tan x=2x\sec x$ (b) $xy\dfrac{dy}{dx}+\dfrac{xy^2}{\tan x}=3y\csc x$ (c) $\tan x\dfrac{dy}{dx}-y\sin x=\sin x$ (d) $(1+\log x)\dfrac{dy}{dx}+\dfrac{y}{x}=e^x$

(6) (a) $\dfrac{dy}{dx}+\dfrac{y(\cos^2 x-\sin^2 x)}{\sin x\cos x}=e^{\cos 2x}$ (b) $\cos x\dfrac{dy}{dx}+\dfrac{y}{\sin x}=\dfrac{1}{\cos^3 x}$ (c) $\dfrac{dy}{dx}+\dfrac{y}{\cot x}=e^{\tan x}.\sec x$

(d) $xdy+\dfrac{y(x+1)}{x}dx=\log x\,dx-\log x\,dy$ (e) $\dfrac{1}{\cos^2 y}dy+\dfrac{\tan y}{x}dx=\cos x\,dx$ (f) $\dfrac{1}{\sqrt{1-y^2}}dy+\dfrac{\sin^{-1}y}{x}dx=\dfrac{dx}{x}$

(7) (a) $\sin y\cos y\dfrac{dy}{dx}+\dfrac{\sin^2 y}{x}=\dfrac{1}{x^3-1}$ (b) $\dfrac{dy}{dx}+\dfrac{xy}{1-x^2}=\dfrac{2x}{\sqrt{1-x^2}}$ (c) $2\dfrac{dy}{dx}+\dfrac{y}{x}=\dfrac{y^2}{x}$ (d) $\dfrac{dy}{dx}-\dfrac{y}{x}=\dfrac{y^2}{x^2}$

(e) $2\dfrac{dy}{dx}-\dfrac{3y}{x+1}=\dfrac{x^2}{y}$ (f) $\dfrac{1}{\sin^2 y}.\dfrac{dy}{dx}+\dfrac{x\cot y}{1+x^2}=1+x$ (g) $y\dfrac{dy}{dx}+\dfrac{y^2}{x}=\sqrt{1+x}$ (h) $\dfrac{1}{x}\dfrac{dx}{dy}+\dfrac{\log x}{1+y}=\dfrac{1}{y}$

(8) (a) $\dfrac{dy}{dx}+\dfrac{y\cos x}{\sqrt{1+\sin x}}=\sin x\cos x$

Answer

(1) (a) $4xy=x^4+4c$ (b) $y=x\log x+xc$ (c) $y=\dfrac{x}{2}\left[\log x-\dfrac{1}{2}\right]+k$ (d) $y=xe^x-1+ce^x$

(2) (a) $y-\cos x-1=ce^{\cos x}$ (b) $x=t^3-3t^2+6t-6+\dfrac{c}{e^t}$ $\left(\text{Put } t=\sqrt{1+y^2}\right)$

(c) $ye^x=\dfrac{xe^{3x}}{3}-\dfrac{e^{3x}}{9}+c$ (d) $xy=-\dfrac{(1-x^2)^{\frac{3}{2}}}{3}+c$

---

(3) (a) $x + y(\log y + 1) = cy^2$ (b) $y + 1 = c(x + 1)$ (c) $2\left[(1+x)^{\frac{3}{2}} + (1+x)^{\frac{1}{2}}\right] + c(1+x)$ (d) $2y - \sec x = 2c\cos x$

(4) (a) $x\log y . \log\left(\frac{c}{x}\right) = 1$ (b) Do yourself (c) $xye^y = e^x(x-1) + c$ (d) $3y = (\log x)^2 + k$ where $k = \dfrac{3c}{\log x}$

(5) (a) $y\cos x = x^2 + c$ (b) $y\sin x = 3\log x + k$ (c) $y + 1 = ce^{\sin x}$ (d) $y(1 + \log x) = e^x + c$

(6) (a) $y\sin 2x + \dfrac{1}{2}e^{\cos 2x} = c$ (b) $y\tan x = \dfrac{1}{4\cos^4 x} + c$ (c) $y = \cos x . e^{\tan x} + c$

(d) $y(x + \log x) = x(\log x - 1) + k$ (e) $x\tan y = x\sin x + \cos x + c$ (f) $\sin^{-1} y = 1 + k$ where $k = \dfrac{c}{x}$

(7) (a) $x^2\sin^2 y = \dfrac{2}{3}\log|x^3 - 1| + k$ (b) $y = \sqrt{1-x^2}\log\left|\dfrac{1}{1-x^2}\right| + k$ where $k = \sqrt{1-x^2}\log|c|$

(c) solution: $-\ \ 2\dfrac{dy}{dx} + \dfrac{y}{x} = \dfrac{y^2}{x}$ or $\dfrac{2}{y^2}\dfrac{dy}{dx} + \dfrac{1}{xy} = \dfrac{1}{x}$ Let $\dfrac{1}{y} = z$ or $-\dfrac{1}{y^2}\dfrac{dy}{dx} = \dfrac{dz}{dx}$

$\therefore\ \ \dfrac{1}{y^2}\dfrac{dy}{dx} = -\dfrac{dz}{dx}$ or $-2\dfrac{dz}{dx} + \dfrac{z}{x} = \dfrac{1}{x}$ or $2\dfrac{dz}{dx} - \dfrac{z}{x} = -\dfrac{1}{x}$ or $\dfrac{dz}{dx} - \dfrac{z}{2x} = -\dfrac{1}{2x}$

$$\therefore\ \ \text{I.F} = e^{-\int\frac{dx}{2x}} = e^{-\frac{1}{2}\log x} = e^{\log(x)^{-\frac{1}{2}}} = x^{-\frac{1}{2}} = \dfrac{1}{\sqrt{x}}$$

The required solution is $-\ \ y.\dfrac{1}{\sqrt{x}} = -\int \dfrac{1}{2x}.\dfrac{1}{\sqrt{x}}dx = -\dfrac{1}{2}\int \dfrac{dx}{x^{\frac{3}{2}}} = -\dfrac{1}{2}\int x^{-\frac{3}{2}}dx = -\dfrac{1}{2}.\dfrac{x^{-\frac{3}{2}+1}}{-\frac{3}{2}+1}$

or $y.\dfrac{1}{\sqrt{x}} = x^{-\frac{1}{2}} + c$ or $y.\dfrac{1}{\sqrt{x}} = \dfrac{1}{\sqrt{x}} + c$ $\therefore\ \dfrac{y}{\sqrt{x}} = \dfrac{1}{\sqrt{x}} + c$ $\therefore\ y = 1 + c\sqrt{x}$ Ans.

(d) $x = y\log\left(\dfrac{c}{x}\right)$ Ans. (Do yourself, same as above question)

(e) $3y^2 = 3(x+1)^3\log(x+1) + 6(x+1)^2 - \dfrac{3}{2}(x+1) + c$ Ans. (Do yourself)

(f) $1 + x^2 + \cot y = -\sqrt{1+x^2}\log\left|x + \sqrt{1+x^2}\right| + k$ (Do yourself)

(g) $x^2 y^2 = \dfrac{2}{7}.(1+x)^3\sqrt{1+x} - \dfrac{4}{5}.(1+x)^2\sqrt{1+x} + \dfrac{2}{3}.(1+x)\sqrt{1+x} + k$ (h) $y(\log x - 1) + \log\left(\dfrac{x}{y}\right) = k$

(8) (a) $y = (1 + \sin x)^{\frac{3}{2}} + \dfrac{1}{2}(1+\sin x)^{\frac{1}{2}} - \dfrac{3}{2}(1+\sin x) + k$ where $k = ce^{-2\sqrt{1+\sin x}}$